# Structure Induced Anelasticity in Iron Intermetallic Compounds and Alloys

by
Igor S. Golovin
Anatoly M. Balagurov

Different anelastic phenomena are discussed in this book with respect to iron-based binary and ternary alloys and intermetallic compounds of $Fe_3Me$ type, where Me are α-stabilizing elements Al, Ga, or Ge. An introduction into anelastic behavior of metallic materials is given, and methods of mechanical spectroscopy and neutron diffraction are introduced for the better understanding of structure-related relaxation and hysteretic phenomena.

To characterize structure and phase transitions - both of the first and second order - in the studied alloys XRD, TEM, SEM, MFM, VSM, PAS, DSC techniques were used. Considerable emphasis is placed on *in situ* neutron diffraction tests that were performed with the same heating and cooling rates as the internal friction measurements. Different types of mechanical spectroscopy techniques were used to study mainly, but not exclusively, Fe-Al, Fe-Ga and Fe-Ge based alloys: from subresonance "low" frequency forced bending and torsion vibrations (0.00001 to 200 Hz) to "high" frequency resonance (above ~200 Hz) free decay bending vibrations.

We discuss (i) thermally activated effects like Snoek-type relaxation, caused by interstitial atom jumps in alloyed ferrite, (ii) Zener relaxation, caused by reorientation of pairs of substitute atoms in iron, (iii) different transient effects due to phase transitions of the first and second order, and (iv) amplitude dependent magneto-mechanical damping; especially with respect to structure, ordering of substitutional solid solution and phase transitions. Special attention is paid to magnetostriction of the alloys - the result of magneto-mechanical elastic coupling.

**Keywords:** Anelasticity, Damping Capacity, Magnetostricition, Structure Transitions, Phase Transitions, Fe-Based Alloys, Intermetallic Compounds, Mechanical Spectroscopy, In Situ Neutron Diffraction

# Structure Induced Anelasticity in Iron Intermetallic Compounds and Alloys

by

**Igor S. Golovin[1] and Anatoly M. Balagurov[2,3]**

[1] National University of Science and Technology "MISIS", Moscow, Russia

[2] Frank Laboratory of Neutron Physics, Joint Institute for Nuclear Research, Russia

[3] M.V. Lomonosov Moscow State University, Moscow, Russia

Published by **Materials Research Forum LLC**
Millersville, PA 17551, USA

Published as part of the book series
**Materials Research Foundations**
Volume 30 (2018)
ISSN 2471-8890 (Print)
ISSN 2471-8904 (Online)

Print ISBN 978-1-945291-64-7
ePDF ISBN 978-1-945291-65-4

This book contains information obtained from authentic and highly regarded sources. Reasonable efforts have been made to publish reliable data and information, but the author and publisher cannot assume responsibility for the validity of all materials or the consequences of their use. The authors and publishers have attempted to trace the copyright holders of all material reproduced in this publication and apologize to copyright holders if permission to publish in this form has not been obtained. If any copyright material has not been acknowledged please write and let us know so we may rectify this in any future reprints.

Distributed worldwide by

**Materials Research Forum LLC**
105 Springdale Lane
Millersville, PA 17551
USA
http://www.mrforum.com

Manufactured in the United States of America
10 9 8 7 6 5 4 3 2 1

# Table of Contents

# Preface

One of the most important physical phenomena in iron alloys and intermetallic compounds under cyclic loading is damping of mechanical vibrations. Damping, or internal friction (IF), is the capacity of any dense material to transform energy of mechanical vibrations into heat by means of different physical processes. Thus, internal friction is the dissipation of mechanical energy inside a material in elastic range of loading. The anelastic behavior may occur due to a wide range of physical processes taking place in the material under the action of elastic stresses. This phenomenon is widely used in solid-state physics, physical metallurgy, materials science to study structural defects and their mobility, transport phenomena, and phase transformations in solids etc. In many cases, the highly sensitive and selective spectra of internal friction (measured as a function of temperature, frequency, and amplitude of vibration: TDIF, FDIF and ADIF, correspondingly) contain unique information that cannot be obtained by other methods. Internal friction spectra may give information not only about defects concentration and distribution but also about their mobility and phase transitions in alloys. In addition to application of internal friction to study structure and behavior of materials, internal friction is a property of metals that is very important for high and low damping materials.

In this book we focus mainly, but not exclusively, on three iron-based systems: Fe-Al, Fe-Ga and Fe-Ge including multicomponent alloys and intermetallic compounds. The intermetallic systems are often considered as functional materials with special properties. Enhanced damping capacity and giant magnetostriction are among those. An enormous research activity was conducted during the last 50 years to develop Fe-Al based alloys for different applications. This system is still under extensive attention of many scientists. After pioneering studies of Fe-Ga alloys mainly by US research groups in the early 2000s, in the last few years a high research activity in this field has taken place in different countries, especially in China. Several new ideas to improve functionality of Galfenol alloys were introduced, including the 'ferromagnetic strain glass' approach, alloying Fe-Ga by rare earth elements, annealing in a magnetic field, formation of special texture by rolling etc. The ability to controllably manipulate magnetic skyrmions, small magnetic whirls with particle-like properties, has recently been demonstrated for Fe-Ga alloys.

First internal friction studies of Fe-Al alloys were performed in the 1960th, while studies of Fe-Ga alloys developed after 2005, and nearly no internal friction studies were carried out on Fe-Ge alloys as of yet. This book intends to present results of our studies of anelasticity in these and similar Fe-based systems. We applied for the first time systematical studies of *in situ* neutron diffractions to characterize phase transitions in these bulk samples to interpret anelastic phenomena measured by different mechanical spectroscopy techniques in the same heating and cooling or annealing regimes. Our book is mainly based on the results of our own studies with some references to relevant findings by other research groups. Some information on related books and conferences is given below.

The International Conference on Internal Friction and Ultrasonic Attenuation in Solids 'ICIFUAS' began in Providence (USA, 1956) and continued in Ithaca (USA, 1961), Manchester (England, 1965), Providence (USA, 1969), Aachen (Germany, 1973), Tokyo (Japan, 1977), Lausanne (Switzerland 1981), Urbana (USA, 1985), Beijing (China, 1989), Rome (Italy, 1993), Poiters (France, 1996), Buenos Aires (Argentina, 1999), and Bilbao (Spain, 2002). In 2005 the conference changed its name to 'International Conference on Internal Friction and Mechanical Spectroscopy, ICIFMS' to incorporate the novel generic name 'mechanical spectroscopy'. Since then conferences have been named ICIFMS, beginning in Kyoto (Japan, 2005), and continued in Perugia (Italy, 2008), Lausanne (Switzerland, 2011), Hefei (China, 2014) and Foz do Iguaçu (Brazil, 2017). The ICIFMS-19 conference will take place at Moscow at National University of Science and Technology MISiS in 2020.

In the 1950s Moscow Institute of Steel and Alloys 'MISiS', now the National University of Science and Technology 'MISiS', was the first educational institution in Moscow to create a Russian research school of anelasticity in metals under the guidance of Prof. B.N. Finkelstein. In the world literature, the anelastic relaxation effect in face centred cubic alloys was named after Boris Finkelstein, who reported it for the first time in 1953. The studies of anelasticity in metallic materials were continued at MISiS by Yu. Piguzov, G. Ashmarin, I. Kekalo, E. Naimi, et al. Research centres on internal friction and mechanical spectroscopy of metals appeared in St. Petersburg (S. Nikanorov, B. Kardashev, S. Kustov), Voronezh (V. Postnikov, B. Darinsky, V. Khonik), Tula (S. Golovin, D. Levin, G. Markova), and some other Russian cities.

The contribution of following basic English-written literature on anelasticity of solids cannot go unnoticed:

- Zener C. Elasticity and anelasticity of metals; University of Chicago Press: Chicago, Illinois, 1948.
- Mason W.P. Physical Acoustics and Properties of Solids; Van Nostrand: Princeton, 1958.
- Nowick A.S., Berry B.S. Anelastic relaxation in crystalline solids. Academic Press, New York, 1972.

- De Batist R. Internal friction of structural defects in crystalline solids. North Holland Publ Comp: Amsterdam, 1972.
- Lakes R.S. Viscoelastic solids. CRC Press: Boca Raton, 1999.
- Schaller R., Fantozzi G., Gremaud G. (Eds.) Mechanical spectroscopy $Q^{-1}$ 2001 with applications to materials science. Trans Tech Publications: Switzerland, 2001.
- Blanter M.S., Golovin I.S., Neuhäuser H., Sinning H.-R. Internal Friction in Metallic Materials. A Handbook. Springer: 2007.
- Ngai K.L. Relaxation and Diffusion in Complex Systems. Springer, 2011.

Our book also focuses on different anelastic effects in iron-based alloys and compounds. The information that can be obtained from internal friction method, sometimes also called 'mechanical spectroscopy', depends on the development of the theory of anelasticity and the knowledge about structure and phase transitions in metallic materials. A specific feature of this book is the close combination of internal friction and *in situ* neutron diffraction studies.

Since the end of the $1940^{th}$ neutron diffraction investigations have been widely used to study atomic structures. Meanwhile, the development of neutron diffraction experimental techniques in the last decade all over the world gives a new opportunity. In contrast with many other diffraction methods, the use of neutron diffraction is often preferred because of its low absorption in the material. This allows us to study the volume effects and to avoid some uncertainties typical for research methods that use surfaces and small volumes of materials. Our *in situ* neutron diffraction studies were used to follow structural transitions in iron alloys with the same heating and cooling regimes as internal friction tests, which helps greatly to interpretation of anelastic phenomena.

This book is intended for students, researchers, and engineers working in solid-state physics, materials science, or mechanical engineering.

The authors are grateful to their colleagues M.S. Blanter, S. Divinski, T. Lograsso, H. Neuhäuser, J. Pons, A. Rivière, H.-R. Sinning, G. Wilde, J. Zhu for long-running cooperation; to V. Cheverikin, I. Chudakov, A. Churyumov, J. Cifre, J. Čížek, L. Dubov, S. Golovina, Ch. Grusewski, S. Jäger, S. Jen, D. Mari, C. Mennerich, A. Pozdniakov, C. Siemers, V. and M. Zadorozhnyy for their invaluable help with different experiments. Neutron diffraction results were obtained with active participation of I. Bobrikov, V. Simkin, S. Sumnikov (experiments), V. Zlokazov, O. Ivanshina (data processing). Special thanks are to recent and former students V. Palacheva, T. Pavlova, T. Pozdova, T. Sazonova, A. Strahl, A. Emdadi, L. Sun, M. Fang, M. Nartey for their enthusiasm and interest for the subject. The experimental results on anelasticity of metals presented in this manuscript are mainly based on our joint publications. We gratefully acknowledge Ms. Valeria Palacheva's enormous input in compiling the list of references.

The authors thank Mr. Thomas Wohlbier (President of Materials Research Forum LLC, USA), who encouraged them to write this book and Ms. Nadya Columbus (President of Nova Science Publishers, Inc.) for her kind permission to use the materials of a chapter written by I.S. Golovin in 2008 for the book 'Intermetallics Research Progress', Ed. Yakov N. Berdovsky. We are grateful to different RFBR grants for financial support of our studies. Last but not least, I.S. Golovin is grateful to National University of Science and Technology MISiS and, in particular, to the Department of Non-Ferrous Materials (Head of Dep. A. Solonin) for providing a favorable and supportive atmosphere in writing this manuscript.

We devote this manuscript to the memory of Professor Stanislav Alexeevich Golovin, one of the first Russian researchers who started to study anelasticity in different metallic materials in the 1950th.

Stanislav A. Golovin
1932-2017

# Chapter 1

# Introduction

Mechanical spectroscopy (MS), referred to as the internal friction (IF) method in earlier literature, offers special opportunities to study elastic and anelastic phenomena in metals and alloys at the atomic level, providing response, e.g., from interstitial atoms, vacancies, substitutional atoms, dislocations, grain and magnetic domains boundaries, phase transformations, etc.

Internal friction (after Charles Coulomb [1]) is the capacity of dense materials to transform energy of mechanical vibrations into heat by different physical processes in the elastic (Hookeian [2]) range of loading. Internal energy dissipation is a similar phenomenon, but in this case the behavior of the materials is not limited by the elastic range of loading. Thus, internal friction is the dissipation of mechanical energy inside a material, caused by its anelasticity – the term used by Clarence Zener [3] to distinguish this phenomena from a broader family of inelastic effects in physics. The anelastic behavior occurs due to a wide range of physical processes in the material under the action of elastic stresses and is widely used in solid state physics, physical metallurgy and materials science to study structural defects and their mobility, transport phenomena, and phase transformations in solids. In many cases the highly sensitive and selective spectra of IF contains unique information that cannot be obtained by other methods. A combination of mechanical spectroscopy with other *in situ* methods provides even more exciting opportunities to study structure, defects of crystalline lattice and their mobility, and, thus, functional properties of modern materials.

Fundamentals of internal friction phenomena and mechanical spectroscopy method are given in several textbooks and monographs: Zener 1948 [3], Mason 1958 [4], Krishtal et al. 1964 [5], Nowick and Berry 1972 [6], De Batist 1972 [7], Postnikov 1974 [8], Lakes 1999 [9], Schaller et al. 2001 [10], Blanter et al. 2007 [11], Ngai [12] and for this reason are considered in this chapter very shortly with respect to studied alloys only.

The variation of the properties of a material depending on time $t$ as a result of a relaxation process is described well by the well known Kohlrausch function [13], which is also frequently called the stretched exponential function:

$$\phi_K(t) = \exp[(-t/\tau)^{1-n}], \tag{1.1}$$

1

where $\tau$ is the relaxation time and $n$ is the parameter of correlation between the elementary acts of the relaxation process ($0 \leq n < 1$).

At $n = 0$ all elementary acts of a relaxation process are independent of one another and the relaxation process is described by a three-element (two springs and dashpot) rheological Zener model [3]. As an example of independent relaxation processes can serve diffusion jumps of interstitial atoms in the crystal lattice of interstitial solid solutions under the effect of an applied cyclic stress. In the case of $n = 0$, the thermally activated relaxation peak of internal friction that corresponds to such a process with a single relaxation time can be evaluated for a standard anelastic solid as a function of the loading frequency.

A mechanical loss peak or a reciprocal quality factor, $Q^{-1}(\omega)$, in the case of a thermally activated relaxation effect with a single relaxation time - no matter which relaxation mechanism is involved - is well-known as described by a Debye equation with respect to frequency ($\omega$):

$$Q^{-1}(\omega) = \Delta \cdot \frac{\omega\tau}{1+(\omega\tau)^2}$$ (1.2a)

and frequency dependent elastic modulus, $E(\omega)$:

$$E(\omega) = E_R \cdot (1 + \Delta \frac{\omega^2\tau^2}{1+\omega^2\tau^2}) = E_U \cdot (1 - \frac{\Delta}{1+\omega^2\tau^2}) \quad ,$$ (1.2b)

where $\tau$ is the relaxation time, $\Delta$ is the relaxation strength, $\omega = 2\pi f$ with $f$ being the frequency of the mechanical vibrations, $E_R$ and $E_U$ are relaxed and unrelaxed modulus.

Several anelastic effects are described by the Debye function of frequency: $\omega\tau/(1+(\omega\tau)^2)$. The relaxation strength ($\Delta$) is individual for each physical mechanism, for example:

1. Thermoelastic relaxation - transverse thermal current between the compressed and dilated sides of a homogeneous and isotropic rectangular beam:

$$Q_T^{-1} = \frac{\alpha^2 ET}{c_\sigma} \cdot \frac{\omega\tau_T}{1+(\omega\tau_T)^2}$$ (1.3a)

where $\alpha$ is the linear thermal expansion coefficient, $E$ is the Young's modulus, $C_\sigma$ is the specific heat capacity at constant stress $\sigma$.

2. <u>Phonon relaxation</u> – recovery of thermal equilibrium of phonon gas:

$$Q_{ph}^{-1} = \frac{\gamma^2 cT}{\rho \upsilon^2} \cdot \frac{\omega\tau_{ph}}{1+(\omega\tau_{ph})^2} ,$$
(1.3b)

where $\upsilon$ is wave speed, $\rho$ is density, $c$ is the specific heat capacity of volume unit; $\gamma$ ($\approx$1) is parameter of lattice anharmonism.

3. <u>Snoek relaxation</u> – diffusion-controlled stress-induced jumps of interstitial atoms in b.c.c. metals:

$$Q_S^{-1} = \frac{\eta C_{AB} V(\lambda_1 - \lambda_2)^2 Mf(\Gamma)}{k_B T_m} \cdot \frac{\omega\tau_{AB}}{1+(\omega\tau_{AB})^2} ,$$
(1.3c)

where $C_0$ is the atomic fraction of interstitial atoms in the solution, $V$ is the volume of one mole of the host metal, $\Gamma$ is orientation parameter. In case of torsional vibration, the parameter $\eta$ = 2/3, and $M$ = $G$, $f(\Gamma)$ = $\Gamma$; for flexural vibration, $\eta$ = 1/9, $M$ = $E$, $F(\Gamma)$ = 1-3$\Gamma$. This results in different orientation dependences for different kinds of deformation.

These were only three examples of Debye type relaxation effects. Grain boundary and Zener relaxation, relaxation due to micro-eddy currents and Bordoni relaxation are other examples. In all these effects two values - the relaxation time, $\tau$ and measuring frequency, $f$ - can be varied in experiments. Consequently, two types of amplitude independent tests can be carried out:

In frequency dependent internal friction tests (FDIF) at a fixed temperature ($\tau$ is a constant in eq. (1.2) for a given temperature) the frequency $f$ is varied over a few orders of magnitude. This method allows direct measurements of $Q^{-1}$ and $E$ spectra *vs.* $\omega\cdot\tau$ as introduced by eqs. (1.2) and shown in Fig. 1.1a. The results can be also presented as a function of linear frequency, $f$ ($f$= $\omega$/2$\pi$, $\tau$ where $\omega$ is angular frequency).

Most of the existing mechanical spectroscopy set-ups (e.g., vibrating reeds, torsion pendula) allows measurements of $Q^{-1}$ as a function of temperature ($T$) but not frequency, i.e. to measure temperature-dependent internal friction (TDIF) (Fig. 1.1b). The temperature dependence (e.g., for jumps of atoms) is typically described by the Arrhenius equation:

$$\tau = \tau_0 \exp(H/k_B T),$$
(1.4)

where $H$ is the activation energy (or enthalpy) of the physical phenomenon which controls the relaxation process.

For a fixed frequency and a single relaxation time, the temperature dependence of $Q^{-1}$ is described by the equation:

$$Q^{-1}(T) = Q_m^{-1} \cosh^{-1}\left\{ \frac{\overline{H}}{k_B}\left( \frac{1}{T} - \frac{1}{T_m} \right) \right\},$$

(1.5)

where $T_m$ and $Q_m^{-1}$ are the peak temperature and height, respectively, $\overline{H}$ is the mean value of activation energy.

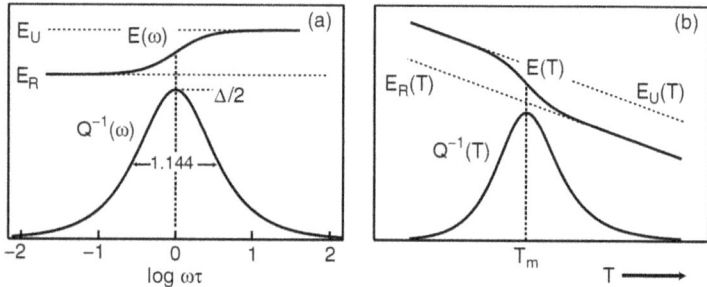

Fig. 1.1 Dynamic modulus E and internal friction $Q^{-1}$ of the standard anelastic solid: (a) as a function of frequency on a log ($\omega\tau$) scale; (b) as a function of temperature at constant frequency. In the latter case, the relaxation-induced step in E(T) is superimposed on the intrinsic temperature dependence of $E_U(T)$ and $E_R(T)$ [11].

A variety of particular distribution functions was developed for the description of different alloys. The most widely used one is the Gaussian distribution of a variable $z = \ln(\tau/\tau_m)$, where $\tau_m$ is the mean relaxation time:

$$\Phi(z) = \frac{\exp\left\{ -(z/\beta_\tau)^2 \right\}}{\beta_\tau \sqrt{\pi}}$$

(1.6)

with $\beta_\tau$ being a parameter characterising the width of the relaxation time distribution, and

$$Q^{-1} = \Delta \int_{-\infty}^{\infty} \Phi(\ln \tau) \cdot \frac{\omega\tau}{1+(\omega\tau)^2} d(\ln \tau)$$

(1.7)

The resulting damping spectrum is broader than a single Debye peak:

$$Q^{-1} = Q_m^{-1} \cosh^{-1}\left\{ \frac{\overline{H}}{k_B r_2(\beta_\tau)}\left(\frac{1}{T} - \frac{1}{T_m}\right) \right\} \qquad (1.8)$$

Here $r_2(\beta_\tau)$ represents the relative peak width, i.e. the peak width with respect to the single Debye peak with $\beta_\tau = 0$.

The relaxation time distribution $(\beta_\tau)$ may originate from distributions in both $H$ and $\tau_0$. In most cases these distributions are interrelated [6]:

$$\beta_\tau = |\beta_{\tau 0} \pm \beta_H/k_B T|, \qquad (1.9)$$

the values $\beta_{\tau 0}$ and $\beta_{\tau H}$ can be obtained by plotting experimental data in axes $\beta_\tau$ vs $1/T$. If $\beta_\tau$ is temperature independent, peak broadening is due to distribution in $\tau_0$, if $\beta_\tau$ is temperature dependent, there is a distribution in activation energy H.

Thus, thermally activated relaxation phenomena are typically described in terms of either frequency or temperature. The above equations are used to evaluate the parameters of an experimental internal friction peak ($\overline{H}$, $T_m$, $Q_m^{-1}$, $\beta_\tau$), i.e. the parameters of a relaxation mechanism, by fitting the experimental curves. Unfortunately, it is not always possible to evaluate all the parameters from available experimental results. The mean values for the activation energy $\overline{H}$ and the pre-exponential factor $\tau_0$ can be calculated relatively easily from the frequency and/or temperature shift of the peak using the Arrhenius equation. Analytical solutions for more complex cases can be found in the literature (e.g. [10, 12]). A collection of experimental data on anelasticity for different metallic materials can be found in [11, 14, 15].

The main thermally activated relaxation effects in annealed iron and iron-based alloys are: the Snoek relaxation (only in alloys with b.c.c. lattice, i.e. with A2 structure), the Finkelshtein-Rosin relaxation (only in alloys with f.c.c. lattice or A1 structure), the Zener relaxation (in alloys with A1, A2 and A3 (h.c.p.) structures) and grain boundary relaxation [16]. The first three types of relaxation are caused by point defects: either interstitial or substitutional atoms. Point defect relaxation means an anelastic relaxation caused by diffusive jumps of point defects in the field of applied stress in the elastic range of loading. In cold worked alloys several additional dislocation related anelastic effects may take place, for example - Bordoni relaxation, Hasiguti relaxation and Snoek-Kê-Köster relaxation. The main condition for anelastic relaxation is that the symmetry of the local elastic distortions, caused by the defects in the crystal lattice, is lower than the symmetry of the lattice itself (so-called selection rule).

The classical *Snoek relaxation*, originally proposed by Snoek [17] to explain the damping due to C in α-Fe, is an anelastic relaxation caused by 'heavy' foreign interstitial atoms (IA) in the *b.c.c.* metals with body centred cubic (*b.c.c.*) lattice. In the case of α-Fe it is observed in C and N interstitial solid solutions. The interstitial atoms are located in the octahedral interstices of the *b.c.c.* metal. The two metal atoms nearest to the interstice (distance $a/2$) move aside; the displacement of the other four atoms, located in a distance of $a/\sqrt{2}$, is about one order of magnitude smaller and has the opposite sign. The lattice distortions around interstitial atoms, i.e. elastic dipoles, are oriented along one axis ($x$, $y$, $z$) and have a tetragonal symmetry. Such an elastic dipole is described by the $\lambda$-tensor with two components $\lambda_1$ and $\lambda_2$, which are not equal to each other. The difference $|\lambda_1 - \lambda_2|$ determines the "elastic dipole strength".

There are three types of interstices corresponding to the three lattice directions ($x$, $y$, $z$). By applying stress, the arrangement of dissolved atoms in some octahedral interstices becomes energetically more favourable. Therefore, the dissolved interstitial atoms will diffuse from one octahedral interstices to the neighboring one just by one stress induced jump for each interstitial atom. Under the alternating stress this "diffusion under stress" of interstitial atoms takes place and leads to energy dissipation of mechanical vibrations, i.e. to the Snoek peak.

Substitutional atoms (SA) dissolved in a host lattice influence the original Snoek relaxation via SA-IA interaction in several coordination shells either by a change in the activation energy of the atomic jump, or by a change of the value ($\lambda_1 - \lambda_2$), i.e. the relaxation strength [18]. Some substitutional atoms may not pronouncedly affect the position and height of the Snoek peak, others may reduce the peak height, lead to the appearance of additional peaks at higher temperature, or completely suppress the Snoek peak. Despite many years of studying Snoek-type relaxation, one can still find many contradicting results in the literature.

Interstitial atoms in face centred cubic (*f.c.c.*) metals occupy octahedral interstitial sites producing the same cubic symmetry of distortions as the host lattice. For this reason an isolated interstitial atom does not contribute to anelastic relaxation in the *f.c.c.* lattice if stress is applied. In case interstitial atoms form pairs in neighboring coordination shells, they produce an elastic dipole and the reorientation of such atom pairs may occur under an external stress leading to anelastic relaxation. This effect was reported for the first time by *Rosin and Finkelshtein* [19] for carbon in *f.c.c.* steels.

A single substitutional atom in a cubic lattice does not produce an elastic dipole with a symmetry lower than that of the host lattice and, consequently, it does not produce any relaxation effect under loading. However, as first shown by Zener [3], the existence of solute next neighbor pairs results in elastic dipole and in a relaxation maximum under cycling stress, called "Zener peak". As pairs of atoms are involved in the relaxation mechanism, for dilute and random solid solutions the peak height, which is proportional to the number of reorienting pairs, should be proportional to the solute concentration squared, $c^2$. The same dependence is also observed for Finkelshtein-Rosin relaxation. This is different to the Snoek relaxation where the peak height is proportional to the concentration of interstitial atoms in the soli solution.

As the atom movements during Zener relaxation resemble those in ordinary diffusion, the rate of relaxation, or relaxation time $\tau$, can in principle be used to estimate the diffusion activation energies $H_D$ at temperatures far below those in ordinary bulk diffusion experiments, because in internal friction the site changes involve near neighbors only. Activation energies for Zener relaxation in different iron based alloys is in the range from 2 to 3 eV [11]. Ordering of substitutional solid solution decreases the Zener peak height by decrease in number of participation atomic pairs, the reorientation of which would mean disordering of the alloy.

A thermally activated relaxation peak is observed in many polycrystalline metals and alloys at about ½ of their melting temperatures. This peak was attributed to grain boundary (GB) sliding [20] and originally explained by the Zener model [21] in terms of the anelastic strain due to grain boundary sliding resulting from a shear stress and slip along the boundary of two adjacent crystals. There were a lot of scientific discussions in the periodic literature about interpretation of this effect in different materials. One of the reasons for these discussions is a huge variety in grain boundary structures in different materials, moreover even in the same material GBs with different orientation contribute differently to applied stress. This latter case was successfully discussed in detail for aluminum bicrystals [22].

A variety of possible grain boundary microstructures, a dependence of grain boundary energies on their orientation, a contribution of segregations of IA or SA at grain boundaries suggest a wide distribution of the relaxation times which may originate from distributions in both $H$ and $\tau_0$ (eq. 8). It is not surprising that the collection of experimental data for iron and iron based alloys [11] also exhibits a wide range for activation energies from 1.6 to 3.7 eV. As a rule of thumb, if both Zener and grain boundary relaxations are recorded in an alloy, the Zener peak locates at low-temperature branch of the GB relaxation which has much bigger relaxation strength.

Another type of relaxation effects may take place due to different phase or structural transitions. Even these effects are thermodynamically driven, i.e. they are thermally activated, they exhibit themselves to be not frequency dependent in the sense that the damping peak temperature does not depend on measuring frequency. Their peak temperature depends on rate of phase or structural (e.g., recrystallization) transition having maximum at the temperature at which the transformation rate has maximum. In contrast with 'true' thermally activated relaxation effects obeyed to Arrhenius law and expressed by Debye equation, these anelastic effects can be called as transient or non-thermally activated. Theory of these transitions invented by L. Landau is given in the book by Nowick and Berry 1972 [6], W. Benoit [23] and G. Fantozzi [24]. Maximum of internal friction for both first order transition and second order (often called lambda, λ-peak) transition according to Landau's theory can be ascribed by equation of the following type:

$$Q_m^{-1}(T_0, T_C) = \frac{\chi^2 A}{J_U \omega},$$
(1.10)

where for the first order transition we use $T_0$ and A = M, and for second order transition we use $T_C$ and A = L, M and L are positive coefficients, $\omega(=2\pi f)$ is frequency, $\chi$ is a coupling coefficient between the internal variable (e.g. ordering parameter) and the mechanical stress, and $J_U$ is the unrelaxed elastic compliance. It is also noted in these Refs [6, 23, 24] that most of experimental internal friction data are not well described by these equations. In this book we will give some opposite results of anelastic effects due to both first and second order phase transitions in intermetallic compounds, which are in agreement with Landau theory.

Anelastic effects for different first order transitions in many metallic materials are well studied. Some empirical coefficients were introduced for a better fit between the theory and experimental observations. If a cyclic stress, $\sigma = \sigma_0 \cos(\omega t)$, is applied to a sample, the internal friction resulting from phase transition ($Q_{Tr}^{-1}$) is:

$$Q_{tr}^{-1} = \frac{1}{\pi J \sigma_0^2} \cdot \oint \sigma d\varepsilon_{an},$$
(1.11)

where J is the elastic compliance, $\sigma_0$ is the oscillation amplitude, $\omega$ (=2πf, f - frequency of mechanical vibrations) is the angular frequency and $\varepsilon_{an}$ is the anelastic part of the deformation during one cycle of vibrations.

The *transient* damping (denoted in the figures in this book as $P_{Tr}$) due to a phase transition, mostly known in literature to accompany the first order martensitic transformations in shape memory alloys, is attributed to the diffusionless transformation

8

rate $\partial n/\partial t$, where $n$ is the amount of the transformed material [25]. This transient peak is also known as the so-called "$\dot{T}$ effect" [26]. The related anelastic strain $d\varepsilon_{an} = k \, (\partial n/\partial t)$ $dt$, which determines the relaxation strength ($\Delta = Q_{Tr}^{-1}$), comes from the lattice deformation when the material is transformed. Delorme and Gobin [27] suggested a "transformation plasticity" approach with the transformation deformation $d\varepsilon_p = k\sigma dn$ to be linearly dependent on the transformed volume fraction and on the applied stress $\sigma$, leading to

$$Q_{Tr}^{-1} = \frac{k}{J} \frac{dn}{dT} \frac{\dot{T}}{\omega},$$
(1.12)

where k is an empirical coefficient.

More recent developments of this model have been discussed by San Juan and Pérez-Sáez: practically all models accept the $Q_{Tr}^{-1} \approx \dot{T} / \omega^n$ dependence [26]. Transient internal friction effects may also accompany the second order transitions, i.e. ordering - disordering reactions, and in this case they are often called as $\lambda$ peaks.

Amplitude dependent internal friction (ADIF) tests allow to study damping capacity of metals and alloys, to distinguish and analyse magnetic and non-magnetic, most often dislocation, damping as a function of amplitude of vibrations. High damping metals, often abbreviated as "Hidamets" or HDM is the biggest and probably most important group of high damping materials widely used in various practical applications (noise and vibration reduction, preventing fatigue problems, increasing the quality of different cutting tools etc.). "HDM should exhibit high damping over a wide frequency and temperature range. In metals, only a limited number of dissipative mechanisms can be used to achieve such performance. Point defect relaxations are not useful in this application because, generally, they give rise to relaxation peaks located in narrow frequency and temperature domains. A damping mechanism is active over a wide frequency range only if it is not thermally activated" R. Schaller (in [29]).

In addition to the obvious demand to have a certain level of strength, the Hidamets must have a stable and powerful source of damping. From the viewpoint of both physics and engineering the damping capacity for different materials should be estimated under homologous conditions at least with respect to temperature: $T_{HD} = n \cdot T_{melt}$, and applied stress $\sigma_{HD} = m \cdot \sigma_{0.2}$, where $T_{melt}$ and $T_{HD}$ are the melting temperature and a certain part of this temperature but not only at room temperature, which is most typical for practical applications. It is desirable for many engineering applications but not obligatory from the viewpoint of physics that room temperature is included the temperature interval where high damping is needed. If damping for different materials is compared at such

9

homologous temperature conditions it is different to engineering Hidamets classifications for room temperature [28].

A specific damping index (SDI) is the quantitative specific damping capacity $\Psi$ ( $\Psi = \Delta W/W$, where $\Delta W$ is the energy absorption during one cycle, and $W$ is the maximum elastic stored energy during the cycle; $Q^{-1} = \Delta W/2\pi W = \Psi/2\pi$) measured by means of a torsion pendulum, when the maximum surface shear stress amplitude is one-tenth of the 0.2 tensile yield strength. This measure is denoted as $\Psi_{0.1\sigma_{YS}}$ or more shortly as $\Psi_{0.1}$, and compared for various materials [28]: Those with $\Psi_{0.1}$ < 1% are called low damping materials, those with $1 < \Psi_{0.1} < 10\%$ medium damping, and those with $\Psi_{0.1} > 10\%$ high damping materials. Four structural mechanisms responsible for high damping are recently being distinguished [29].

Magnetomechanical damping is one of four main sources for high damping in metallic materials: (materials with 1 - thermoelastic martensite, 2 - easy movable magnetic domains, 3 - high heterogeneity, and 4 - easy movable dislocations). Total damping in ferromagnetic alloys is composed of non-magnetic (dislocation, grain boundaries, point defects) and magneto-mechanical damping: $Q^{-1} = Q_d^{-1} + Q_h^{-1}$. The magneto-mechanical damping in ferromagnetic materials has its source in a stress-driven irreversible movement of the magnetic Domain Walls (DWs). The DW motion also leads in ferromagnetic materials to the effect of magnetostriction, discovered by J.P. Joule in 1842 (Phil.Mag. 1847). Magneto-mechanical damping in Fe alloys at ambient temperatures is typically much higher as compared with a non-magnetic one, and it is used as one of the main sources for high damping materials. High damping capacity in that case is determined by nonreversible movement of 90° DWs. The dissipated energy accompanying the DWs motion depends on domain walls mobility and density. The mobility of DW is controlled by the following factors:

- Structure and size of magnetic domains and domain boundaries, i.e. magnetic parameters of alloys including their saturation magnetostriction;

- Structure of the crystalline lattice in which the DWs move;

- Interaction between the DWs and different imperfections of the crystalline lattice.

Classical review of elastic energy dissipation in ferromagnets was done in [30] from the point of view of the physics of magnetism. Phenomenologically the energy loss $\Delta W$ due to magnetic domains for a vibration stress ($\sigma$) below some critical stress is well described usually given by the Rayleigh law: $\Delta W \sim \sigma^3$, and damping is proportional to the stress applied: $Q_h^{-1} = \Delta W/2\pi W \propto \lambda^3 E \sigma$, where $\lambda$ is magnitostriction [31]. With increase in stress, the dissipated energy saturates ($\Delta W_S$), and damping decreases with increase in applied stress: $Q_h^{-1} = \Delta W_S / 2\pi W = E \cdot \Delta W_S / \pi \sigma^2$. Thus, the amplitude dependence of internal friction $Q_h^{-1}(\sigma\ or\ \varepsilon)$ has a maximum.

The maximal damping $Q_{h.max}^{-1}$ due to the magneto-mechanical hysteresis is proportional to $\lambda_S E/\sigma_i$, where $\lambda_S$ is the saturation magnetostriction, E is Young's modulus, and $\sigma_i$ is the average internal stress opposing domain boundary motion. For a Maxwell distribution of internal stresses, the value of maximal hysteretic internal friction ($Q_{h.max}^{-1}$) was described by Smith and Birchak [31] as:

$$Q_{h.max}^{-1} = 0.34\ k\ \lambda_S\ E\ /\ \pi\ \sigma_i \approx 0.25\ k\ \lambda_S\ E\ /\ \pi\ \sigma_{max}, \tag{1.13}$$

where $\sigma_i$ is the average residual internal stress in a material, $\sigma_{max}$ is a stress at which ADIF curve has a maximum ($\sigma_m \approx 0.7\sigma_i$), k = 1 is a constant characteristic of the shape of the hysteresis loop and E is Young's modulus. Note that values of damping should correlate with values of saturation magnetostriction strain $\lambda_S$. For iron $\lambda_S$ is about 10 ppm: Al, Ge and specifically Ga increase $\lambda_S$ in binary Fe alloys up to 400 ppm. Giant magnetostriction is known for rare-earth elements and their alloys. The most famous composition is Terfenol-D ($Tb_{0.3}Dy_{0.7}Fe_2$), which can exibite a strain ~0.1% at room temperature. However, the rare-earth based magnetostrictive materials are expensive and brittle. Thus, cheaper 3d transition metal based alloys such as Fe-Al and, in particular, Fe-Ga may offer an attractive opportunity for formation of both high magnetostriction and high damping capacity.

## References

[1] Coulomb, C.A. Recherches Theoretiques et Experimentales sur la Force de Torsion et sur l'elasticite Des Fils Metals;Academie Royale des Sciences: Paris, France, 1784 (in French).

[2] Hooke, Robert. Lectures De Potentia Restitutiva, or of Springs, Explaining the Power of Springing Bodies. Oxford, 1678.

[3] Zener, C. Elasticity and anelasticity of metals; University of Chicago Press: Chicago, Illinois, 1948.

[4] Mason, W. P. Physical Acoustics and Properties of Solids; Van Nostrand: Princeton, 1958.

[5] Krishtal, M. A., Pigusov, Yu. V., Golovin, S. A. Internal friction in metals and alloys (in Russian); Metallurgizdat: Moscow, 1964.

[6]     Nowick, A. S., Berry, B. S. Anelastic relaxation in crystalline solids; Academic Press, New York, 1972.

[7]     De Batist, R. Internal friction of structural defects in crystalline solids; North Holland Publ Comp: Amsterdam, 1972.

[8]     Postnikov, V.S. Internal Friction in Metals; Metallurgiya: Moscow, 1974.

[9]     Lakes, R.S. Viscoelastic solids; CRC Press: Boca Raton, 1999.

[10]    Schaller, R., Fantozzi, G., Gremaud, G. (Eds.) Mechanical spectroscopy Q–1 2001 with applications to materials science; Trans Tech Publications: Switzerland, 2001.

[11]    Blanter, M.S., Golovin, I.S., Neuhäuser, H., Sinning, H.-R. Internal Friction in Metallic Materials. A Handbook; Springer: 2007.

[12]    Ngai K.L. Relaxation and Diffusion in Complex Systems. Springer, 2011. https://doi.org/10.1007/978-1-4419-7649-9

[13]    Kohlrausch R. Theorie des elektrischen Rückstandes in der Leidener Flasche; R. Phys. Chem., 91 (1854) 179-214.

[14]    Smithells, K. J. Metals Reference Book: Elastic Properties and Damping Capacity; 5th Ed.; London, Boston, 1976, 975-1006.

[15]    Blanter, M.S., Pigusov, Yu.V. (Eds.) Internal friction method in physical metallurgy researches (in Russian); Metallurgiya: Moscow, 1991.

[16]    Blanter M.S., Golovin I.S. Internal Friction. In Encyclopedia of Iron, Steel and Their Alloys. Edited by Rafael Colás, George E. Totten. Taylor and Francis: New York, 1852-1870

[17]    Snoek, J.L. Effect of small quantities of carbon and nitrogen on the elastic and plastic properties of iron. Physica VIII 1941, 8(7), 711–733. https://doi.org/10.1016/S0031-8914(41)90517-7

[18]    Koiwa, M. Theory of the Snoek effect in ternary b.c.c. alloys I. General theory, Philos. Mag. 24 (1971) 81-106, 107-122. https://doi.org/10.1080/14786437108216426

[19]    Rosin, K.M., Finkelshtein, B.N. Study of phase transformations by internal friction method. Dok Akad Nauk SSSR. 91(4) (1953) 811–814 (in Russian).

[20]    Kê, T.S. Experimental evidence of the viscous behavior of grain boundaries in metals, Phys. Rev. 71, 533-546 (1947). https://doi.org/10.1103/PhysRev.71.533

[21]    Zener, C. Theory of the elasticity of polycrystals with viscous grain boundaries, Phys. Rev. 60, 12, 906-908 (1941). https://doi.org/10.1103/PhysRev.60.906

[22]    Shi, Y.; Cui, P.; Kong, Q.P.; Jiang, W.B.; Winning, M. Internal friction peak in bicrystals with different misorientations, Phys. Rev. B 71, 6, 060101(R) (2005).

[23]    Benoit, W. 'Thermodynamics of 2nd and 3rd order transitions' in Schaller R., Fantozzi G., Gremaud G. (Eds.) Mechanical Spectroscopy Q-1 2001 with Applications to Materials Science. Trans Tech Publications, Switzerland. 2001, 341-360.

[24]    Fantozzi, G. 'Ferroelectricity' in Schaller R., Fantozzi G., Gremaud G. (Eds.) Mechanical Spectroscopy Q-1 2001 with Applications to Materials Science. Trans Tech Publications, Switzerland. 2001, 361-381.

[25]    Scheil, E., Müller, J. Die Dompfung mechanischer Schwingungen wdhrend der Martensitbildung in Eisen-Nickel-Legierungen (1956) Arch Eisenhutt 27: 801-805. https://doi.org/10.1002/srin.195602984

[26]    San Juan, J.; Perez-Saez, R.P. 'Transitory effects' in Schaller R., Fantozzi G., Gremaud G. (Eds.) Mechanical Spectroscopy Q-1 2001 with Applications to Materials Science. Trans Tech Publications, Switzerland. 2001, 416-437.

[27]    Delorme, J.F., Gobin, P.F. Metaux-Corrosion-Ind 1973, 573:185–222

[28]    James, D. W. High damping metals for engineering applications. Mater Sci Eng 1969, 4, 1-8. https://doi.org/10.1016/0025-5416(69)90033-0

[29]    Golovin, I. S. Damping Mechanisms in High Damping Materials. Key Engineering Materials 2006, 319, 225-230. https://doi.org/10.4028/www.scientific.net/KEM.319.225

[30]    Cochardt, A., "Magnetoelastic Internal Friction". In "Magnetic Properties of Metals and Alloys". Pergamon Press, NY , pp. 328-363, (1958).

[31]    Smith, G.W.; Birchak, J.R. Effect of Internal Stress Distribution on Magnetomechanical Damping. J. Appl. Phys. 39 (1968) 2311-2315.

Chapter 2

# Materials and Methods

**Contents**

### 2.1 Materials

Several groups of materials are discussed in this book. Some of them are "true" intermetallic compounds of $A_3B$-type (I), while the rest of them are either alloys on the basis of these compounds (II) or alloys with compositions in which the amount of alloying elements is not sufficient for forming of an intermetallic compound (III) in iron but they are helpful for better understanding of the physical origin of acting anelastic mechanisms:

    I) $Fe_3Al$, $Fe_3Ga$, $Fe_3Ge$, $Fe_3Si$;

    II) $Fe_3(Al,Me)$ and $(Fe,Me)_3Al$, $Fe_3(Ga,Me)$ and $(Fe,Me)_3Ga$ alloys, where Me stays for Cr, Si, Ge, Mn, Co, Ni, Tb etc.;

    III) Fe-Al, Fe-Ga, Fe-Ge, Fe-Cr, Fe-Si based disordered and ordered alloys.

The second (II) group consists mainly of the following ternary Fe-Al and Fe-Ga based alloys:

    1) strongly carbide forming elements like Ti, Nb, Zr, Ta in Fe-Al were added to trap carbon, some of these elements enhance the yield strength by producing Laves phases;

    2) elements providing increased ductility and strength in Fe-Al like Cr, which is also a carbide forming element;

    3) not strongly carbides forming elements which enhance the tendency of ordering in Fe-Al (Si);

4) elements (Mn, Ge, Co) which do not strongly affect the interstitial carbon concentration in solid solution and may affect Fe-Al ordering in different ways;

5) Tb which increases magnetostriction of Fe-Ga alloys;

6) Al, Co, Ni which change phase diagram and influence on properties of Fe-Ga alloys.

*Fe-Al:* In the Fe-rich corner of the Fe–Al phase diagram there are three phases, namely disordered *b.c.c.* A2 (*Im3m*) phase, *b.c.c.*-born ordered $D0_3$ (*Fm3m*) and B2 (*Pm3m*) phases. In the $D0_3$ ordered ($Fe_3Al$) structure there are three types of sublattices with sites denoted as 4a, 4b, and 8c in Wyckoff's notation. In the binary $D0_3$ structure the 4a positions are occupied by Al atoms, while 4b and 8c positions are occupied by Fe atoms; in B2 lattice Fe and Al atoms are distributed on 4a and 4b sites. In ternary alloys each or all of these sublattices can be occupied by Me atoms. In case of the $D0_3$ structure the 4b sublattice, i.e. Fe antisite positions, are preferably occupied by Me atoms; this structure type is called $L2_1$. At higher temperatures or higher Al concentration the $D0_3$-to-B2 transformation takes place. The Curie temperature $T_C$ for a given alloy is different in the A2, B2, and $D0_3$ phases.

Alloying Fe-Al by a "third" element (Me) changes the parameters of order and the temperatures of phase transformation: Si improves the $D0_3$ order and thus increases the transition temperature $T_O$ from $D0_3$ to B2. Nb and Zr have the same effect on $T_O$, and in addition they produce Laves phases. Co stabilises the B2 phase. Cr and Mn change $T_O$ only slightly up to certain concentrations, adding Ge in significant amounts produces $L2_1$ order and at lower concentrations increases $T_O$. Interstitial atoms, which in the case of Fe–Al-based alloys are mainly carbon atoms, occupy octahedral interstice positions in *b.c.c.* iron and also in derivatives from the *b.c.c.* lattice, i.e. in $D0_3$, B2 or $L2_1$ lattices. The interaction between $Me_1$-$Me_2$ and C-Me atoms and its influence on the parameters of anelasticity (e.g., the Snoek-type and Zener relaxation) is one of the main subjects of this work. The third (III) group of studied alloys with amount of alloying elements from 3 to 13 at.% was used to clarify these effects in disordered alloys.

*Fe-Ge, Fe-Si:* Different types of order take place in Fe-Ge alloys with Ge < 40 at.%: namely there are the B2 ($\alpha_1$), $D0_3$ ($\alpha_2$), and $L1_2$ ($\epsilon'$) cubic phases, and the $D0_{19}$ ($\epsilon$) and $B8_1$ ($\beta$) hexagonal phases. Several Fe-Si alloys with B2 and $D0_3$ order are also studied. Tendency to ordering increases in these systems from left to right: Fe-Al, Fe-Ge, Fe-Si [32].

*Fe-Ga:* In the last decade, the Fe-Ga 'Galfenol' alloys have been the focus of a lot of attention due to their good mechanical properties and high magnetostriction in low saturation magnetic fields [33] and have the potential to be widely used in magnetostrictive actuators and sensor devices. The long standing question is how the addition of non-magnetic Ga atoms into *b.c.c.* iron can create nearly ten-fold increase in magnetostriction. It is believed that the increase in magnetostriction for Fe-Ga alloys is because of the preferential (110) Ga-Ga pairing or nano sized clustering in the disordered body centered cubic (*b.c.c.*) structure. It has been found that dependence of the tetragonal magnetostriction with Ga content, '$\lambda_{100}$ *vs* %Ga' exhibits two peaks near 19 at.% of $265 \times 10^{-6}$ and 27 at.% of $235 \times 10^{-6}$ [34]. However, the nature of the tetragonal nanodomains and how they are coupled to the matrix remains unclear. Doping of Fe-Ga alloys by rare earth elements additionally increases their magnetostriction. Though having a lower magnetostriction as compared with Terfenol-D, Fe-Ga alloys have the advantage of a very low coercive field of ~100 Oe together with low hysteresis and high temperature stability, an desirable property for most applications.

Recently, Prof. X. Ren at Xi'an Jiaotong University has demonstrated that a large magnetostriction (800 ppm) triggered by a low saturation field (0.8 kOe) can be achieved in Fe-Pd alloys in ferromagnetic strain glass composition range [Phys. Rev. Lett. 119, 125701 (2017)], which bears an unique two-phase nanostructure with nanosized frozen strain domains (tetragonal) embedded in the ferromagnetic matrix. These nanosized strain nanodomains (so-called 'strain glass' state) are "alive" and able to grow under external magnetic field stimuli, resulting in a desired high magnetostriction at low magnetic field, which fulfils the virgin region of available magnetostrictive materials. High magnetostriction of Fe-Ga may be related to a ferromagnetic strain glass and it may become possible to design high-performance magnetostrictive alloys using the properties of ferromagnetic strain glass.

Fe-Ga alloys, being known by their high magnetostriction, may also be good candidates for damping applications. Magnetic domains with sharp 90° and 180° domain walls, which are typical for soft-magnetic materials with positive anisotropy energy [35], are the source for high damping properties even at low amplitudes of vibrations [36]. According to the Smith and Birchak theory maximal damping at amplitude dependent internal friction curves is proportional to magnetostriction constant ($\lambda$) of ferromagnetic materials (eq. (11)). The values of $\lambda$ in Fe-Al alloys are significantly higher than those in $\alpha$-Fe (e.g., for Fe-16Al: $\lambda_{100} = 85 \times 10^{-6}$, $\lambda_{111} = -2 \times 10^{-6}$, $\lambda_{polycr} \approx 35 \times 10^{-6}$) but, in turn, they are lower than those in Fe-Ga alloys (e.g., for Fe-17Ga: $\lambda_{100} = 207 \times 10\text{-}6$, $\lambda_{111} = -12 \times 10\text{-}6$, $\lambda_{polycr} \approx 76 \times 10\text{-}6$) [36]. Compositions with around 19%Ga have reported strains of up to 400 ppm along <100> direction with low saturation fields of several hundred Oersteds [34].

Fe-Ga alloys are known by their ordering of Ga atoms in *b.c.c.* iron: the type of order depends on temperature and concentration of Ga atoms [37, 38]. Ordering decreases mobility of magnetic domain walls and dislocations, leads to low ductility [39] and decreases damping capacity [40, 41]. In contrast to systematic study of magnetic characteristics, very little is yet known on internal friction in Fe-Ga alloys.

Taking into account that the Fe-Ga alloys structure is rather similar to Fe-Al alloys: the ground-state electron configurations of nonmagnetic elements Al and Ga are $1s^2 2s^2 2p^6 3s^2 3p^1$ and $1s^2 2s^2 2p^6 3s^2 3p^6 3d^{10} 4s^2 4p^1$, it is not surprising that Fe-Ga and Fe-Al alloys have several common features of magnetic and anelastic effects. Both Al and Ga can enhance the magnetostriction of *b.c.c.* iron giving a magnetoelastic contribution to damping capacity of these alloys. Atomic ordering in both systems decreases damping due to additional pinning of magnetic domain walls at antiphase boundaries [42].

The formation of the equilibrium face-centered cubic (*f.c.c.*)-based L1$_2$ ordered phase below ~620°C (according to the *equilibrium* Fe-Ga diagrams) is slow and in most cases the decomposition and ordering develop in accordance with the *metastable* phase diagram and at room temperature it is presented by a mixture of A2 and D0$_3$ phases [43]. Formation of transient nonstoichiometric nanosized B2 phase may take place prior to nearly equilibrium D0$_3$ phase [44]. Quenching suppresses the formation of D0$_3$ structure in favor of a disordered supersaturated A2 structure and creates freeze-in vacancies. Similar situation takes place with Fe-Ge alloys – they need long term annealing to reach their equilibrium state. The situation with ordering type and kinetics in ternary Fe-Ga-Al alloys is not clear enough yet. Based on the metastable Fe-Ga diagram, D0$_3$ ordering may also take place in the ternary composition.

Most of the Fe-Al, Fe-Al-(Si, Ge, Cr, Co, Mn), Fe-Si, Fe-Ge alloys were produced at the Institute for Physics of Condensed Matter at the Technical University of Braunschweig by induction melting of 99.98% Fe, 99.999% Al and addition of third elements under argon atmosphere in a vacuum induction furnace (the group of Prof. H. Neuhäuser). A small amount of carbon (typically from 0.005 to 0.04 %) was added in order to attain a sufficiently developed Snoek peak to study carbon related effects in some of these alloys. That is why it would be more correct to speak about multi-component Fe-Al-(Me)-C alloys, however, for simplicity the indication of carbon is omitted everywhere below.

Several Fe-(10-12)%Al alloys were produced at the Moscow Central Research Institute of Iron and Steel Industry (Dr. I.B. Chudakov). The Fe-Al-Me (Me = Ti, Nb, Zr, Ta) and most of Fe-Al-Si alloys were produced at the Max-Planck-Institut für Eisenforschung GmbH, Düsseldorf (Dr. F. Stein): details of the compositions, structures and production procedure are given in the cited papers. The nominal compositions of all alloys used in this research and the compositions were determined by inductively coupled plasma optical emission spectroscopy (ICPOES analysis) and for carbon by combustion to $CO_2$.

The amount of nitrogen in our alloys was at least two orders of magnitude lower than that of carbon, and therefore neglected in further consideration.

Fe-Ga, Fe-Ga-Tb, Fe-Ga-Al alloys were produced at NUST MISiS, Moscow, by directional solidification using pure Fe and Ga by induction melting under protection of high-purity inert argon gas using an Indutherm MC–20V mini furnace (Dr. A. Churyumov, A. Bazlov). Using energy dispersive spectroscopy, the chemical compositions of the cast buttons were measured with accuracy within ±0.2%. Several Fe-20%(Ga+Al) alloys were provided by Prof. J. Zhu (Beijing), and FeCo-26Ga alloy by Prof. S. Jen (Taipei).

## 2.2 Methods

### 2.2.1 Mechanical spectroscopy

Mechanical spectroscopy is a research method of measurements of anelastic and elastic properties of solids as a function of frequency, amplitude or temperature. With ascending frequency, the experimental techniques of mechanical spectroscopy are generally divided into four groups [11]: quasi-static, subresonance, resonance and wave-propagation methods. While measuring different quantities and response functions, they all can be used to determine internal friction of metallic materials, preferably under vacuum to avoid unwanted aerodynamic losses. More details about the following techniques can be found in the books [4-10]. In this section we introduce briefly some equipment used in our study.

*"High" frequency resonance free-decay tests:*

1.  Damping $Q^{-1}$ and modulus change (shear modulus $G$ or Young's modulus $E$, both are $\sim f\,2$, where $f$ is the resonance frequency of flexural vibrations) were measured in three vibrating reed set-ups at the Institut für Werkstoffe and Institut für Physik der Kondensierten Materie at Technical University of Braunschweig (Germany) designed by Profs. H.-R. Sinning and H. Neuhäuser. These vibrating-reed set-ups, working in the range from 200 Hz to 3 kHz, all in the temperature range from 80 to 900 K, have either optical detection of the vibrations (Fig. 2.1) or electrostatic registration;

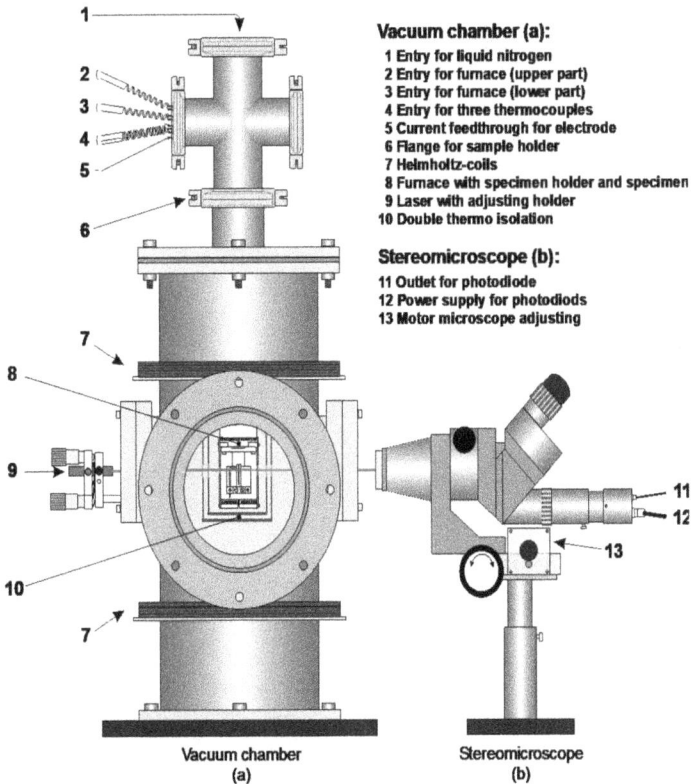

**Vacuum chamber (a):**
1 Entry for liquid nitrogen
2 Entry for furnace (upper part)
3 Entry for furnace (lower part)
4 Entry for three thermocouples
5 Current feedthrough for electrode
6 Flange for sample holder
7 Helmholtz-coils
8 Furnace with specimen holder and specimen
9 Laser with adjusting holder
10 Double thermo isolation

**Stereomicroscope (b):**
11 Outlet for photodiode
12 Power supply for photodiods
13 Motor microscope adjusting

Vacuum chamber
(a)

Stereomicroscope
(b)

*Fig. 2.1 Vibrating reed (Technical University of Braunschweig, setup is developed by Prof. H. Neuhäuser).*

*"Low" frequency subresonance forced vibrations:*

2) Inverted torsion pendula using free decay vibrations in the frequency range from 0.5 to 3 Hz at Tula State University (Russia) (Fig. 2.2) carried out in saturated magnetic field 2400 A/m in the temperature range below 650 °C;

3) Dynamical Mechanical Analyzers Q800 TA Instruments (tests were carried out at NUST MISiS Moscow with assistance of Drs. V. and M. Zadorozhnyy), at UIB Spain with assistance of Mr. J. Cifre, and DMA Metravib (Mr. R. Barbin). The DMA Q800, based on a combined motor and transducer design, uses advanced, non-

contact, linear motors to control stress and measures strain with an optical encoder. The sample is clamped at both ends and either flexed in the middle (dual cantilever) – Fig. 2.3.

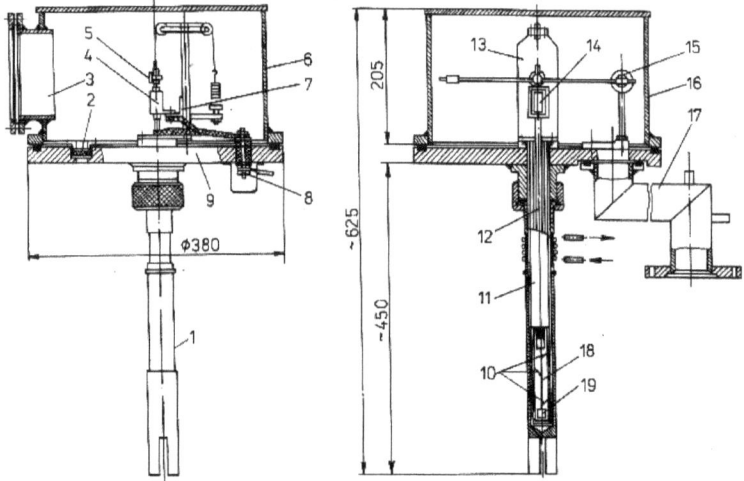

Fig. 2.2 Inverted torsion pendulum PKM-TPI (Tula State University, constructed by S. Arkhangelsky and S. Golovin [257]): 1 - lower cover, 2,8 – let and outlet, 3 – window from quartz glass, 4, 14 – damper, 5 – mirror, 6, 16 upper cover "bell", 7 – insulator, 9 – bottom of bell, 10 – thermocouple, 11 – internal pipe, 12 – torsion mandrel, 13 – stand, 15 – magnetos, 17 – to vacuum system, 18 - specimen, 19 – jaw.

Dynamical mechanical analyser DMA Q800 TA Instruments operates in the frequency range between 0.01 and 200 Hz but it has not sufficient stiffness for tests of metalic materials at frequencies above 30 Hz. Measurements at frequencies below 0.1 Hz are time consuming: one experimental point is measured by averaging tanφ over seven cycles of loading, and already at f = 0.1 Hz it takes more than 1 min. This is critical for temperature dependent tests with a typical heating and cooling rate of 1 K/min because at low frequencies density of experimental points decreases drastically.

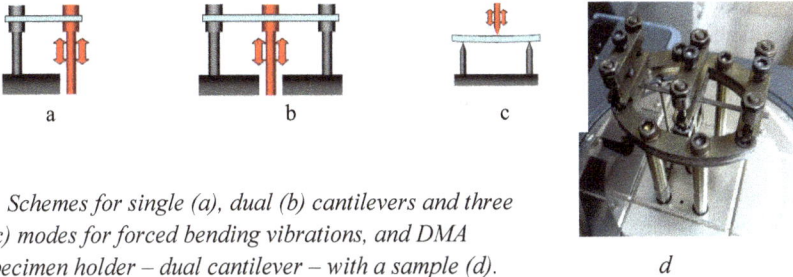

*Fig. 2.3 Schemes for single (a), dual (b) cantilevers and three points (c) modes for forced bending vibrations, and DMA Q800 specimen holder – dual cantilever – with a sample (d).*

d

Testing of anelastic properties was carried out using forced *bending* vibrations in the temperature range from 0 to 600°C. The internal friction was measured as $\tan\varphi$ ($=Q^{-1}$), where $\varphi$ is the phase lag between applied cyclic stress, $\sigma$, and resulting strain, $\varepsilon$, $\sigma = \sigma_0\cos(t)$ and $\varepsilon = \varepsilon_0\cos(\omega t+\varphi)$; $\varphi = 2\pi f$, and $\varepsilon_0 = 5\times10^{-5}$. These measurements were conducted as a function of temperature between 0 and 600°C, using forced bending vibrations in the range between 0.1 and 30 Hz with $\varepsilon_0 = 5\times10^{-5}$ with a heating and cooling rate of 1 or 2 K/min (temperature dependent IF and elastic modulus: TDIF and TDEM). Also measurements were conducted as a function of amplitude between $4\times10^{-6}$ and $2\times10^{-3}$ at the frequency of 3 Hz (amplitude dependent IF and elastic modulus: ADIF and ADEM, correspondingly).

Acoem DMA 25/50 Metravib (Fig. 2.4, Lyon) was used mainly for three points bending tests to study the influence of static (pre-loading) stress on damping capacity of the samples.

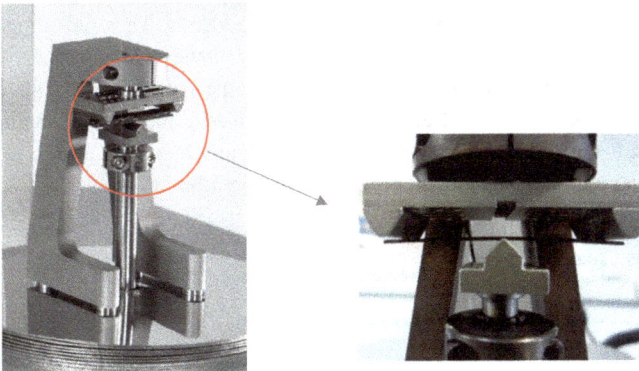

*Fig. 2.4 Acoem DMA 25/50 Metravib specimen holder for three points bending.*

The difference in 'how damping is calculated in the case of free decay and forced vibrations' determines, as a rule of thumb, some advantages and disadvantages of forced and free-decay vibrations methods. For free decay method the equipment is relatively simple and the number of vibrations in a fixed interval of amplitudes (e.g., between $A_i$ and $A_{i+n}$, Fig. 2.5a) is counted to calculate logarithmic decrement $\delta = \ln(A_i/A_{i+1}) = (n^{-1} \times \ln(A_i/A_{i+n}))$. The more vibrations take place in a chosen interval of amplitudes, the higher is the accuracy: low damping materials can be measured with high precision. In contrast, in many cases high damping materials cannot be measured well in free decay mode without special software. Nowadays, the use of modern electronics helps to overcome this problem.

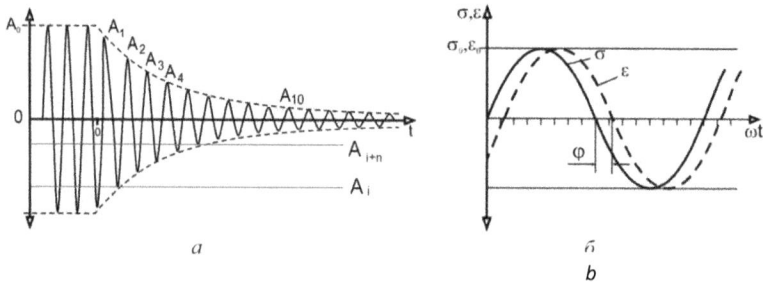

*Fig. 2.5. Schemes for free-decay (a) and forced (b) vibrations.*

For forced vibrations the phase lag ($\varphi$) between applied cyclic stress ($\sigma$) and resulting strain ($\varepsilon$) is measured (Fig. 2.5b). If this value is low, the accuracy of measurements decreases and scattering of experimental points increases. Thus, this mode is better for materials with relatively high damping. A disadvantage of many DMA setups, which were originally designed for soft materials, is insufficient stiffness. This often leads to problems with estimations of absolute values of elastic moduli of metallic materials if samples are not thin enough. For that reason in this book we do not always provide absolute values for elastic moduli but present data in arbitrary units.

For relatively low damping the following relation between different afore-mentioned methods is usually accepted

$$Q^{-1} = tg\varphi = \delta/\pi = \Delta W/2\pi W, \tag{2.1}$$

while this ratio does not work well if damping is high.

For specimen characterisation the following methods were applied: TEM (transmission electron microscopy), SEM (scanning electron microscopy), LOM (light optical microscopy), XRD (X-ray diffraction), DSC (differential scanning calorimetry), positron annihilation spectroscopy (PAS), vibrating sample magnetometry (VSM). Neutron diffraction has, complementary to the afore-mentioned methods, several advantages for the study of Fe-based alloys. First, a neutron diffraction experiment could easily be arranged in an in situ and real-time mode with a sufficient temporal resolution. Second, high penetration of thermal neutrons allows analyzing bulk effects unlike XRD. Finally, in neutron experiments, the data on the magnetic structure of a material can be obtained together with the structural information.

### 2.2.2 Neutron diffraction

Neutron diffraction studies of atomic and magnetic structures of alloys began in the late $1940^{th}$, and their efficiency is well known for a long time (e.g., [45, 46]). Substantially, the efficiency of neutron diffraction studies is determined by remarkable differences between interactions of neutrons and X-rays or synchrotron radiation with the matter. In the case of metallic alloys the ability to distinguish between neighboring elements in the periodic table and to analyze magnetic structure of the materials, a large penetration depth and a large cross-section of the neutron beam are particularly important. These features of neutron diffraction allow to study the volumetric properties of alloys and to avoid some uncertainties associated with alloys non-uniform and coarse-crystalline structure. Neutron diffraction can be effectively used to study microstructural characteristics of metallic alloys: the average dimensions of coherently scattering atomic domains and the static fluctuations of the unit cell parameters (microdeformations), as well as to measure the residual or dynamic internal stresses in the bulk materials to the depth of several centimeters.

For diffraction studies neutrons with energies of about 0.025 eV (slow neutrons) are used. Their speed is around 2200 m/s, and the de Broglie wavelength $\lambda$ is about 1.8 Å. Scattering of slow neutrons is usually characterized by the coherent scattering length $b$, which is a constant, independent of the neutron's energy. The scattering lengths and other important characteristics of interaction of slow neutrons with isotopes of elements are listed in [47]. For some transition metals the coherent scattering lengths are: $b_{Al} = 3.45$, $b_{Ti} = 3.44$, $b_V = -0.38$, $b_{Cr} = 3.64$, $b_{Mn} = -3.73$, $b_{Fe} = 9.45$, $b_{Co} = 2.49$, $b_{Ni} = 10.30$, $b_{Cu} = 7.72$, all in Fermi units ($=10^{-13}$ cm). The difference between these values is significant, that provides the necessary contrast to refine the occupancies of the atoms in the unit cell in the case of ordered alloys.

Additionally to the nuclear interaction there exists a magnetic dipole interaction between the neutron magnetic moment and the magnetic moment of an atom, which is used for analysis of crystal magnetic structure. The magnetic coherent scattering length depends on neutron energy and for magnetic moment of 1 $\mu_B$ it is equal to 5.39 fm, which is comparable with the majority of the nuclear scattering lengths.

To the disadvantages of the neutron diffraction technique one can refer small, in comparison with the X-ray and synchrotron radiation, intensity of neutron beams and the possible activation of samples. Often, even after a brief exposure to the neutron beam, a sample has to endure for several days or even weeks for the disappearance of the induced activity.

**Neutron structural experiment.** The experimental setup for neutron diffraction studies is known as *neutron diffractometer*. Its main units are the neutron source, systems forming a neutron beam on a crystal, detectors, electronics for data acquisition and control of the experiment. They also include devices specifying the conditions under which a crystal is studied: temperature, pressure, magnetic field, etc. At present, two types of research neutron sources are used for obtaining beams of slow neutrons: *steady state* and *pulsed source.* The former include numerous nuclear reactors, the second type is formed with, still a few, sources based on proton accelerators (Spallation Neutron Sources, SNS). Research work is going on also at two "non-standard" sources: SINQ at PSI (Switzerland), which is the SNS source, but not pulsed, and IBR-2 at JINR (Dubna, Russia), which is a nuclear pulsed reactor. A detailed review of the situation with the research neutron sources around the world and in Russia, as well as analysis of their European development strategy are described in the review [48].

In accordance with these two types of neutron sources there are two alternative, complementary, and at the same time more or less equivalent modes of neutron diffraction experiments. At a steady state source neutrons with fixed (monochromatic) wavelength are used and the neutron diffraction pattern is a function of the scattering angle. Neutron diffraction studies, examples of which are given in this book, were performed at pulsed neutron source with a polychromatic ("white") beam using the time-of-flight (TOF) technique. Correspondingly, the neutron diffraction pattern is a function of the neutron energy or wavelength. This type of instruments is usually called as TOF-diffractometer. Its general arrangement is shown in Fig. 2.7: neutrons from pulsed source after moderation down to thermal energy fly some path (usually between 10 to 30 meters), and then they are scattered by the sample and registered by detector or by several detectors placed at fixed scattering angles. The diffraction patterns after successive source pulses are added together to accumulate the necessary statistics.

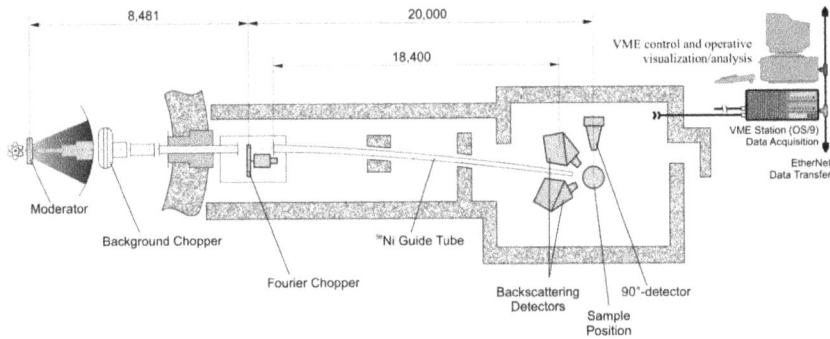

*Fig. 2.7 The layout of TOF-diffractometer HRFD at the IBR-2 pulsed reactor. The flight path between moderator and sample position is equal to 28.48 m. At HRFD a special device – fast Fourier chopper – is used for improving $\Delta d/d$ resolution. Neutron diffraction data included in this book were obtained with this instrument.*

For neutron sources with a long ($\Delta t_0 > 300$ µs) pulse (IBR-2 reactor in Dubna and a new European spallation source, which is under construction in Sweden) the pulse must be reduced to an acceptable width, which is achieved by using fast Fourier or Fermi choppers. Fourier-chopper is a multislit disk (rotor) with a regular arrangement of slits of the same width. A large number of slits dramatically increase its transmission (25% instead of 1%), but requires specific, correlation, technique for analysis of diffraction data.

The world's first Fourier diffractometer HRFD (High-Resolution-Fourier-Diffractometer) at a pulsed neutron source was constructed at the IBR-2 reactor in Dubna. The HRFD combines both high neutron flux at the sample, $\sim 10^7$ n/cm2/s, and high resolution, which is close to $\Delta d/d = 0.001$ over a wide range of $d$-spacings. One of the HRFD features is the possibility to switch easily between high-resolution and high-intensity modes of operation. In the second case the correlation analysis is not used and diffraction data can be measured in a very short (about 1 minute) intervals, which is important for *in-situ* studies of changes in both crystal structure and microstructure.

In the case of neutron diffraction on polycrystalline materials for the intensity of the diffraction scattering the following general expression can be written:

$$I(d_i) = C \sum_{hkl} j_{hkl} |F_{hkl}|^2 Q_L \cdot \varphi(d_{hkl} - d_i) + I_b(d_i), \tag{2.2}$$

where C is normalized constant, $I_b$ is intensity of the incoherent background, $d$ is interplanar distance or $d$-spacing, $Q_L$ is the Lorentz factor, $j$ is the multiplicity factor, $\varphi$ is a function for describing of the diffraction peak profile. The value $F_{hkl}$ in (2.2) is called as *structure factor* and depends on atomic coordinates in the following way:

$$F_{hkl} = \sum_j b_j \exp[2\pi i(hx_j + ky_j + lz_j)] \cdot \exp[-W_j(d_{hkl})], \qquad (2.3)$$

where $b_j$ is the coherent scattering length, $(x, y, z)_j$ are coordinates of the $j$-th atom in the unit cell. The second exponent in eq.(2.3) called as thermal factor or *Debye-Waller factor*. The $W_j$ values are characteristics of thermal motion of particular atom, they depend on $(hkl)$ and on crystal temperature.

These equations are the basis for determining the atomic structure of the polycrystalline substance, and for verifying their modification under composition or environmental conditions changes, and in the course of phase transitions. Currently, as the conventional mathematics for diffraction data processing the Rietveld method is used. A detailed description of this method, which includes the features of refinement of neutron data, is containing in the Ref. [49]. There are several software packages, in which the Rietveld method is implemented for the treatment of neutron diffraction data. The most popular are the packages FullProf [50], GSAS [51] and MRIA [52]; the last was specially developed for the analysis of the data obtained with a Fourier diffractometer. The example of the Rietveld refinement pattern for Fe-27Ga alloy is shown in Fig. 2.8.

The special feature of the structural analysis of ordered alloys is the need to determine the structure factors of *superstructure* diffraction peaks, intensities of which depend on the difference between the coherent scattering lengths of ordered atoms. For instance, in the $Fe_3Al$ alloy, as it was already mentioned, three possible states with varying degrees of ordering are possible: completely disordered (A2 phase), partially ordered (B2 phase) and fully ordered ($D0_3$ phase) states. In this case the main (fundamental) peaks have the Miller indices $(h, k, l)$, so that they are even and their sum $(h + k + l)$ is a multiple of 4. For them the structural factor is proportional to the sum of the coherent scattering lengths, i.e., $F_{hkl} \sim (3b_{Fe} + b_{Al})$. Besides, there are two types of superstructural peaks, with even $(h, k, l)$ and odd value $(h + k + l)/2$, allowed to phase $D0_3$ and B2, and with all odd $(h, k, l)$, allowed for $D0_3$ only. For both of them $F \sim (b_{Fe} - b_{Al})$, and since the peak intensity $I$ is proportional to $|F_{hkl}|^2$, they are much weaker than the main peaks.

The difference in the intensities of the main and superstructural diffraction peaks will be even greater for the compositions based on iron, in which the second atom has the $b$-

value closer to $b_{Fe}$. For instance, in the $Fe_3Ga$ compound, $b_{Ga}=7.29$ fm and, respectively, the intensity of superlattice peaks will be several times weaker than in $Fe_3Al$.

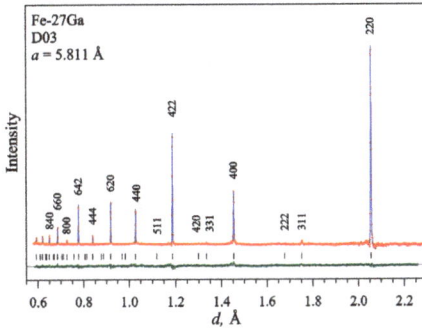

*Fig. 2.8   Diffraction pattern of Fe-27Ga powder sample measured at HRFD and processed by the Rietveld method. Experimental points, calculated and difference lines are shown. The vertical bars indicate calculated peak positions (D0$_3$ phase, a = 5.811 Å). The difference between experimental and calculated intensities fluctuates around zero line, which confirms correctness of the structural model. Small negative deeps at the peaks profiles are associated with special features of the high-resolution correlation mode of data acquisition at HRFD.*

**Real-time neutron diffraction.** Advanced neutron diffractometers, acting on high-flux neutron sources, can provide the counting rate of about 105 n/s and more. So high intensity allows to study structural phase transitions in alloys in real-time mode. This means that some transition processes in crystalline material can be traced by using neutron diffraction with the measuring time, $t_s$, for a whole diffraction pattern (temporal resolution), which is much smaller than the characteristic time, $\tau$, of the investigating processes.

In diffraction experiment the available characteristic time scale depends on a sufficient level of statistics, which can be gathered in the time $t_s$. This value depends on the neutron flux at the sample position, solid angle of a detector system, and scattering capacity of a sample. At HRFD instrument when the iron-based alloys with the size of 4×4×8 mm were studied the measuring time $t_s = 1$ minute was used that was absolutely enough for tracing phase transformations during heating or cooling. Examples of such studies are given in the relevant sections of the book, and in reviews [48, 53].

27

**Neutron diffraction analysis of alloy microstructure.** In diffraction experiment several characteristics of the alloy microstructure can be analyzed: the average size of coherently scattering domains (CSD), crystallographic texture, the internal micro and macrostresses. Microstructure of the alloy materials is often the subject of a special study, because it directly affects their specific physical or physicochemical properties, which are especially important in engineering and technological applications.

To determine the crystallographic texture it is necessary to conduct an experiment on the dedicated texture diffractometer to perform rather complicated data analysis [54]. The stress state arising from static or dynamic deformation of the material, which modifies the lattice parameters in volumes covering a large number of crystallites, called as macrostress, leads to a shift of the diffraction peaks from the nominal positions. Microstresses arise in the volume of individual crystallites as a consequence of imperfect internal structure, i.e. the presence of point or volume defects, the composition inhomogeneities, block structure, etc. With the arising stress field in this case usually only peak broadening occurs without shifting their position.

To analyze the effects that lead to an increase in the width of the diffraction peaks the classic Williamson-Hall and Warren-Averbuch methods and more advanced WPPM (Whole Powder Pattern Modeling) one are used. Their current description is contained in a collection of articles [55]. For the joint estimation of both size ($L$) and strain ($\varepsilon = \Delta d/d$) effects in an alloy the most simple and clear Williamson-Hall method can be used as the first step. For the particular case of the TOF-diffractometer the dependence of the peak width $\Delta d$ (integral or FWHM) on d-spacing can be expressed as:

$$(\Delta d)^2 = C_1 + (C_2 + C_3) \cdot d^2 + C_4 \cdot d^4, \tag{2.4}$$

where the constants $C_1$ and $C_2$ are related with resolution function of the diffractometer, $C_3 = (2\varepsilon)^2$, $C_4 = (1/L)^2$. This relation implies that $(\Delta d)^2$ is a linear function of $d^2$ in the absence of size effect and is a parabolic function for finite CSD size. If the widths of the diffraction peaks are measured in a large enough d-spacing range and resolution function is taken into account then the value of both $\varepsilon$ and $L$ can be determined reliably. For example, the HRFD resolution allows determining microstrains in crystallites at the level of $\varepsilon \approx 0.0008$ and more, and the average CSD size at the level of $L \approx 350$ nm and less.

For alloys an anisotropic broadening of the diffraction peaks is often observed, which in crystals with cubic lattice in the frame of Williamson-Hall method can be accounted, as shown in [56], by introducing into the equation (2.4) dislocation-related anisotropy factor $\Gamma$:

$$(\Delta d)^2 = C_1 + C_2 \cdot d^2 + C_3(A - B\Gamma) \cdot d^2 + C_4 \cdot d^4, \tag{2.5}$$

where $\Gamma = (h^2 k^2 + h^2 l^2 + k^2 l^2)/(h^2 + k^2 + l^2)^2$, and A and B are constants dependent on the density and the relative content of the edge and screw dislocations. If the size effect is absent ($C_4 = 0$) and the resolution function is subtracted, the formula (2.5) is written as:

$$(\Delta d_\varepsilon)^2 = (K_1 - K_2\Gamma) \cdot d^2, \tag{2.6}$$

where new constants $K_1$ and $K_2$ are introduced instead of products $C_3 \cdot A$ and $C_3 \cdot B$. By using the peaks with different $\Gamma$ factors, which range from 0 to 1/3, and selecting coefficients $K_1$ and $K_2$, it is possible to minimize the deviation of $(\Delta d_\varepsilon)^2$ values from the linear dependence on $d^2$. If the peak broadening is really connected with the deformation anisotropy, the $[(\Delta d_\varepsilon)^2 + K_2\Gamma d^2]$ value as a function of $d^2$ should be approximated by a straight line passing through the origin of coordinates. How it revealed itself in the experiment and calculations is shown in the relevant sections of the book.

Different mechanical spectroscopy techniques, shortly discussed above, with different methods of structural investigations, in particular with *in-situ* neutron diffraction, allowed us to study anelastic phenomena listed in the introduction with respect to temperature, frequency and amplitude of vibrations.

## References

[32]  Glezer, A.M.; Molotilov, B. V. Ordering and Deformation of Iron Alloys (in Russian). Metallurgiya: Moscow, 1984.

[33]  Clark, A.E.; Restorff, J.B.; Wun-Fogle, M.; Lograsso, T.A.; Schlagel D.L. Magnetostrictive properties of body-centered cubic Fe-Ga and Fe-Ga-Al alloys. IEEE Trans. Magn., 36, No. 5, (2000) 3238. https://doi.org/10.1109/20.908752

[34]  Clark, A.E.; Hathaway, K.B.; Wun-Fogle, M.; Restorff, J.B.; Lograsso, T.A.; Keppens, V.M.; Petculescu, G.; Taylor, R.A. Extraordinary magnetoelasticity and lattice softening in bcc Fe-Ga alloys. J. Appl. Phys., 93, No. 10, (2003) 8621. https://doi.org/10.1063/1.1540130

[35]  Mudivarthi, C.; Na, S.-M.; Schaefer, R.; Laver, M.; Wuttig, M.; Flatau, A.B. Magnetic domain observations in Fe-Ga alloys. J.Magnetism and Magnetic Materials, 322 (2010) 2023–2026. https://doi.org/10.1016/j.jmmm.2010.01.027

[36] Ishimoto, M.; Numakura, H.; Wuttig, M. Magnetoelastic damping in Fe-Ga solid-solution alloys. MSE(A), 442 (2006) 195-198. https://doi.org/10.1016/j.msea.2006.02.215

[37] Kubaschewski, O.: Iron-Binary Phase Diagrams, Springer-Verlag, Berlin 1982.

[38] Massalski, T.B.: Binary Alloy Phase Diagrams, 2nd ed., ASM, OH (1990) 1740.

[39] Li, J.; Gao, X.; Zhu, J. et al. Ductility enhancement and magnetostriction of polycrystalline Fe-Ga based alloys. J. All. Comp., 484 (2009) 203-206. https://doi.org/10.1016/j.jallcom.2009.03.008

[40] Golovin, I.S.; Palacheva, V.V.; Zadorozhnyy, V.Yu.; Zhu, J.; Jiang, H.; Cirfe, J.; Lograsso, T.A. Influence of composition and heat treatment on damping and magnetostrictive properties of Fe-18%(Ga+Al) alloys. Acta Materialia, 78 (2014) 93–102. https://doi.org/10.1016/j.actamat.2014.05.044

[41] Golovin, I.S. Anelasticity of Fe-Ga based alloys. Materials and Design, 88 (2015) 577-587. https://doi.org/10.1016/j.matdes.2015.08.160

[42] Udovenko, V.A.; Tishaev, S.I.; Chudakov, I.B. Magnetic domain structure and damping in alloys of the FeAl system. Physics Doklady, 38 (1993) 168-176.

[43] Ikeda O, Kainuma R, Ohinuma I et al. Phase equilibria and stability of ordered b.c.c. phases in the Fe-rich portion of the Fe-Ga system. J. All. Comp., 347 (2002) 198. https://doi.org/10.1016/S0925-8388(02)00791-0

[44] Boisse, J.; Zapolsky, H.; Khachaturyan, A.G. Atomic-scale modeling of nanostructure formation in Fe–Ga alloys with giant magnetostriction: Cascade ordering and decomposition. Acta Mat., 59 (2011) 2656–2668. https://doi.org/10.1016/j.actamat.2011.01.002

[45] Shull, C.G.; Siegel, S. "Neutron Diffraction Studies of Order-Disorder in Alloys" Phys. Rev. 75 (1949) 1008. https://doi.org/10.1103/PhysRev.75.1008

[46] Bacon, G.E.; Cowlam, N. "A study of some alloys of gamma-manganese by neutron diffraction" J. Phys. C: Solid State Phys. 3 (1970) 675. https://doi.org/10.1088/0022-3719/3/3/023

[47] Sears, V.F. "Neutron scattering length and cross sections" Neutron News, 3 (1992) 26. https://doi.org/10.1080/10448639208218770

[48] Aksenov, V.L.; Balagurov, A.M. "Neutron diffraction on pulsed sources" Physics-Uspekhi 59 (2016) 279-303. https://doi.org/10.3367/UFNe.0186.201603e.0293

[49]  "The Rietveld Method", Ed. R.A. Young, Oxford Univ. Press, 1993.

[50]  Rodríguez-Carvajal, J. "Recent advances in magnetic structure determination by neutron powder diffraction" Physica B, 192 (1993) 55. https://doi.org/10.1016/0921-4526(93)90108-I

[51]  Larson, A.C.; Von Dreele, R.B. "General structure analysis system" Los Alamos National Laboratory, Report No. LAUR. 86–748 (2004).

[52]  Zlokazov, V.B.; Chernyshev, V.V. "MRIA - a program for a full profile analysis of powder multiphase neutron-diffraction time-of-flight (direct and Fourier) spectra" J. Appl. Cryst., 25 (1992) 447. https://doi.org/10.1107/S0021889891013122

[53]  Aksenov, V.L.; Balagurov, A.M. "Neutron time-of-flight diffractometry" Physics-Uspekhi, 39 (1996) 897-924. https://doi.org/10.1070/PU1996v039n09ABEH000169

[54]  Popa, N. "Microstructural Properties: Texture and Macrostress Effects" Chapter 12 in "Powder Diffraction. Theory and Practice" Ed-s R.E. Dinnebier, S.J.L. Billinge, RSC Publishing, Cambridge, 2008.

[55]  "Modern Diffraction Methods" Eds E.J. Mittemeijer, U. Welzel, Wiley-VCH Verlag & Co. KGaA, 2013.

[56]  Ungár, T. "Dislocation model of strain anisotropy" Powder Diffraction 23 (2008) 125-132. https://doi.org/10.1154/1.2918549

Chapter 3

# Structure and Anelasticity of Fe-Al, Fe-Ga and Fe-Ge Intermetallic Compounds

## Contents

### 3.1    Equilibrium and non-equilibrium structures of Fe$_3$Me - type alloys

*Equilibrium* structures of Fe-Al, Fe-Ga and Fe-Ge alloys are known from existing phase diagrams. They are presented in Fig. 3.1(a,b,c), correspondingly. Al, Ga and Ge are α-stabilizing alloying elements in iron. Also they are known to be not carbide forming elements in iron apart from some special cases, e.g. κ-carbides in Fe-Al.

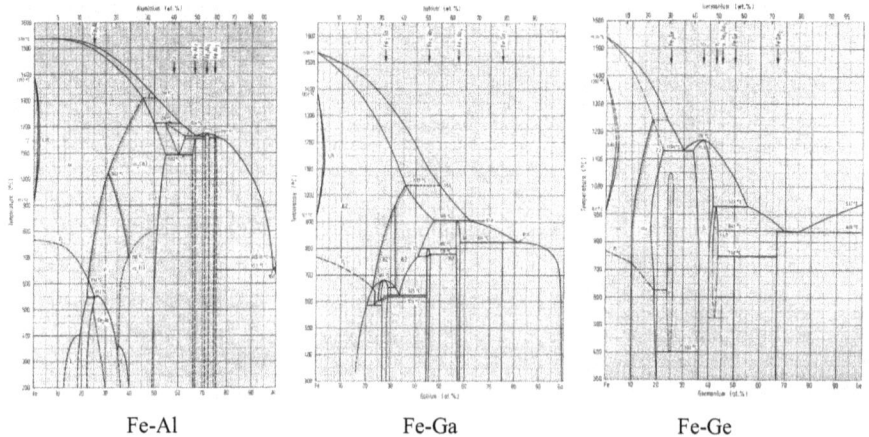

|  Fe-Al  |  Fe-Ga  |  Fe-Ge  |

*Fig. 3.1 Equilibrium phase diagrams Fe-Al, Fe-Ga, and Fe-Ge [37].*

According to the phase diagrams and our studies, the following phases are observed in these alloys at different temperatures in the concentration range below roughly 30 at.% (we will not consider alloys with higher concentration of alloying elements):

- the **A1** (for all three systems) has an γ-Fe-type structure with Fe and Me = Al, Ga, Ge atoms randomly distributed, sp. gr. $Fm3m$;
- the **A2** (for all three systems) has an α-Fe-type structure with Fe and Me = Al, Ga, Ge atoms randomly distributed, sp. gr. $Im3m$;
- the **A3** (for Fe-Ge) has Mg-type structure with randomly distributed atoms, sp. gr. $P6_3/mmc$
- the **B2** (for all three alloys) has a CsCl-type structure with Fe and Me atoms partially ordered, sp. gr. $Pm3m$;
- the **D0₃** (for all three alloys) has a $BiF_3$-type structure with Fe and Me atoms partially ordered, sp. gr. $Fm3m$;
- the **D0₁₉** (for Fe-Ga and Fe-Ge alloys) has a $MgCd_3$-type structure with Fe and Ga/Ge atoms partially ordered, sp. gr. $P6_3/mmc$;
- the **L1₂** (for Fe-Ga and Fe-Ge alloys) has a $Cu_3Au$-type structure with Fe and Ga/Ge atoms partially ordered, sp. gr. $Pm3m$;
- the **B8₁** (for Fe-Ge alloy) has a NiAs-type structure with Fe and Ge atoms partially ordered, sp. gr. $P6_3/mmc$.

The structure of alloys is not always described by a equilibrium phase diagram. In most cases treatment of the samples, such as casting, quenching in different media (water, air, oil etc.), leads to significant structural deviations from equilibrium diagrams and it plays a very important role in practice. Heating, cooling or isothermal annealing of Fe-based alloys leads to significant changes in their structure and, consequently, in their properties.

For example, the formation of the equilibrium face-centered cubic (*f.c.c.*)-based L1₂ ordered phase in Fe-Ga alloys below 620°C (according to the *equilibrium* diagrams [37]) is rather slow and in most cases the decomposition and ordering develop in accordance with the *metastable* phase diagram (Fig. 3.2) and at room temperature it is presented by mixture of A2 and D0₃ phases [43]. Formation of transient nonstoichiometric nanosized B2 phase may take place prior to nearly equilibrium D0₃ phase [44]. Quenching suppresses the formation of D0₃ structure in favor of a disordered supersaturated A2 structure and creates freeze-in vacancies. This mainly relates to a sample surface, which its bulk structure is at least partly ordered. Low temperature annealing may produce a two-phase mixture of A2 + D0₃ even on the surface of alloys with 15-25% Ga.

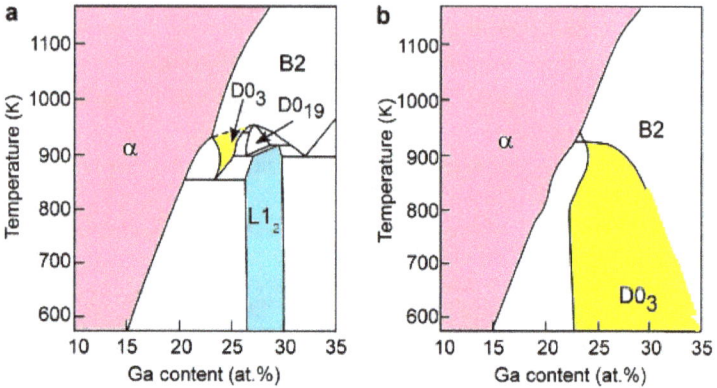

*Fig. 3.2. Sketches of equilibrium (a) [37] and metastable (b) [43] Fe-Ga diagrams.*

Similar situation - fixation of high temperature phases at room temperature by quenching may take place in other studied systems. An example of structures of as cast alloys (Fig. 3.3) and phase transitions (Fig. 3.4) towards phase equilibrium in binary systems for three different compositions roughly corresponding to $Fe_3Me$ (where Me is Al, Ga or Ge) compounds are considered below for instant heating and cooling [57].

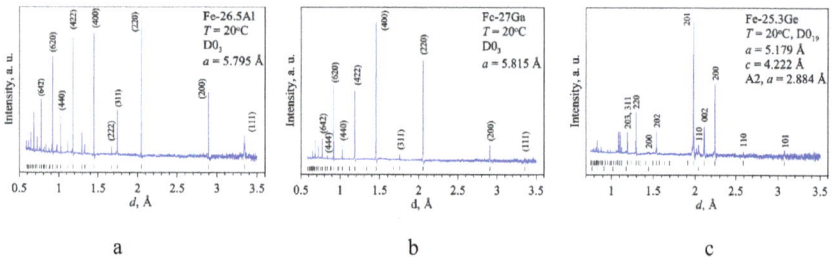

*Fig. 3.3 Examples of the neutron diffraction patterns of the Fe-26.5Al (a), Fe-27.4Ga (b), and Fe-25.3Ge (c) samples measured at room temperature in high-resolution mode ($\Delta d/d \approx 0.001$). The vertical bars indicate calculated peak positions of the $D0_3$ phase at (a) and (b), and $D0_{19} + A2$ phases at (c). The Miller indices of several first peaks are shown.*

The initial state after direct solidification for the studied bulk samples is characterized by $DO_3$ structure in Fe-26.5Al (Fig. 3.3a), by $DO_3$ structure in Fe-27.0Ga (Fig. 3.3b), and by $DO_{19}$ (with a little amount of A2) structure in Fe-25.3Ge (Fig. 3.3c).

In-situ 3D neutron diffraction patterns for the same Fe-26.5Al, Fe-27.4Ga, and Fe-25.3Ge alloys at continuous heating with 2K/min are presented in Fig. 3.4. Upon continuous heating, several phase transitions occur according to the neutron diffraction (Fig. 3.4 a, c, e):

- In **Fe-26.5Al** alloy only second order transitions – "ordering (of B2 and $DO_3$ types) $\leftrightarrow$ disordering" take place. At heating, the recorded sequence of phase transitions for *water quenched* from 900°C sample is: B2 $\rightarrow$ $DO_3$ $\rightarrow$ B2 $\rightarrow$ A2 (Fig. 3.4a), whereas in *cast* alloy it is: $DO_3$ $\rightarrow$ B2 $\rightarrow$ A2 (this result is not presented in the Figure), i.e. as cast alloy gets higher ordering due to slower cooling compared with water quenching. The $DO_3$ and B2 ordered phases appear and disappear as a result of Al atoms ordering in *b.c.c.* Fe-Al (A2) solid solution. Ordering needs some time: that is why the appearance of the $DO_3$ and B2 ordered phases depends on the heating or cooling rate. According to neutron diffraction, water quenching from 900°C prevents $DO_3$ ordering but does not allow to preserve A2 structure, and the sample has the B2 structure at room temperature. Further analysis of the experimental data shows that a major part of the sample at room temperature is partially B2 ordered, whereas the completely ordered $DO_3$ phase fills only small (hundreds of angstroms) domains [58]. The characteristic size of the $DO_3$ phase clusters increases after slow heating and cooling by almost a factor of four. This result follows from significantly different dependences of the widths of the main and superstructure diffraction peaks on the interplanar distance.

- In **Fe-27.0Ga**, Fe-27.4Ga and Fe-27.8Ga alloys (denoted below and in the figures as Fe-27Ga), the recorded sequence of first order phase transitions upon heating is $DO_3$ $\rightarrow$ $L1_2$ $\rightarrow$ $DO_{19}$ $\rightarrow$ B2/A2 (depending on the heating rate: A2 for 2 K/min or B2 for 1 K/min). A second order A2 $\rightarrow$ $DO_3$ transition precedes the sequence of first order transitions if only a surface of the directly solidified or quenched samples is analysed by XRD [59] and position annihilation [60] studies. The range of the $DO_3$ $\rightarrow$ $L1_2$ transition depends on heating rate (Fig. 3.5) as shown by in situ neutron diffraction.

- In **Fe-25Ge** alloy, the non-equilibrium ordered $DO_{19}$ phase recorded at room temperature turns to the A3 disordered phase at heating (Fig. 3.4e). In contrast with the equilibrium phase diagram we did not record neutron diffraction from mixture of A2 + B8₁ phases below 400 °C and L1₂ phase above 400 °C at the chosen test conditions. Some additional results of the XRD study of Fe-27Ge sample are discussed below.

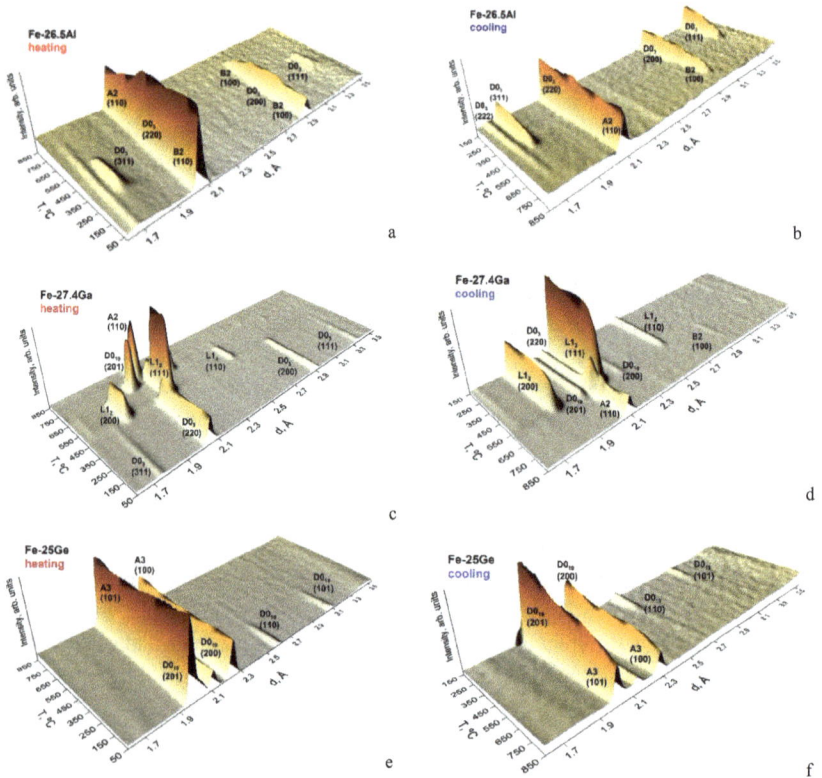

*Fig. 3.4 The 3D visualization of in-situ neutron diffraction pattern evolution for Fe-26.5Al (a, b), Fe-27.4Ga (c, d), and Fe-25.3Ge (e, f) samples at continuous heating (a, c, d) and cooling from 850°C (b, d, f).*

Upon continuous cooling from 850 °C the following phase transitions occur according to neutron diffraction (Fig. 3.4b, d, f):

- In Fe-27Al alloy the A2 → B2 → D0₃ transitions take place (Fig. 3.4b);

- In Fe-27Ga alloy the recorded sequence of first order phase transitions upon cooling is A2(B2) → D0$_{19}$ (limited amount) + L1$_2$ (dominating phase) and at the same time a small amount of remaining A2 phase gets D0$_3$ ordering (Fig. 3.4d);
- In Fe-25Ge alloy the A3 → D0$_{19}$ transition takes place (Fig. 3.4f).

Prior to the discussion of relaxation phenomena in these alloys let us summarise structural and magnetic transitions in the studied alloys in Fig. 3.6. Phase transitions in the studied alloys at heating from RT to 850 °C according to in-situ neutron diffraction are presented in Fig. 3.6 (a,b,c) and their magnetisation in Fig. 3.6 (d,e,f).

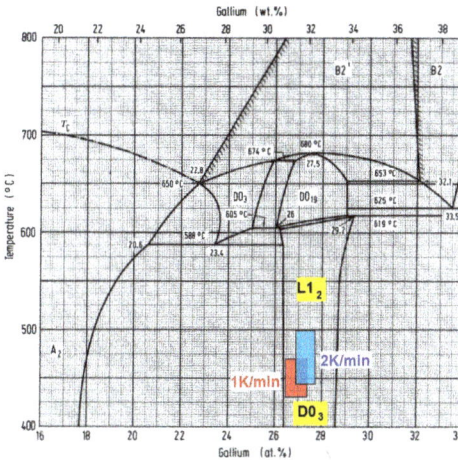

*Fig. 3.5. Fe-27Ga-type composition: from metastable D0$_3$ to stable L1$_2$ phase.*

The magnetisation curve in Fe-Al is affected by the D0$_3$ to B2 transition. This results in two "Curie points" known since 1935 [61]. The data extrapolation from the D0$_3$ range gives the Curie temperature in the vicinity of 530 °C (T$_{C\ (D03)}$), whereas the Curie point at higher temperatures represents some contribution of high-temperature phase, in agreement with the Fe-Al phase diagram. A decrease of the magnetisation in Fe-27Ga sample heated up to about 475 °C (Fig. 3.4d) is related to a decrease of the ferromagnetism of the D0$_3$ phase [62]. However, direct observation of this dependence down up to zero magnetisation (i.e. to the Curie temperature) is not possible as the D0$_3$ phase itself transforms to the magnetic L1$_2$ phase. An increase in magnetisation above

37

~475 °C is due to formation the the $L1_2$ phase. Finally, magnetisation decreases above 610 °C due to the $L1_2 \rightarrow D0_{19}$ transition.

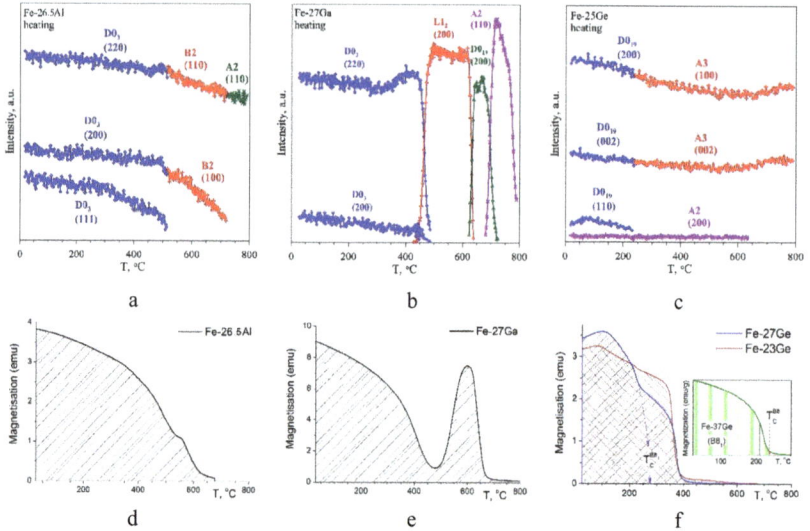

*Fig. 3.6 Phase transitions at heating with a rate of ~2.2 K/min shown as intensity changes of the characteristic diffraction peaks (a, b, c) and magnetisation (d, e, f) in the Fe-26.5Al (a, d), Fe-27Ga (b, e) and Fe-27Ge (c, f) samples. The Miller indices of the peaks are indicated nearby the curves (a, b, c). In Fig. 3.4f magnetisation curves for Fe-23Ga and Fe-37Ga are shown additionally.*

The results of the *in-situ* neutron diffraction studies for the Fe-25.3Ge composition are significantly different to the equilibrium Fe-Ga phase diagram in contrast with Fe-Al and Fe-Ga alloys reported earlier, which fits well to the corresponding diagrams. In contrast with the equilibrium phase diagram, we recorded neither mixture of $A2 + B8_1$ phases below 400 °C nor $L1_2$ phase above 400 °C in the chosen test conditions. The only $D0_{19} \leftrightarrow A3$ transition is proven. Our results evidences that time needed to reach the equilibrium state in the Fe-Ge alloy is significantly longer than that for Fe-Al and Fe-Ga.

A two-steps decrease of the magnetisation in the Fe-27Ge sample is presented in Fig. 3.6f. The Curie temperature extrapolated to $T_C \approx 260$ °C most probably corresponds to the $B8_1$ phase. The presence of the $B8_1$ phase in the structure of the Fe-27Ge alloy is confirmed by XRD study in contrast with Fe-25Ge where we have not resolved it. The $\beta$ phase (B8$_1$) is thermodynamically stable in a wide range of temperatures. The Curie temperatures for the $\beta$ phase measured in Fe-37Ge (see inset in Fig. 3.6f) is $T_C = 230\,°C$ [6363]. The step-like decrease of magnetisation around 370-390 °C is associated with the Curie temperature of the D0$_{19}$ hexagonal phase, which is a dominating phase in the structure of the studied alloy. Yamamoto [64] suggests the Curie temperature for the $\varepsilon$ phase to be $T_C = 382\,°C$, and Drijver [65] $T_C = 367\,°C$ (both for Fe-25Ge, D0$_{19}$ structure), which reasonably complies with that temperature from Fig. 3.6f.

Thus, using *in-situ* neutron scattering we have defined ranges of phase transitions in the bulk samples of Fe-Al, Fe-Ga and Fe-Ge alloys with 25-27 at.% of alloying elements under continuous heating and cooling. In case of Fe-Al and Fe-Ga alloys, our data in general support the existing phase diagrams; whereas heating and cooling tests of Fe-25Ge do not demonstrate low temperature transitions suggested by equilibrium diagram. The contribution of these phase transitions to anelastic effects is discussed in the next chapters. Influence of additional alloying on these transitions is also discussed below.

### 3.2 Neutron diffraction study of atomic ordering in Fe$_3$Al-type alloys

The ordering of atoms in a stoichiometric alloy with the composition of Fe$_3$Al (Fe-25Al) is known since the 1930s. Already in [66] it was established that during the crystallization of Fe$_3$Al, an ordered structure, now called D0$_3$, is formed. In this structure the atoms are distributed over 16 possible Wyckoff positions in the cubic space group $Fm3m$ ($a \approx 5.834$ Å), and, as follows from an X-ray analysis [66], Al atoms occupy mainly (~92%) four of these positions. If Fe$_3$Al alloy is heated, the second-order structural transitions occur: D0$_3 \rightarrow$ B2 $\rightarrow$ A2, where B2 is a partially ordered ($Pm3m$), and A2 completely disordered ($Im3m$) phase.

The presence of atomic ordering determines the appearance in the diffraction patterns of superstructure ($s$) peaks the intensities of which are:

$$I_s \sim |F_s|^2 \sim \xi^2 \cdot (b_{Fe} - b_{Al})^2, \tag{3.1}$$

where $F$ represents structure factor, $b_{Fe}$ and $b_{Al}$ are atomic form-factors for X-ray diffraction or coherent scattering lengths for neutron diffraction, and $\xi$ is a degree of Fe and Al ordering over crystallographic positions ($0 \leq \xi \leq 1$). Intensities of fundamental ($f$) peaks in diffraction patterns are:

$$I_f \sim |F_f|^2 \sim (3b_{Fe} + b_{Al})^2. \tag{3.2}$$

The so called contrast R = $F_s/F_f$ at complete ordering is 0.19 for neutrons ($b_{Fe}$ = 9.45, $b_{Al}$ = 3.45, in $10^{-13}$ cm units), which is quite acceptable for reliable registration of superstructure peaks.

An important feature of the Bragg diffraction patterns of $Fe_3Al$ is the specific modulation of the widths of the diffraction peaks: they are large for superstructure peaks and relatively small for fundamental reflections. For the first time, a similar phenomenon was observed for the $Cu_3Au$ compound [67], in which atomic ordering also occurs under certain conditions. This effect was explained using the model of antiphase domains (APDs) [68], i.e. the neighboring regions with the same arrangement of atoms in the unit cell but shifted to each other by a part of lattice translation vector. The waves diffracted by these ordered domains are coherent and shifted in phase, which determined their name.

The APDs model suggests that the width of the fundamental peaks is determined by the finite, coherently scattering domain size $L_{coh}$ (CSDS) regardless of their partition onto antiphase domains, whereas the width of superstructure peaks is determined by the average size of individual antiphase domains. Accordingly, if the domains with ordered atomic structures are connected by a coherent antiphase boundary, APB, and their dimensions are smaller than $L_{coh}$, the superstructure peaks will be broadened because of the size effect. This model became generally accepted, and it was expounded and used in numerous articles, monographs and textbooks (e.g., paper [69] and books [70, 71]).

The appearance of antiphase domains in ordered alloys has been justified theoretically and confirmed by various observations and calculations. Most of these observations were made using transmission electron microscopy (TEM). For example, in Ref. [72] this method was used to determine the critical behavior of antiphase boundaries in $Fe_3Al$ at a temperature transition between the B2 and $D0_3$ structures. From the obtained dependence of $l_c$ on $(T_O - T)/T_O$, where $T_O$ is the disorder-order transition temperature, it follows that the B2 $\leftrightarrow$ $D0_3$ reaction is the second-order transition, and that, near $T_O$ (at $T_O - T <$ 10 K), the correlation length can exceed 100 Å (up to 400 Å). Diffraction of a coherent X-ray beam was used along with electron microscopy to visualize APB in $Fe_{65}Al_{35}$ in the B2 structural phase [73]. On the one hand, this method confirmed, on the whole, the results of the TEM observations, but at the same time allowed the registration of the intensity modulation in the superstructure peak (001) predicted by the APDs model.

A detailed analysis of the TEM data regarding the formation of the APDs and the motion of the APB in the course of the B2 $\rightarrow$ $D0_3$ transition was performed in a review [74]. The performed 1D and 2D simulations based on the minimization of the thermodynamical energy coincided well with the experimental data. According to these data, if the initial B2 phase is isothermally annealed, the disperse D0$_3$ domains of a small size appear, grow and merge if there is no phase shift between them or they form APDs if this shift takes place. In parallel with an increase in size and a change in the shape of domains, the degree of atomic ordering in them gradually increases. With prolonged annealing, the B2 regions gradually disappear, becoming APB between ordered domains. The final stage of the process is the alignment of the APB due to the disappearance of small domains that are absorbed by domains with the opposite ordering phase.

A question about universality of this model of the APD formation still remains open. The attainability of the final equilibrium state significantly depends upon the conditions of sample preparation, the duration and temperature of its annealing, and its specific microstructure. Often, TEM analysis is conducted with samples whose state is far from the equilibrium. In the literature, one can easily find a lot of TEM pictures that clearly do not correspond to the state with the APDs separated by narrow flat APB. A typical example of such an intermediate state is shown in Fig. 3.7 [75], where the regions of the ordered D0$_3$ phase of an arbitrary size and shape are clearly visible. Interlayers between them also have an arbitrary thickness and shape. Nevertheless, even in this case it is customary to describe the regions of the ordered phase as antiphase domains, although if the interlayers between adjacent regions are large, it is impossible to say whether their atomic structures are in phase or antiphase. The concept of a matrix of a disordered or less ordered phase with dispersed clusters of an ordered phase is more appropriate for the situation shown in Fig. 3.7.

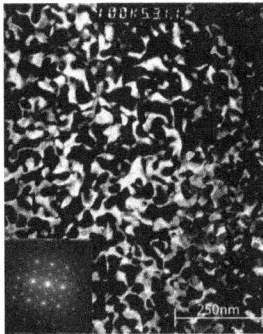

*Fig. 3.7 Dark-field transmission electron microscope image of the structure of Fe-26Al in annealed at 300 °C state: D0$_3$ domains (in white) immerged into B2 matrix (in black) [75].*

Interpretation of nonstandard profiles of the diffraction peaks in ordered alloys based on the APDs model suggested in [68] is not universal. Already its author, A.J.C. Wilson, pointed out that the model based on the 'foam structure' proposed by W.L. Bragg in [76] can also explain the observed effects. Moreover, the Bragg's model looks more realistic than Wilson's assumption of antiphase domains of equal volume with flat boundaries. In some papers, both assumptions were used to explain the widths of the fundamental and superlattice peaks. For example, both models were used to interpret XRD diffraction spectra in differently quenched Fe-Al alloys with different Al content [77]. In Fe-Al alloys with ~20 at. % Al nano-sized dispersed inclusions of the ordered $D0_3$ phase with dimensions from 20 to 50 Å in the disordered $A2$ matrix were observed by the TEM method [78, 79]. A similar structure (nano-dispersed $D0_3$ clusters in $A2$ matrix) was identified for the Fe-19Ga composition using high-resolution transmission electron microscopy (HRTEM) [80]. Moreover, in that paper, this structure was used to explain the enhanced magnetostriction in Galfenol alloys. The existence of such nano-sized regions also follows from calculations based on the Atomic Density Field (ADF) theory [81] carried out in [82] for the Fe-(15-20)Ga compositions.

Thus, the model of dispersed clusters should be revisited using experimental results obtained with the help of modern techniques. This is especially true for analysis of non-equilibrium states including those gradually formed during isothermal annealing. Our first results [58], acquired at a high-resolution neutron diffractometer for as-cast polycrystalline Fe-27Al, suggest that the main volume of the sample is the partially ordered B2 phase. Whereas the nearly completely ordered $D0_3$ phase (the terms "partially" and "completely" ordered phases are clarified below) of hundreds Å in size is coherently incorporated into the B2 matrix. Similar result was obtained for an almost equilibrium state of the sample after slow cooling from 950 °C, with the only difference being in the characteristic size of the $D0_3$ clusters which increased from 230 Å to 880 Å.

A slightly off-stoichiometric alloy with a nominal composition Fe-26.5Al was produced using directional solidification with 99.99% pure Fe and Al by induction melting under protective high-purity argon gas. A single crystal of the same composition Fe-26.5Al of 4×4×4 mm size was cut out from a large single crystal 4×8×100 mm, which had been grown in Institute for Metal Physics and Nuclear Solid State Physics, Technical University of Braunschweig (Prof. Neuhäuser's group). These two samples are denoted as Fe-27Al-poly and Fe-27Al-single below.

In the stoichiometric $Fe_3Al$ composition, the completely ordered - if the thermal disorder is not taken into account – the $D0_3$ phase is an equilibrium phase at room temperature with iron atoms in positions $(4a)$ - $(0,0,0)$ and $(8c)$ - $(1/4,1/4,1/4)$ and Al in position $(4b)$ - $(1/2,1/2,1/2)$ of the $Fm3m$ space group. In our slightly non-stoichiometric $Fe_{2.77}Al$ (Fe-

26.5Al) composition, both Fe and Al can be found in one of the positions. For example, we can assume that in (8c) there are 7.76Fe + 0.24Al, whereas (4a) and (4b) are completely filled with Fe and Al, respectively. In the cubic cell of the B2 phase, the atoms are in the positions (1a) - (0,0,0) and (1b) - (1/2,1/2,1/2). A complete B2 ordering is possible only for the $Fe_{0.5}Al_{0.5}$ composition. For alloys $Fe_{1-\delta}Al_\delta$ ($\delta < 0.5$) the maximal ordering takes place if Fe atoms completely fill the (1a) position, while the (1b) position is filled by both Fe or Al atoms with probabilities (1 - 2$\delta$) and 2$\delta$, respectively, i.e. 0.47 and 0.53 for $\delta = 0.265$. In the disordered A2 phase, the Fe and Al atoms occupy both the (0,0,0) and (1/2,1/2,1/2) positions with probabilities of 0.735 and 0.265. Thus, we will refer below to $D0_3$ as nearly ordered, B2 as partially ordered, and A2 as disordered phases.

Neutron diffraction patterns were measured with HRFD instrument (see section 2.2.2.) With back-scattering detectors (2θ = 152°) the HRFD $d$-spacing range is (0.6 - 3.6) Å for high-resolution and up to 4.5 Å for high-intensity modes, thus all fundamental and superstructure diffraction peaks of Fe-27Al with low Miller indices can be measured. Examples of high-resolution diffraction patterns of Fe-27Al-poly and Fe-27Al-single are shown in Fig. 3.8.

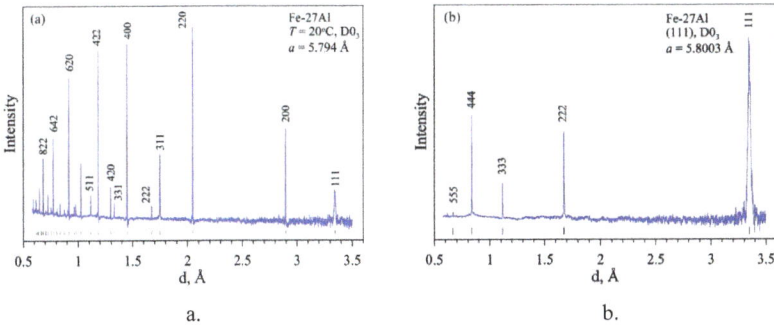

Fig. 3.8 Diffraction patterns of Fe-27Al-poly (a) and Fe-27Al-single crystals, (111) plane (b). Superstructure reflections are the peaks with odd Miller indices. The vertical bars indicate the calculated peak positions.

With the as-cast Fe-27Al polycrystalline sample a heating (up to 900 °C) and subsequent cooling (down to 20 °C) procedure with a constant rate of 2.2 K/min and $t_s$ = 1 min was carried out. Before heating and after cooling to room temperature, the high-resolution diffraction patterns were measured. Heating-and-cooling cycle took about 15 hours and led to the annealing of the sample and bringing its structure to the equilibrium state. Accordingly, the high-resolution data were obtained in two states: in as cast state followed by long-term natural aging (>10 years for single crystal), referred to as 'weakly non-equilibrium', and after heating and cooling with an instant rate of 2.2 K/min, referred to as 'nearly equilibrium'.

With a Fe-27Al-single sample, high-resolution diffraction patterns for the (111), (110), (311) and (511) planes were measured. Then, orders of reflections from the (111) plane were measured in real-time mode with $t_s = 1$ min in the temperature range from 20 to 85 °C with a heating rate of 2.2 K/min and then with fast cooling (~20 K/min).

As it was discussed in the section 2.2.2 for the particular case of the TOF-diffractometer the dependence of the peak width Δd (integral or FWHM) on d-spacing can be expressed as:

$$(\Delta d)^2 = C_1 + (C_2 + C_3) \cdot d^2 + C_4 \cdot d^4,$$ (3.3)

where the constants $C_1$ and $C_2$ are related with resolution function of the diffractometer, $C_3 = (2\varepsilon)^2$, $\varepsilon$ is a static variance of unit cell parameters $C_4 = (1/L_{coh})^2$. The $(\Delta d)^2$ on $d^2$ is dependence is linear if the size effect is absent (large coherently scattering domain size, $L_{coh} > 3500$ Å for HRFD) and parabolic otherwise. More details according this analysis can be found in [83].

A schematic representation of both models of atomic ordering is given in Fig. 3.9. Both patterns represent some intermediate non-equilibrium states. The principal difference between them is that the ordered D0$_3$ domains *are* or *are not* separated by a coherent antiphase boundary (APB).

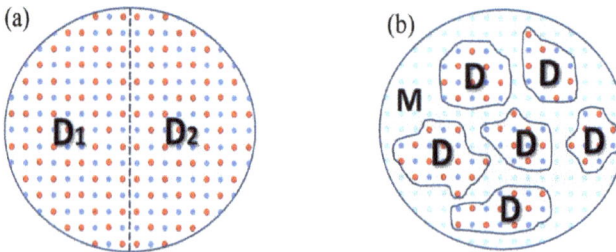

Fig. 3.9 (a) Pair of antiphase domains (D$_1$/D$_2$) with an ordered atomic structure. Straight lines denote coherent (plane) antiphase boundaries. (b) Dispersedly distributed ordered clusters in a disordered matrix.

For monochromatic beam diffraction by the APDs in $Fe_3Al$, the relation for the widths of $s$-peaks was obtained [77]. Rewriting it for the TOF-diffractometer, one can obtain:

$$\Delta d = d^2/L_D \cdot \beta_{hkl}, \qquad (3.4)$$

where $L_D$ is an average domain size. From the Scherrer equation and, correspondingly, from the conventional size contribution in (3.4), this formula is distinguished by the presence of an additional factor: $\beta_{hkl} = (h + k + l)/(h^2 + k^2 + l^2)^{1/2}$ (all indices are positive and odd integers), varying in the range from 1 to 1.73. In the contribution of domains size to the fundamental peaks, the factor $\beta_{hkl}$ is absent ($\beta_{hkl} = 1$), and instead of $L_D$, the average size of the coherently scattering domain size must be used .

Thus, in the simplest case of diffraction by two neighboring APDs with similar shape and size separated by a plain coherent APB, the contribution to the width of the fundamental peaks is $\Delta d_f \sim (d^2/2L_D)$, while for the superstructural peaks $\Delta d_s \sim (d^2/L_D) \cdot \beta_{hkl}$. For example the (111) peak is ~3.5 times broader than fundamental peaks.

From dispersed cluster structure, two types of diffraction peaks are formed: intense (fundamental) peaks, the width of which is determined by CSDS of the matrix, $\Delta d_f \sim (d^2/L_M)$, and relatively weak superstructure peaks, the width of which is determined by the average cluster size of the ordered phase, $\Delta d_s \sim (d^2/L_C)$. The obvious differences with the APDs model are that, in the case of dispersed clusters, there is no dependence of the width of $s$-peaks on the Miller indices and that the ratio between $\Delta d_f$ and $\Delta d_s$ can be arbitrary.

The neutron diffraction patterns measured from Fe-27Al-poly and Fe-27Al-single crystal samples in the initial state are fully indexed in accordance with the structure of the $D0_3$ phase ($Fm3m$, $a \approx 5.834$ Å). Upon heating and subsequent cooling, the $D0_3$, B2, and A2 phases are recorded (Fig. 3.10). In contrast with results for as quenched sample (Fig. 3.4a), the results presented in Fig. 3.10a correspond to slow cooled sample.

By analogy with the analysis carried out in Ref. [74], it can be assumed that dispersed clusters of the partially ordered phase B2 appear and grow up on cooling from the high-temperature disordered A2 phase. If this structural state is achieved, the A2 is the matrix which gives the fundamental (narrow) diffraction peaks and the B2 phase gives superstructure (with increased width) peaks. If further ordering takes place, the A2 structure transforms into B2 phase in which clusters of the $D0_3$ phase appears. Thus, the B2 becomes the matrix giving the $f$-peaks, while broadened $s$-peaks are from the $D0_3$ clusters.

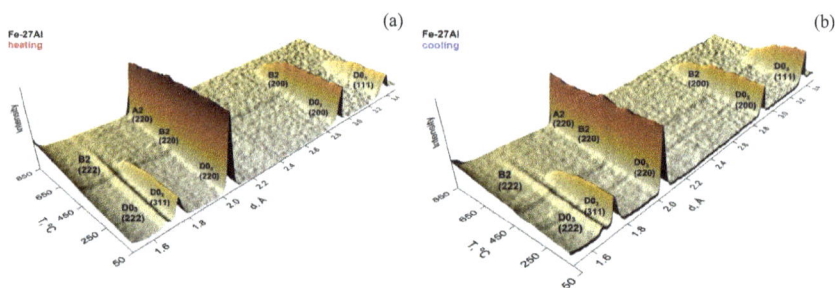

*Fig. 3.10 3D visualization of structural phase transitions in Fe-27Al-poly at heating (a) up to 850 °C with 2.2 K/min followed by cooling down to 50 °C (b). The peak indices are given for the $DO_3$ phase, which is characterized by (111) and (311) lines. In the B2 phase, these peaks disappear, at high temperatures, only (220) line of the A2 phase is visible.*

**High-resolution diffraction:** the $(\Delta d)^2$ over $d^2$ dependences (Williamson-Hall plot), measured in the initial state of the Fe-27Al polycrystal alloy and after its heating-cooling are shown in Fig. 3.11. The widths of peaks with the even Miller indices, allowed for the A2 and B2 structural phases, fit the linear dependence well. On the contrary, the widths of superstructure peaks with odd Miller indices, which are allowed only in the $DO_3$ phase, differ sharply in magnitude. More over, the $(\Delta d)^2$ on $d^2$ dependences for them are parabolic. Another important fact is that the widths of the both *f*- and *s*-peaks becames narrower by a factor of 2.5 after heating-cooling cycle. The $(\Delta d)^2$ on $d^2$ dependences are smooth for both the fundamental and superstructural peaks. In contrast, an attempt to take into account the $\beta_{hkl}$ factor for the widths of *s*-peaks leads to a sharp violation of their regular dependence on $d^2$, i.e. the APB model does not explain our experimental data.

Thus, the diffraction data for the as-cast Fe-27Al polycrystalline sample in both states - "weakly non-equilibrium" (before heating) and "nearly equilibrium" (after heating and cooling) - are in good agreement with the model of mesoscopic $DO_3$ clusters randomly distributed in the B2 matrix. The average $DO_3$ cluster sizes are $L_C = 230 \pm 20$ Å before heating and $L_C = 880 \pm 22$ Å after heating and cooling, respectively. The fundamental diffraction peaks are only slightly broadened in comparison with the contribution from the resolution function. For them, a parabolic dependence on $d^2$ is not observed and this means that the coherently scattering domain size of the matrix exceeds the HRFD limit $L_M > 3500$ Å. A decrease in the widths of the fundamental peaks after heating and cooling is the result of a decrease in microstrains in crystallites. The calculation of these values leads to $\varepsilon \approx 8\times10^{-4}$ and $3\times10^{-4}$ for the initial and final states.

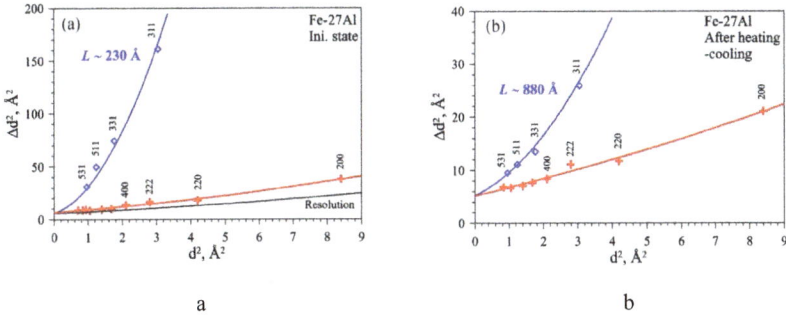

*Fig. 3.11  The $(\Delta d)^2$ vs $d^2$ dependences (Williamson-Hall plot) for Fe-27Al polycrystal in the initial (before heating) state (a) and after heating and cooling cycle (b). The $(\Delta d)^2$ values are multiplied by $10^6$. Statistical errors of experimental points are about symbol size. Bottom line corresponds to the diffractometer resolution function measured with a standard sample.*

For the Fe-27Al single crystal, the measured $(\Delta d)^2$ on $d^2$ dependences are shown in Fig. 3.12: both dependences can be fitted by parabolas - for fundamental peaks it corresponds to the CSDS $L_C \approx 920$ Å and for superstructural peaks it corresponts to $L_C \approx 400$ Å. Similar situation takes place in Fe-Al-Cr allows, see Chapter 5.

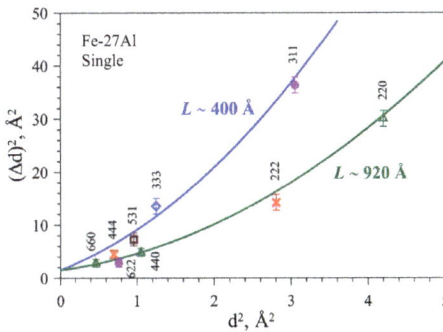

*Fig. 3.12  The same as in Fig. 3.11 but for Fe-27Al single crystal (initial, before heating state) for all measured directions. The widths of the peaks with even Miller indices fit into a curve corresponding to the coherently scattering domain size $L \sim 920$ Å. The $(\Delta d)^2$ values are multiplied by $10^6$.*

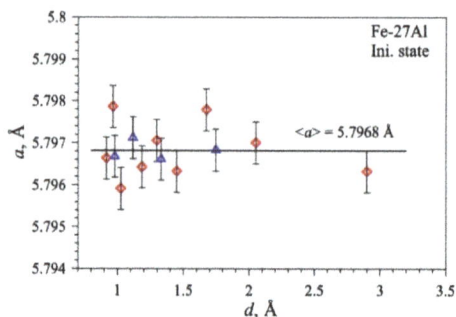

*Fig. 3.13 Unit cell parameter determined from the positions of individual diffraction peaks for the B2 (doubled value) and D0$_3$ structural phases. The average value of the parameter is the same with a relative accuracy of ~10$^{-5}$ for the f-peaks (red diamonds) and s-peaks (blue triangles).*

In all these cases (Fe-Al single and polycrystals, and Fe-Al-Cr), calculation of parameters of the unit cells according to the positions of the fundamental and superstructural diffraction peaks proves with very high accuracy their correspondence (Fig. 3.13) (relative difference of parameters is at the level of 10$^{-5}$), i.e. a more ordered phase (D0$_3$ or B2) is embedded into the matrix of a less ordered phase (B2 or A2) with a very high degree of coherence.

***In situ* real-time temperature scanning:** Temperature dependences of the intensities of the characteristic peaks and of the atomic volume (the volume of the unit cell per atom) for the Fe-27Al polycrystalline samples are reported in Ref. [58]. At heating and cooling these quantities vary continuously indicating that all the D0$_3$ ↔ B2 ↔ A2 transitions are typical order-disorder transitions of the second order. For the B2 phase, it was shown that the degree of atomic ordering in this phase is close to the maximum possible value for the Fe-26.5Al composition at temperatures lower than 500 °C. In the Fig. 3.14 the temperature dependences of intensities and widths of several orders of reflection from the (111) crystallographic plane of Fe-27Al single crystal are shown. Upon heating, the superstructure (111) peak, allowed only in the D0$_3$ phase, disappears first, then the peak (222) allowed in the D0$_3$ and B2 phases disappears, while the (444) peak, allowed in the disordered A2 phase, is present in the entire temperature range. The widths of even orders (f-peaks) are practically unchanged on heating, while the width of the (111) superstructure peak decreases drastically.

If the analysis is limited by the presence of diffraction peaks in the patterns, one could conclude that the structure of studied Fe-27Al samples in the initial state is the $D0_3$ phase, and after heating and cooling the structural transformations $D0_3 \rightarrow B2 \rightarrow A2$ take place as predicted by the phase diagram (see, for instance, the handbook [11]). However, a deeper analysis of the experimental data shows that in the initial state the volume of the samples has a partially ordered B2 phase, in which the 'islands' of the completely ordered $D0_3$ phase occupy only mesoscopic regions (hundreds of Å in size), being dispersed in the B2 matrix. After heating and cooling at the rate ~2 K/min, the size (the mean value) of the $D0_3$ clusters increases almost 4-fold (Fig. 3.15).

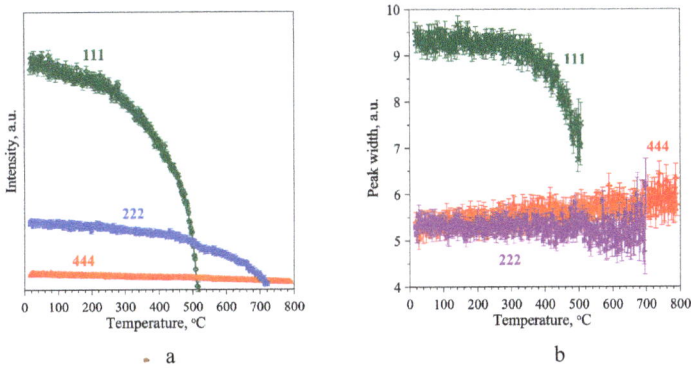

*Fig. 3.14 Intensities (a) and widths (b) of the orders of reflection from the (111) plane of Fe-27Al single crystal measured upon heating. Experimental points are shown with statistical uncertainties.*

Upon continuous heating in Fe-27Al single crystal sample the width of $s$-peaks above the temperature of about 350°C decreases rapidly. This is the result of an increase in the size of ordered clusters. Taking into account the resolution function of HRFD and using the well-known Scherrer equation, it is possible to calculate the change of cluster size with temperature. For Fe-27Al single crystal, this dependence is shown in Fig. 3.15a. At first glance, the increase in the size of the clusters contradicts the observed decrease in the intensities of the (111) (for $T > 500°C$) $s$-peaks. However, this contradiction disappears if we assume that the process of decreasing the atomic order degree in clusters starts simultaneously with heating.

49

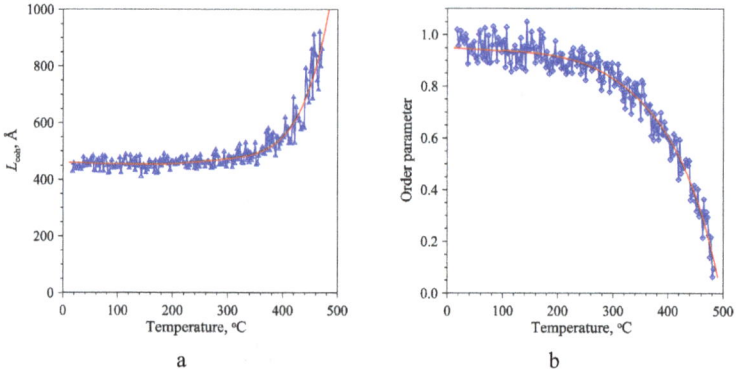

*Fig. 3.15 The temperature dependences of the characteristic cluster size of the ordered DO₃ phase in Fe-27Al single crystal determined from the width of the diffraction peak (111) (a) and the degree of atomic ordering (b) in these clusters determined from the ratio between the intensity and width of the peak (111).*

The decrease in order degree with increase in temperature can be calculated. For the superstructural peaks of the $DO_3$ clusters, intensity of the peaks $I_s \sim L_C^3 \cdot \xi^2$, where $L_C$ is their average size and $\xi$ is the degree of ordering. Since the width of the superstructural peaks is $W \sim 1/L_C$, then $\xi \sim (I_s \cdot W^3)^{1/2}$. Thus, making a correction for the Debye-Waller factor, we obtain:

$$\xi \sim (I_s \cdot W^3)^{1/2} \cdot \exp[B(T)/2d^2], \tag{3.5}$$

where $B(T)$ is the thermal factor with the Debye temperature, $\Theta_D \approx 410$ K for Fe-based alloys. The temperature dependence of the degree of order $\xi(T)$ is shown in Fig. 3.14b with a normalization of 0.95 at 20 °C. It can be seen that the main changes in the ordering factor inside clusters begin above 300 °C.

Thus, neutron diffraction results on polycrystalline and single crystal Fe-26.5Al allow to make several important conclusions. The initial state of the as cast and long-term naturally aged Fe-26.5Al compositions can be described as a partially ordered B2 matrix with dispersedly distributed $DO_3$ clusters. The characteristic sizes of the clusters are on the mesoscopic level, ranging from ~200 to ~400 Å. Upon slow heating and subsequent cooling, the intensities and widths of the diffraction peaks change in a regular way, showing the dependences typical for order-disorder structural transitions: $DO_3 \rightarrow B2 \rightarrow$

$A2 \rightarrow B2 \rightarrow D0_3$. These standard notations for structural transitions in Fe-Al based alloys should be interpreted as conditional, since some of them occupy only part of the volume (from 8% to 60%) of the sample. In spite of the fact that one can find a lot of evidence in favor of the cluster model in the literature [74, 77, 78, 79, 80, 82], the direct confirmation of the cluster-like structure in bulk samples is obtained in this study for the first time. The unusual behavior of the widths for fundamental and superstructure diffraction peaks observed in ordered metal alloys fits well to the described above cluster model.

It should be noted that a successful analysis of the microstructure of Fe-27Al alloys was carried out because of the large difference between the coherent scattering lengths of iron and aluminum and, consequently, the relatively high intensity of superstructure peaks. It seems that the discussed effect of clustering exists for other compositions as well, for instance for Fe-27Ga alloy. However, lower contrast, i.e. small intensity of superstructure peaks prevent determination the width of diffraction peaks with necessary precision. That is why the cluster like character of the ordered microstructure is not discussed for other types of alloys.

### 3.3 Tailoring magnetostriction in intrinsic Fe-27Ga alloy by isothermal annealing

Magnetostrictive materials play an increasingly important role in applications ranging from active vibration control, controlled surface deployment, actuators, damping devices, linear motors, positioning devices, and energy harvesting to stress and torque sensing [84,85,86,87,88,89]. Galfenol exhibits rather high magnetostriction (350 ppm) under very low magnetic fields of 100 Oe (8 kA/m) [90]. This alloy also shows a very low hysteresis [86, 91] and demonstrates a high tensile strength (500 MPa) [39, 92, 93, 94] and limited variation in magnetomechanical properties for temperatures between 20 and 80 °C [34, 95]. Fe-Ga alloys possess a high permeability ($\mu_r > 100$) [33], a high Curie temperature ($T_C > 650\,°C$) [96], and are corrosion resistant [97]. Such unique engineering capabilities make galfenol a sustainable alternative to conventional Terfenol-D that suffers extremely from brittleness and low yield stress under shock and tensile loads [87, 98, 99].

Various structures including disordered *b.c.c.* A2, ordered *b.c.c.* B2 and $D0_3$, ordered *f.c.c.* $L1_2$, and hcp $D0_{19}$ can be obtained when the amount of Ga is ~27 (~26–29) at.%. The interplay between the metastable $D0_3$ and the stable $L1_2$ phases in as-cast Fe-Ga alloys plays a crucial role for the formation of functional properties of Galfenol. Both phases are ferromagnetic but they have different physical characteristics. For example, they show magnetostriction of different signs, different coercive forces, and magnetization [100, 101]. Recently, a potential route has been suggested to obtain zero-magnetostriction in Fe-Ga alloys by establishing a controlled ratio between the $D0_3$ and $L1_2$ phases, which is essential for the design of soft magnets [101, 102]. For Fe-Ga alloys, samples with A2 and $D0_3$ structures possess positive magnetostriction; whereas samples with $L1_2$ and $D0_{19}$ structures exhibit negative magnetostriction. Compared to the *b.c.c.* and hcp phases, the $L1_2$ phase has a larger magnetization [100, 104, 103]. Following this structure-property relation, one could achieve an intrinsic Fe-Ga composite containing both, a *b.c.c.* phase with positive magnetostriction and an *f.c.c.* phase with negative magnetostriction [101, 102].

The kinetics of the $D0_3$ to $L1_2$ transition in Fe-27Ga under constant heating- and cooling conditions was discussed in the previous section. It was shown that during heating with a rate of 2 K/min, the $D0_3$ to $L1_2$ transition occur in the temperature range of about 50 °C [104]. Obviously, it is nearly impossible in practice to produce controlled amounts of the co-existing phases using linear heating. In order to solve the problem, i.e., to create an alloy with a given ratio between $D0_3$ and $L1_2$ phases, we decided to use isothermal annealing at temperatures slightly below the interval at which this transition occurs during continuous heating. To study the phase transitions, several techniques were involved: scanning electron microscopy equipped with EBSD, *in situ* XRD, and neutron diffraction.

Another important problem intensively discussed in the literature is the mechanism of the transition from the ordered *b.c.c.* ($D0_3$) to the ordered *f.c.c.* ($L1_2$) phase. Several atomic mechanisms were proposed:

- A diffusionless displacive *b.c.c.* → *f.c.c.* reconstruction was proposed by A.G. Khachaturyan [82]: in that model, the atomic order inherited from the $D0_3$ phase corresponds to a transformed tetragonal ($D0_{22}$ structure) lattice and the corresponding Bain strain transforms the *b.c.c.* lattice further into an fct lattice bringing Fe and Ga atoms to their new positions in an *f.c.c.*-based lattice without an atomic interchange between these sites.

- In the work by Yin-Chih Lin and Chien-Feng Lin [105], an $Fe_{73}Ga_{27}$ alloy was as-quenched and aged at 700 °C for 24 h, the $D0_3$ nanoclusters underwent a phase transformation to an intermediate tetragonal phase ($L1_0$ - like martensite) via Bain distortion, and finally $L1_2$ structures precipitated, as demonstrated by TEM and XRD analyses. The $L1_0$ - like martensite and the $L1_2$ structures in the aged $Fe_{73}Ga_{27}$ alloy drastically decreased the magnetostriction from positive to negative values, as also confirmed by experimental magnetostriction measurements.

- Mingming Li et al. [106] found that according to the analysis of the results of *in situ* heating, XRD and TEM measurements, the structure transitions of $D0_3 \rightarrow A1 \rightarrow L1_2$ were identified in the temperature range of 20-500 °C, which correspond exactly to the changes of Young's modulus in as cast and directionally solidified samples.

- Balagurov et al [83] suggested that the $D0_3 \rightarrow L1_2$ transition in bulk and powder-shaped Fe-27Ga alloys is of the order-disorder-order type, i.e., an intermediate disorder state exists between the two order states.

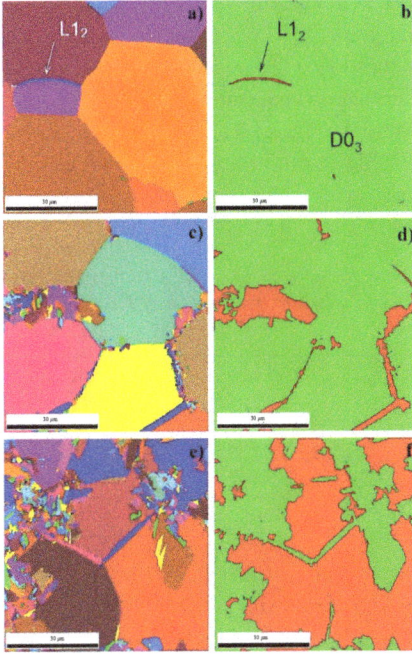

*Fig. 3.16 The microstructure (left) and EBSD analysis (right column) of the Fe-27Ga in the annealed states: 400 °C, 200 min (a, b), 350 min (c, d) and 600 min (e, f). For the EBSD micrographs (right column) the green color corresponds to the b.c.c. – and the red color to the f.c.c. phase. The scale bars in all the figures correspond to 30 μm (V. Palacheva).*

Thus, in this section, we focus on: (i) the type and the kinetics of the $D0_3 \rightarrow L1_2$ transition, and (ii) on a new pathway is elucidated to adjust an intrinsic composite structure with a controlled ratio of b.c.c. and f.c.c. phases and, consequently, with controlled magnetostriction, allowing to tune the magnetostriction in accordance with the requirements of a given specific application. The phenomenon of a concomitant transformation of the microstructure and the evolution o f ordering is discussed below using different characterization techniques: scanning electron and magnetic forced microscopy, XRD, and neutron diffraction [107]. It is a pleasure for author to underline that similar and very interesting results were published recently by Chinese group [102, 108, 109].

The electron back-scatter diffraction was used to identify the b.c.c. (b.c.c.-derivative $D0_3$) and f.c.c. (f.c.c.-derivative $L1_2$) phases and to provide a measure of the residual strains within the grains. The results of energy-dispersive X-ray spectroscopy (EDX) for Fe and Ga show that the distribution of chemical elements in the sample is sufficiently homogeneous for all samples. Table 3.1 summarizes the information about the samples, heat treatment, and the amount of the f.c.c. and b.c.c. phases identified by several methods.

The microstructure of the alloy in the annealed state (400 °C, 200 min) consists of almost equiaxed grains of the b.c.c. phase (Fig. 3.16a). The results of EBSD are shown in Fig. 3.15b: the amount of the f.c.c. phase is 0.4 %. It should be noted that given the EBSD measurements, we cannot distinguish between A2 and $DO_3$ as well as between A1 and $L1_2$ phases. As neutron diffraction data confirms $DO_3$ and $L1_2$ ordering of the A2 and A1 phases, we can use these designations. The new $L1_2$ phase nucleates at grain boundaries and grows into the body of the *b.c.c.* grains. The grain size was estimated using the EBSD maps. The average grain size for the bcc grains is 77.6 μm, for the *f.c.c.* ones it amounts to 8.8 μm. The *f.c.c.* grain size does not follow a Gaussian distribution because it is mostly formed by heterogeneous nucleation at grain boundaries. The microstructure of the alloy annealed at 400 °C for 350 min is shown in Fig. 3.16c. The average grain size for the *b.c.c.* grains is 70.2 μm. The result of EBSD shown in Fig. 3.16d quantified the amount of the *f.c.c.* phase as 9.9 % (area fraction).

*Table 3.1. Influence of annealing time at 400 °C on the $DO_3$ to $L1_2$ phase transition. Note that the area fractions are given for the EBSD data whereas volume fractions are listed for XRD and neutron diffraction data.*

| Name | Annealing time at 400 °C, min | *f.c.c.* phase, % (EBSD) | *f.c.c.* phase, % (XRD) | *f.c.c.* phase, % (ND) |
|---|---|---|---|---|
| Sample n.1 | 200 | 0.4 | 1.3 | 4 |
| Sample n.2 | 350 | 9.9 | 6.7 | 18 |
| Sample n.3 | 600 | 61.4 | 65.2 | 69 |

The microstructures of the samples annealed at 400 °C for 600 min show that the *f.c.c.* phase ($L1_2$) grows from the grain boundaries to the center of the *b.c.c.* grains (Fig. 3.16e). The average grain size of the *b.c.c.* phase amounts to 23.5 μm, the average grain size of the *f.c.c.* phase is 9.9 μm. The amount of the *f.c.c.* phase is 61.4 %. The largest *b.c.c.* grains were observed to disappear first after annealing. After annealing for 600 min, the average size of the *b.c.c.* phase decreases from 77.6 to 23.5 μm and the average size of the *f.c.c.* phase increases slightly from 8.8 to 9.9 μm.

The MFM facility used for this study is equipped with an optical microscope with a magnification of $300^x$. The topography analyses revealed that the *b.c.c.*-derivative phase (A2 or $DO_3$) had darker colour than the *f.c.c.* (A1 or $L1_2$) phase. Given just the results of the MFM analysis, we cannot distinguish between A2 or $DO_3$ or between A1 or $L1_2$ phases: based on our neutron diffraction results for room temperature, these phases are designated in figures as $DO_3$ and $L1_2$.

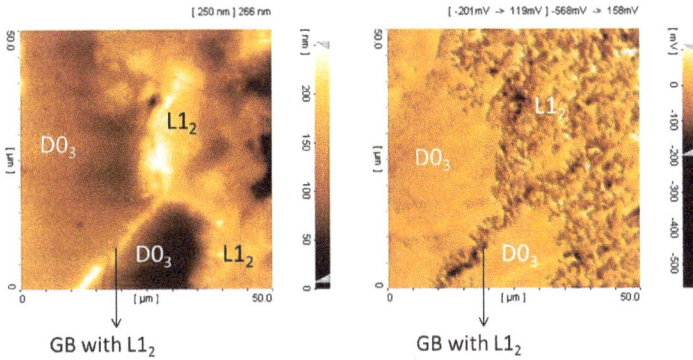

*Fig. 3.17. Topography (left) and MFM (right) images of an Fe-27Ga sample annealed at 400 °C for 600 min showing A2/D0₃ (darker resolution) and A1/L1₂ (brighter resolution) phases (A. Emdadi).*

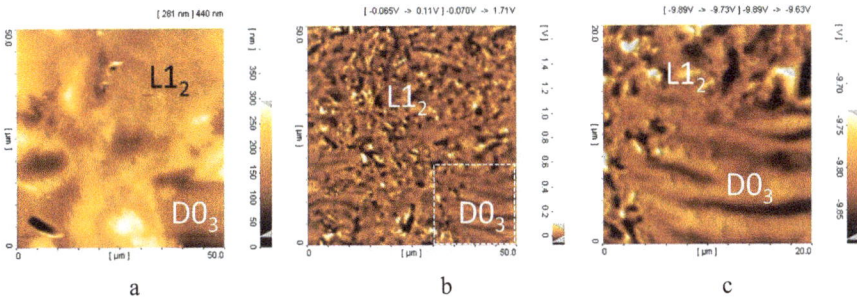

*Fig. 3.18. Topography (a) and MFM (b) images of Fe-27Ga sample showing A2/D0₃ (darker resolution) and A1/L1₂ (brighter resolution) phases; the enlargement of the marked area (white rectangle) in Fig. 3.18b is shown in Fig. 3.18c (A. Emdadi).*

Fig. 3.17 shows the topography (left) and the MFM images (right) of the Fe-27Ga sample annealed at 400 °C for 600 min. Several grains with $D0_3$ and $L1_2$ structure and a grain boundary between two *b.c.c.* grains with an elongated fringe of the *f.c.c.* ($L1_2$) phase that nucleated at the grain boundary are visible. Thus, the *f.c.c.*-derivative $L1_2$ phase nucleates at the grain boundaries between the grains with the $D0_3$ structure and continues to grow inside the *b.c.c.* grains. The magnetic domain structure for $D0_3$ and $L1_2$ phases are rather different: the $L1_2$ phase contains random and irregular magnetic domains; whereas the $D0_3$ phase has plate-like magnetic domains with a distinct magnetic substructure. The

55

stronger magnetic contrast of the $L1_2$ phase regions in the magnetic force image can be attributed to the larger magnetization of the $L1_2$ phase compared with the $D0_3$ matrix [103]. Figs. 3.18 a and b display the topography and the MFM images in a different region of the studied sample. Fig. 3.18c shows an enlarged region of a marked area (white rectangle) of Fig. 13.8b. It clearly shows irregular maze-like magnetic domains in the $L1_2$ phase and a large (width: 3-5 μm, length: about 40 μm) well-aligned stripe domain in the $D0_3$ phase. Such regular but smaller stripe domains were already observed for the A2 phase in a Fe-19Ga alloy [110,111,112,113]. It is known [84, 114, 115] that the domain walls in well-aligned stripe domains are more easily movable compared with irregular maze-like magnetic domains.

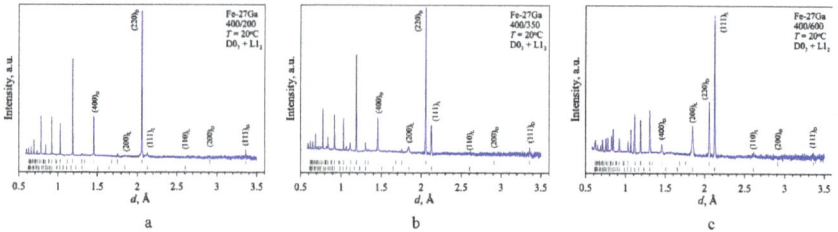

Fig. 3.19 Neutron diffraction patterns of the Fe-27Ga samples after annealing for 200 (a), 350 (b) and 600 min (c) at 400 °C, measured with high resolution at room temperature.

In contrast with SCAN-EBSD and XRD, neutron diffraction (ND) characterizes the bulk structure of the samples. Fig. 3.19 exhibits neutron diffraction patterns for the same three samples. All three samples are characterized by two phase structures with different ratios of the metastable $D0_3$ and the stable $L1_2$ phases in agreement with the EBSD and XRD results. The calculated peak positions for $D0_3$ (upper row) and $L1_2$ (lower row) are given at the bottom of each figure. In contrast to the as-cast or as-quenched states of the samples in which the structures of surface and bulk are different due to the differences of the cooling rate, a similar structure is recorded by both XRD and ND in the annealed samples. Based on high-resolution neutron diffraction data, the amount of the $D0_3$ phase can be roughly determined as 96% (200 min annealing), 82% (350 min), and 31% (600 min) (Table 3.1). Table exhibits the volume fractions of the f.c.c. phase in the sample annealed at 400 °C for different times obtained by using different methods: EBSD, XRD, and ND. By increasing the annealing time at 400°C from 200 min to 600 min, the volume fraction of the f.c.c. phase increases significantly. The results obtained by different methods are in a good agreement with each other.

The widths of the diffraction peaks in the samples n.2 and n.3 are appreciably different for the coexisting phases and they vary irregularly from peak to peak, i.e., the peak width strongly depends on a particular set of the Miller indices (Fig. 3.20). This effect can be analyzed using the Williamson–Hall relation (see Section 2.2.2.).

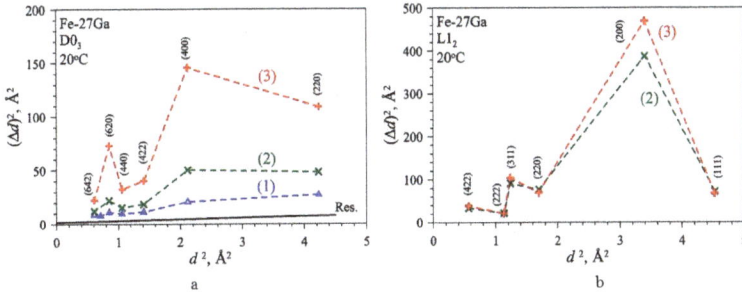

a          b

Fig. 3.20  The $(\Delta d)^2$ over $d^2$ dependences (Williamson–Hall plots) for the $D0_3$ phase for samples (1) - (3) (on the left) and for the $L1_2$ phase for the (2) - (3) samples (on the right). The $(\Delta d)^2$ values are multiplied by $10^6$. The statistical errors of the experimental data points are about the same size as the symbols. The bottom line in the left figure corresponds to the diffractometer resolution function measured with a standard sample. In sample (1), the amount of the $L1_2$ phase was very small, and thus, the peak widths were not determined. Only for the $D0_3$ phase in sample (1), microstresses were evaluated and were found to be almost isotropic, which is manifested in the linear dependence of $(\Delta d)^2$ on $d^2$. The Miller indices of the diffraction peaks are also indicated.

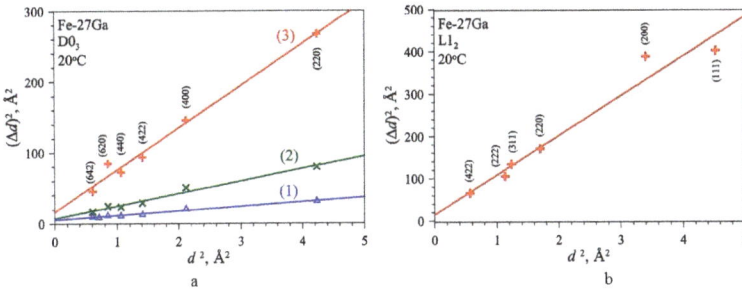

a          b

Fig. 3.21  The same experimental points as in Fig. 3.19 after correction on the dislocation anisotropy factor. For the $L1_2$ phase, the points only for sample (3) are shown, for sample (2), the result is similar. The $(\Delta d)^2$ values are multiplied by $10^6$. Statistical errors of the experimental points are about the symbol size. The Miller indices of the diffraction peaks are indicated.

After the correction, the $(\Delta d)^2$ over $d^2$ dependences become linear with a good accuracy, as it is shown in Fig. 3.21, which means that the initial anisotropy was, indeed, connected to micro-stresses. The increase of the volume fraction of the $L1_2$ phase from sample (2) to sample (3) leads to an increase of the stress mainly in the $D0_3$ phase (about three times), while the stress in the $L1_2$ phase remains on the same level, which is relatively high from the very beginning of the appearance of the $L1_2$ phase. At the $D0_3 \rightarrow L1_2$ transition, the micro-strain increases to $\varepsilon \approx 0.0021$ in sample (2) and to $\varepsilon \approx 0.0039$ in sample (3) from the initial value $\varepsilon \approx 0.0013$ in the as-cast sample. In spite of the fact that these stresses are very local, it is possible to estimate roughly the local micro-stresses taking the value of the Young's modulus for the Fe-27Ga alloy to be about 100 GPa: $\sigma \approx E \times \varepsilon \approx 400$ MPa. These are very high local stresses but they are still below the yield stress of the alloy [116], which leads to the internal friction transient effect discussed in [117].

Thorough analyses of the phase transitions at continuous heating shows that when the phase transition from *b.c.c.* to *f.c.c.* occurs, the $D0_3$ phase transforms to a disordered A2 phase that further transforms into an A1 phase that finally becomes $L1_2$ ordered. Isothermal annealing allows analysing this effect in more detail. Sample n.1 (previously annealed at 400°C for 200 min, the microstructure is given in Fig. 3.15 a, b) was rapidly heated (25 K/min to reach 410 °C) and then kept at 410 °C during the next 600 min. The entire sequence of the $D0_3 \rightarrow L1_2$ transformation is shown in as 3D plot and several selected diffraction patterns measured during this annealing time are shown in Fig. 3.22.

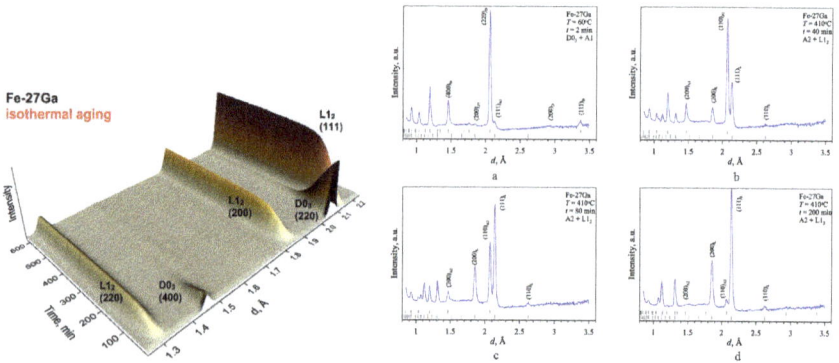

*Fig. 3.22 Left: A 3D visualization of the neutron diffraction pattern evolution of the Fe-27Ga sample no.1 pre-annealed for 200 min at 400 °C and subsequently measured upon fast heating up to 410 °C and exposed at this temperature for 600 min. Right: Neutron diffraction patterns (medium resolution) of sample no.1 measured at the very beginning of the annealing process (a) and at 410°C 40 min (b), 80 min (c) and 100 min (d) after the thermal processing was started.*

The $D0_3 \rightarrow L1_2$ transformation is nearly completed during the first ~100 minutes of exposure at 410°C without any noticeable signal at later times. Fig. 3.23a confirms the existence of the $D0_3$ phase in the initial state (a small superlattice (111) peak is clearly seen); whereas the next patterns do not show any $D0_3$ superlattice reflexions indicating a transformation of the $D0_3$ phase into the disordered A2 state, $D0_3 \rightarrow A2$. In contrast, the superlattice peak (110), which corresponds to the ordered $L1_2$ phase, is clearly visible in Fig. 3.22b, c, and d but not in Fig. 3.22a. Thus, it is verified that the transformation between the two ordered phases ($D0_3 \rightarrow L1_2$) proceeds via the disordered transient phases/states at the early stages of the transition: $D0_3 \rightarrow A2 \rightarrow A1 \rightarrow L1_2$. We did not mark these transitions in the 3D figure in order to simplify it, as this sequence of transitions is considered in detail in the next figures.

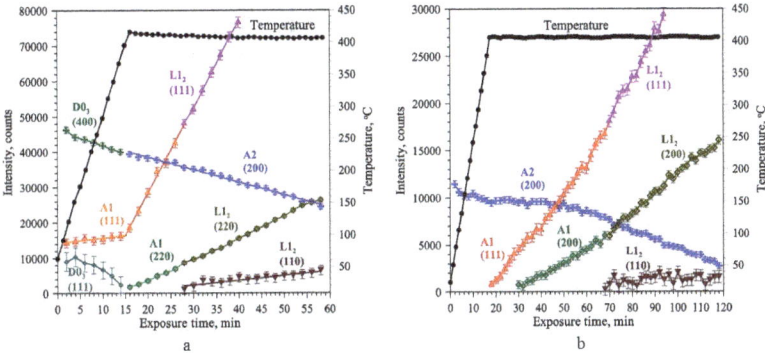

*Fig. 3.23 Comparison of the sequence of the phase transitions for sample n.1 (left) and for the as-cast state (right) in the course of rapid heating followed by annealing at 410 °C. The intensities (left scale) of several main and superlattice peaks are shown as functions of the annealing time. The dependence of the sample temperature on time is also shown (black line, right scale).*

Details of the $D0_3 \rightarrow A2 \rightarrow A1 \rightarrow L1_2$ transition can be seen more clearly in Fig. 3.23a, where for some particular diffraction peaks intensity changes are shown for the first 60 minutes of heating and annealing. The intensity of the superlattice $(111)_{D03}$ peak decreases rapidly and the peak disappears when the temperature reaches 410°C indicating the $D0_3 \rightarrow A2$ transformation $((400)_{D03} \rightarrow (200)_{A2})$. During the interval from 15 to 28 minutes of the thermal treatment, two disordered phases A1 and A2 coexist. Further, the formation of the ordered $L1_2$ phase is started $((111)_{A1} \rightarrow (111)_{L12})$. It means that under the conditions of the annealing treatment performed here, the ordered $D0_3$ state is completely converted via the two disordered A2 and A1 states into the ordered $L1_2$ phase.

The entire phase transformation sequence can be summarized as follows:

(i) along with $D0_3$, a small amount of the $L1_2$ phase is present in the sample, i.e., the nucleation time for the $L1_2$ phase at 400 °C is about 200 min,

(ii) the $D0_3$ to $L1_2$ transition is completed during next 200 min at 410 °C, and

(iii) the $D0_3$ to $L1_2$ transition is accompanied by disordering of the $D0_3$ to A2 phase, followed by a transition from A2 to A1 and completed by the final ordering of the A1 phase to form the observed $L1_2$ structure: $D0_3 \rightarrow A2 \rightarrow A1 \rightarrow L1_2$.

In order to compare this sequence of the phase transitions and their kinetics, the as cast sample was also studied under the same conditions of heating and annealing (Fig. 3.23b). As expected, pre-annealing of the sample at 400 °C for 200 min enhances the kinetics of the phase transition due to the nucleation of the $L1_2$ phase (see Table 3) during this pre-treatment. In case of the as cast sample, the appearance of the A1 phase at 410 °C takes about 7-8 min, and after about 70 min of annealing at 410 °C the $L1_2$ ordered phase forms from the disordered A1 phase.

The initial state of sample no.1 is $D0_3$ and, additionally, a small amount of the $L1_2$ phase is present (Table 3.1). The transition $D0_3 \rightarrow A2$ is completed after 15 minutes of heating. The ordered $L1_2$ state (A1 $\rightarrow L1_2$ transition) appears after 28 minutes of heating followed by isothermal annealing at 400 °C. The initial state of the as cast sample consists of the disordered A2 phase with some weak sign of $D0_3$ ordering. Upon heating, the A2 phase transforms into the disordered A1 phase after 20 - 30 minutes of heating followed by annealing. The ordered $L1_2$ state appears after 70 min or after 54 min of isothermal annealing at 400 °C.

The kinetics of the *b.c.c.* to *f.c.c.* transformation is compared for the sample previously annealed for 200 min at 400 °C no.1 and the as-cast sample in Fig. 3.24. The pre-annealed sample contains small fractions of the $L1_2$ phase (up to 4%). The as cast sample at the moment of heating to 410 °C does not show any content of the *f.c.c.*-derivative phases. In result, the *b.c.c.*-to-*f.c.c.* transition in the pre-annealed sample starts from the very beginning of the isothermal annealing, whereas in the as-cast sample, an incubation period was observed. Thus, the time difference of the kinetics of the isothermal phase transition between these two samples is about 20-25 min.

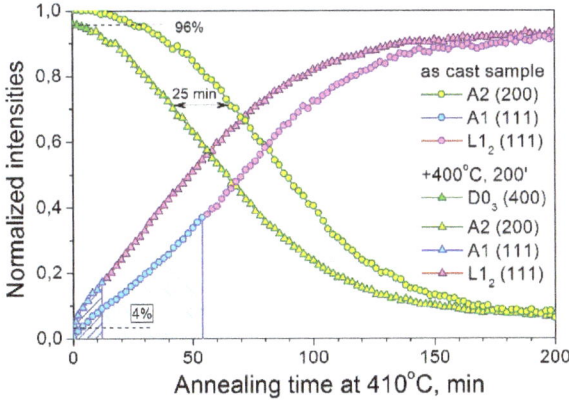

*Fig. 3.24 Comparison of the kinetics of the phase transitions in sample no.1 and the as cast samples in the course of rapid heating and subsequent annealing at 410 °C.*

Fig. 3.25a illustrates the room temperature magnetostriction measurements on the Fe-27Ga samples in different states measured up to a maximum magnetic field value of 40 kA/m. In this paper, the parallel and perpendicular magnetostriction ($\lambda_\parallel$ and $\lambda\perp$) refers to the strain measured along and perpendicular to the field direction, respectively. Using $\lambda_S$ = 2/3 ($\lambda_\parallel$ - $\lambda\perp$) [118], the saturation magnetostriction ($\lambda_S$) of the samples are calculated as: 105 ppm, 86 ppm, and 52 ppm at the field of 40 kA/m, respectively, for the as-cast material and for the samples annealed at 400 °C for 350 min and 600 min. The as-cast material contains practically 100% of the $D0_3$ phase and, thus, shows the highest saturation magnetostriction, whereas annealing at 400 °C leads to a decrease of the saturation magnetostriction by ~20% and ~50%, respectively, for the annealing times of 350 min and 600 min measured at the field of 40 kA/m. The formation of the $L1_2$ phase during annealing reduces the magnetostriction of the sample, since $L1_2$ has a negative magnetostriction in contrast to the positive magnetostriction of the $D0_3$ phase. It was observed that the longer the annealing time, the higher is the volume fraction of the $L1_2$ phase in the sample; and consequently, the lower is the saturation magnetostriction.

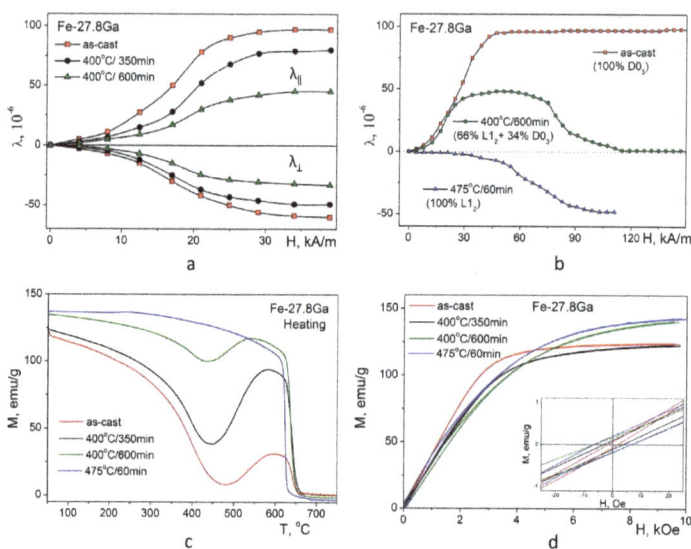

*Fig. 3.25 Parallel (positive) and perpendicular (negative) magnetostriction for the Fe-27Ga alloy in different states: as cast (black squares), annealed at 400 °C for 350 min (red circles) and 600 min (olive triangles) (a). Magnetostriction for Fe-27Ga alloy in different states: 475 °C, 60 min (blue triangles), 400 °C, 600 min (olive circles), and as cast (red squares) (b), magnetization vs. temperature curves for the Fe-27Ga samples in different states (c), and magnetization hysteresis loops of the studied samples at room temperature (d). The inset shows the coercivity $H_C$ derived from the magnetization loops measured at the lower fields of -15 to +150 Oe (A. Emdadi, E. Zanaeva).*

The measurements of magnetostriction at higher magnetic fields are shown in Fig. 3.25b. Upon increasing the field strength, the axial magnetostriction of the as-cast alloy remains unchanged after reaching a maximum value at H ≈ 50 kA/m. The saturation magnetostriction is positive (100 ppm) for the as-cast state with a dominant $D0_3$ structure, whereas it is negative (-50 ppm) after annealing at 475 °C for 60 min, since the material consists mainly of the $L1_2$ phase. The observed difference of the sign of magnetostriction between the *b.c.c.* (A2 or $D0_3$) and the *f.c.c.* (A1 or $L1_2$) phases is consistent with the previous work [100, 101, 119]. For the sample annealed at 400 °C for 600 min, which contains a phase mixture roughly of 66% *f.c.c.* + 34% *b.c.c.*, a magnetostrictive behavior with zero magnetostriction is recorded. With increasing the magnetic field strength, the magnetic domains in the $D0_3$ phase first move along the magnetic field and the axial magnetostriction approaches a saturation value when all the magnetic domain wall motions and the rotation of magnetizations take place [101]. When the magnetic field is increased to its critical value, the domain wall motions and magnetization rotations of the $L1_2$ phase lead to a significant reduction of magnetostriction down to zero. Thus, the sample first expands and then contracts along the field direction.

The changes of the sign of magnetostriction are due to the net effect of the domain wall motions and magnetization rotations in materials with an intrinsic composite microstructure consisting of the $D0_3$ and $L1_2$ phases [101]. Under lower magnetic fields, the positive magnetostriction of the $D0_3$ phase dominates. It is not completely compensated by the negative magnetostriction of the $L1_2$ phase. When the magnetic field increases to a certain level (~50 kA/m), the contributions of the two phases completely compensate each other. With a further increase of the magnetic field, the domain wall motion and magnetization rotation of the $L1_2$ phase with stronger magneto-crystalline anisotropy leads to a reduction of the magnetostriction to zero. In summary, the preferable domain wall motion and magnetization rotation of the positive magnetostriction component ($D0_3$) at lower magnetic fields result in an elongation up to a maximum value, meanwhile the contribution from the negative magnetostriction component ($L1_2$) at higher magnetic fields leads to contraction till zero.

The different ratio between the *b.c.c.*-derivative (A2 or $D0_3$) and *f.c.c.*-derivative (A1 or $L1_2$) phases present in the samples that are produced by different heat treatment procedures alters the temperature dependence of the magnetization and the magnetic properties at room temperature. Fig. 3.25c shows the magnetization vs. temperature, M(T), curves for the Fe-27Ga samples in different states: as-cast and annealed at 400 °C for 350 and 600 min, respectively. For the as-cast sample, the M(T) curve shows a decrease in magnetization close to 8 emu/g up to ~480 °C. According to [120], this temperature corresponds to the presence of a two phase structure (50 % $D0_3$ and 50 % $L1_2$) after heating with a heating rate of 2 K/min. Taking into account that the heating rate applied in the present VSM tests is 6 K/min, we can be sure that the decrease of the magnetization is not due to the disappearance of the $D0_3$ phase but it corresponds to the Curie point for the $D0_3$ phase. Further heating leads to an increase in the magnetization due to the appearance and growth of the $L1_2$ ferromagnetic phase, at T~600 °C the phase transition is completed and the magnetic properties of the material are determined by the ferromagnetic *f.c.c.* $L1_2$ phase that possesses a high specific magnetization. The higher saturation magnetization at ~600 °C is due to a marked increase of the magnetic moment per Fe atom due to the change in the atomic environment [103]. Upon further heating, the magnetization starts to decrease sharply, which agrees well with the structural change from the $L1_2$ phase to the $D0_{19}$ phase, as proved by the neutron diffraction studies [106, 120, 121]. The zero magnetization at ~700 °C corresponds to the transition from the hcp $D0_{19}$ to the non-magnetic ordered *b.c.c.* B2 phase. For the sample annealed at 400 °C for 350 min, the magnetization decreases at about 450 °C to a level above the as-cast sample, since this sample already contains 6.7-18% $L1_2$ phase as estimated by different methods (Table 3.3). The sample annealed at 400 °C for 600 min exhibits only a small decrease in the magnetozation at 450 °C as it already contains 61-69% $L1_2$ phase at room temperature, and only a little amount of the *b.c.c.*-derivative (A2 or $D0_3$) phase transforms to a *f.c.c.*-derivative (A1 or $L1_2$) phase. Finally, the magnetization of the as-cast Fe-27Ga sample additionally annealed for one hour at 475 °C with 100% $L1_2$ structure as confirmed by XRD has a very low dependence of the magnetization on temperature. In equilibrium state, i.e. having the $L1_2$ structure, Fe-27Ga alloy exhibit highly thermal-stable magnetization over a wide temperature range but not advanced and negative values of magnetostriction.

The magnetic properties (saturation field and magnetization as well as the coersive force) at room temperature are considerably influenced by the amount of phases present in the given material's state. Fig. 3.25d displays the room temperature magnetization loops for the as-cast and annealed samples. The inset shows the coercivity, $H_C$, derived from the magnetization loops measured at a low field of -15 to +150 Oe. The saturation magnetization, $M_S$, and the coercivity increase from 126 emu/g and 1.61 Oe for the as-cast sample to 137 emu/g and 3.17 Oe for the sample annealed at 400 °C for 350 min, and a further increase to 151 emu/g and 6.45 Oe for the sample annealed at 400 °C for 600 min. The longer the annealing time at 400 °C, the higher the volume fraction of the $L1_2$ phase, and subsequently, the larger the values of the $M_S$ and $H_C$ are. The magnetization of the sample annealed at 400 °C for 600 min saturates at a much higher magnetic field (~10 kOe), demonstrating that the $L1_2$ phase possesses a higher magnetocrystalline anisotropy than that of the $D0_3$ phase. Consequently, magnetostriction for the sample with the $L1_2$ phase as the main constituent phase also saturates under a much higher magnetic field than that of the as-cast sample (Fig. 3.25b).

In summary, a change of the ratio between *b.c.c.*-derivative (A2 or $D0_3$) and *f.c.c.*-derivative (A1 or $L1_2$) phases obtained by different heat treatment procedures alters both the magnetic properties and the magnetostrictive behavior of the samples. The saturation magnetostriction is positive (~100 ppm) for the as-cast state with a dominant $D0_3$ structure, whereas it is negative (~ -50 ppm) after annealing at 475 °C for 60 min, since the material then consists mainly of the $L1_2$ phase. Zero magnetostriction is recorded for the sample annealed at 400 °C for 600 min with a phase mixture of 66% *f.c.c.* + 34% *b.c.c.* The formation of the *f.c.c.* phases affects not only the temperature dependence of the magnetization but also the magnetic properties at room temperature. With an increase in the volume fraction of the $L1_2$ phase, the values of $M_S$ and $H_C$ increase. The higher magnetocrystalline anisotropy of the $L1_2$ phase, as compared to that of the $D0_3$ phase, causes the magnetostriction and the magnetization to saturate under a much higher magnetic field in the sample annealed at 400 °C for 600 min than that of the as-cast material.

We have also underline that the phase transition from an ordered *b.c.c.*-derivative $D0_3$ phase to an *f.c.c.*-derivative ordered $L1_2$ phase first leads to disordering of the $D0_3$ phase to obtain an A2 structure followed by an A2 to A1 transition with final A1 phase ordering to achieve the $L1_2$ structure. This sequence ($D0_3 \rightarrow A2 \rightarrow A1 \rightarrow L1_2$) is proven here for the first time for the Fe-27Ga alloy. This sequence amends results recently reported by [83] and [106].

### 3.4   Temperature dependent anelastic effects in Fe$_3$Me - type alloys

In this section we overview of anelastic effects in Fe$_3$Me-type Fe-Al, Fe-Ga and Fe-Ge alloys as measured at forced bending vibrations (DMA Q800) between 0 and 600 °C.

Temperature Dependent Internal Friction (TDIF) curves measured using six frequencies from 0.1 to 30 Hz at continuous heating and cooling with a rate of 2 K/min are presented in Fig. 3.26. Several internal friction peaks can be distinguished at the TDIF curves: those of them which are thermally activated, i.e. their temperature increases with an increase in measuring frequency, are denoted as the P1, P2 and P3 peaks (in Fe-27.1Ge only one peak, denoted as P, exists: TDIF data for Fe-25.3Ge are practically the same with Fe-27.1Ge). Non-thermally activated but phase transition peaks are denoted as $P_{Tr}$: their temperatures (~550 °C for Fe-27Al and ~470 °C for Fe-27Ga) do not depend on measuring frequency in contrast with their height.

Two different types of relaxation processes – thermally activated (denoted as P1, P2, P3 and P effects) and structural induced anelasticity ($P_{Tr}$) are recorded and discussed with respect to the structure in the studied alloys.

**Fe-Al:** The P1 and P2 internal friction effects belong to the $D0_3$ ferromagnetic phase, whereas the P3 peak may belong either to the $D0_3$ ferromagnetic or B2 paramagnetic phase dependently on the measuring frequency. This case – shift of a peak temperature from one phase to another will be discussed later. The $D0_3 \leftrightarrow B2$ transition is accompanied with transient (or lambda) peak both at heating and cooling.

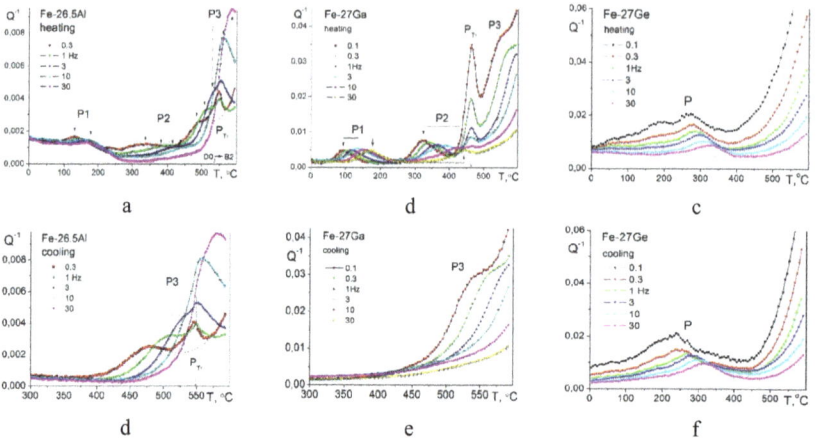

Fig. 3.26 Temperature dependent internal friction for $Fe_3Al$ (a, d), $Fe_3Ga$ (b, e) and $Fe_3Ge$ (c, f) at heating (a, b, c) and cooling (d, e, f). DMA Q800 TA Instruments, frequencies of forced bending vibrations are shown in figures.

TEM pictures for Fe-26.5Al samples after water quenching from 1000 °C and selected annealing regimes confirm their $D0_3$ structure (Fig. 3.27): the $D0_3$ domains are observed directly after quenching and their size increases with an increase in the annealing temperature and time unless the $D0_3 \rightarrow B2$ transition takes place at around 550 °C.

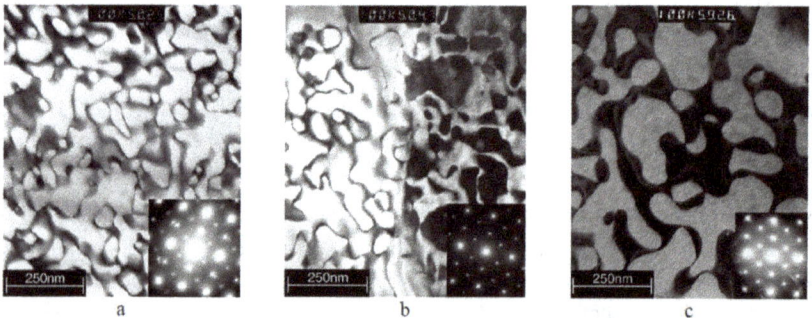

Fig. 3.27 TEM study of the $D0_3$ domains [110] (111) in Fe-26.5Al alloy (dark field): a) water-quenched from 1000°C, b) plus 8 hrs at 200 °C, c) plus 48 hrs ageing at 400 °C (Ch. Grusewski).

**Fe-Ga:** Thus, the P1 and P2 internal friction thermally activated effects belong to the $D0_3$ ferromagnetic phase, whereas the $P_{Tr}$ effect at about 465°C corresponds to the $D0_3$ to $L1_2$ phase transition, consequently the P3 peak belongs to the $L1_2$ phase. The $D0_3 \rightarrow L1_2$ first order phase transitions in Fe-Ga at 450-500 °C leads to a noticeable difference in atomic unit cell volumes of these phases giving rise to transient peak $P_{Tr}$.

*The P1 peak* is observed in Fe-Al and in Fe-Ga alloys (Fig. 3.26) at heating only. This peak is significantly broader than the Debye peak with one relaxation time. The mean values of the activation energy are between 1.03 to 1.09 eV. This type of relaxation mechanism in Fe-Al alloys was discovered in 1962 by Fischbach [122] and later investigated by Hren [123], Tanaka [124], Koiwa [18] and Golovin [125, 126] and it was identified as the carbon Snoek-type relaxation, as well as the similar anelastic effect in Fe-Ga alloys [127, 128]. It means that the origin of this peak is stress-induced diffusion of interstitial atoms (carbon atoms) in the solid solution. Originally, the mechanism of such relaxation in the interstitial solid solution of α-Fe was proposed by J. Snoek [129] in 1941. In case of Fe-Al or Fe-Ga alloys, one can speak about a Snoek-type mechanism, which extends the ordinary Snoek mechanism to jumps of interstitial carbon atoms under stress in a *substitutional solid solution*. Diffusional jumps of C atoms from one octahedral to another octahedral interstices in the lattice of Fe-Al and Fe-Ga alloys are influenced by the presence of substitute atoms (Al and Ga) in the crystalline lattice. Al and Ga atoms generate local compressive stresses in the lattice and increase the lattice parameter of iron. Al atoms (effective Goldschmidt atomic radius equal to 0.143 nm) and Ga (0.135 nm) atoms are larger than Fe (0.128 nm) atoms. Thus, their 'elastic' interaction with C atoms increases the activation energy for C-atom jumps as compared with 'pure' α-Fe (with the activation energy between 0.83 [130, 131] and 0.87 eV [132]).

The carbon Snoek-type peak is observed neither in Fe-25Ge nor in Fe-27Ge compositions. Contrary to Fe-Al and Fe-Ga alloys with their *b.c.c.* or *b.c.c.-derivative* (B2 and $D0_3$) structures, this peak is crystallographically forbidden in the *hexagonal* $D0_{19}$ phase dominating the structure of these Fe-Ge alloys.

*The P2 peak* (Fig. 3.26) is also observed only at heating. The mean value of the activation energy is ranged between 1.37 to 1.91 eV. In some earlier papers this peak for Fe-Al alloys was denoted as the X-peak. In case of Fe-27Ga, this peak – if measured at a higher limit of used frequencies – overlaps with the $P_{Tr}$ peak (discussed below). A similar effect in Fe-Al alloys was explained earlier either as re-orientation of complexes of vacancies or of a complex carbon atom – vacancy. This relaxation effect has been repeatedly observed in Fe-Al alloys since 1997 by different authors. Further hypotheses have been introduced to explain this peak: Damson et al. explained this anelastic effect by "vacancy – Fe atom" pair reorientation [133], and later by a four jump reorientation of divacancy complexes in the iron sublattice $v_{Fe}$-$v_{Fe}$ [134]. A similar peak in an Fe-30%Al *single crystal* [134] ($\approx$ 525 °C, $\approx$20 kHz, roughly H $\approx$ 1.55 eV) was explained by $D0_3$-B2 transformation. Based on the fact that the P2 peak in Fe-Al is sensitive to both the content of carbon in solid solution and of vacancies, this peak was suggested [135,136,137] to be a carbon-and-vacancy related effect in the Fe-Al-$v$-C system ($v$ stands for vacancies), where the binding energy between C and $v$ in solid solution ($H_B^{C-v}$) of about 0.4-0.5 eV corresponds to the difference in activation energies between the P2 and P1 peaks: $H_{P2} = H_{P1} + H_B^{C-v}$. The P2 peak in Fe-Ga alloys, which is rather similar to that of in Fe-Al, is less studied. The temperature range where it is observed is close to the temperature at which nearly all quenched-in vacancies annihilate. None of these hypotheses about the nature of P2 effect has been finally confirmed to date.

***The P3 peak*** is observed in Fe-Al and Fe-Ga alloys at heating, cooling and in subsequent heating-cooling tests. The activation energy of the P3 peak in Fe-27Al alloy according to this study varies from 2.4 to 2.8 eV and typical values for $\tau_0$ range from $10^{-17}$ to $10^{-19}$ s in agreement with values of H from 2.5 to 3.2 eV and of $\tau_0$ from $10^{-18}$ to $10^{-20}$ s given in [138]. These activation parameters may correspond to grain boundary relaxation or to Zener relaxation in Fe-based alloys.

In order to exclude possible grain boundary (GB) contribution to relaxation processes in the studied temperature - frequency range, we have performed tests of two single crystals: Fe-26Al and Fe-17Ga. Presence of the P3 peak in single crystals confirms the presence of this peak in single crystal and consequently this test excludes grain boundary mechanism for the P3 anelastic effect. Thus, this effect corresponds to Zener relaxation.

The Zener type relaxation in Fe-Al was reported in 1960 by Shyne and Sinnot [139] and then by different authors using temperature- [122, 123, 140, 141] and frequency [142] dependent internal friction tests. The Zener relaxation in Fe-Ga was probably for the first time reported in [127] using both temperature- and frequency dependent tests. Zener peak in Fe-26.5Al overlaps with a transient frequency independent $\lambda$-type [6] IF peak at 550 °C both at heating and cooling. Zener peak in Fe-27Ga is observed at higher temperatures compared with a transient frequency independent IF peak which takes place only at heating at about 460 °C.

The activation energy of the Zener peak observed in Fe-27Al single and polycrystals at heating, cooling and in subsequent heating-cooling tests varies from 2.4 to 2.8 eV and typical values for $\tau_0$ range from $10^{-17}$ to $10^{-19}$ s in agreement with values of H from 2.5 to 3.2 eV and of $\tau_0$ from $10^{-18}$ to $10^{-20}$ s given in Ref. [138]. Activation parameters for the Zener relaxation peak in Fe-17Ga single crystal are at heating $H_Z = 2.54$ eV and $\tau_0 = 2\times10^{-18}$ s, and at cooling $H_Z = 2.50$ eV and $\tau_0 = 3\times10^{-18}$ s.

***The P peak*** is observed in Fe-Ge alloy at heating (Fig. 3.26c), cooling (Fig. 3.26f) and in subsequent heating-cooling tests. Very few internal friction studies were performed as yet for Fe-Ge alloys. Except for paper [141], where Snoek peak in Fe-3%Ge was reported in *b.c.c.* structure, a thermally activated effects were recorded in the Fe-15 wt. %Ge (~12 at.%Ge) alloy which belongs to the $\alpha_1$ (B2) structure with the following parameters $H = 2.05$ eV and $\tau_0 = 10^{-13.5}$ s and it was classified as a Zener peak [143]. Similar peaks were later recorded in Fe-12Ge, -19Ge and -22Ge alloys [144]: authors underlined possible contribution of vacancies to this effect. Finally, the IF peak in Fe-27Ga with the activation energy and the pre-exponential frequency factor estimated using high frequency IF were reported [63]: $H = 1.78$ eV and $\tau_0 = 2 \cdot 10^{-17}$ s, respectively.

The values of activation energies in Fe-27Ge from 2 to 2.5 eV, as well as the stability of the peak height and position with respect to heating/cooling experiments rather supports that this peak can be classified as a Zener peak in the $D0_{19}$ structure. Thus, the peak is caused most probably by reorientation of pairs of substitutional atoms (Ge in Fe) under the applied stress. It is notable that there are no pronounced phase transitions except $D0_{19}$ ordering of hexagonal A3 phase in the Fe-25Ge alloy according to neutron diffraction. At the same time temperature dependence of magnetisation in Fe-27Ge (Fig. 3.26f) suggests phase transition at about 400 °C ($A2(D0_3) + B8_1 \rightarrow L1_2$ according to equilibrium diagram) and Curie temperature of one of participating phase ($B8_1$ according data in inset to Fig. 3.6f) at about 250 °C. The $D0_{19} + B8_1$ structure in Fe-27Ga is also confirmed by XRD. These transitions take place in the same temperature with the P peak and decrease accuracy of evaluation of activation parameters of the P peak.

Some tendency for the P3 peak height as well as internal friction background decrease with an increase in quenching temperature can be explained by increase in vacancy concentration. At cooling the peak narrowing takes place. Two hypotheses can be considered: (i) quenching increases the vacancy concentration and decreases the degree of order in the studied Fe-25Ge and Fe-27Ge alloys: both factors should increase the Zener peak height; (ii) the relaxation (Zener) peak can overlap with a phase transition peak (from undercooled $D0_{19}$ to $L1_2$ phase or from $A2(D0_3) + B8_1 \rightarrow L1_2$). This latter hypothesis is partly supported DSC tests [43]. In both cases, a clear distinction between different contributions to the P peak in Fe-Ge was not possible.

***The $P_{Tr}$ peak*** may accompany phase transitions of the first and the second order.

In contrast with 'true' thermally activated relaxation effects that obey the Debye and Arrhenius laws, this anelastic peak can be referred to as transient effect related to phase transitions. Theory of these transition effects invented by L. Landau is given in the book by Nowick and Berry 1972 [6], W. Benoit [23] and G. Fantozzi [24]. Maximum of internal friction for both first-order transition and second-order (often called lambda, $\lambda$) transition according to Landau's theory can be described by eq. (1.10). In Refs [6-8, 23-24], the authors noted that most of the experimental internal friction data are not well described by these equations. Here we provide opposing results, which are in agreement with Landau's theory.

The peak position (temperature) does not depend on measuring frequency; whereas the peak height is roughly proportional to inverse frequency of measurements. The *transient* damping, well known in literature to accompany the shear diffusionless transformations, is attributed to the transformation rate $\partial n/\partial t$, where $n$ is the amount of transformed material. This effect is also known as the so-called "$\dot{T}$ effect".

This transient effect in Fe-27Ga was reported in [59] and explained by a partly diffusionless transition from $DO_3$ to $L1_2$ phase [145] in agreement with Khachaturyan and Viehland's calculations [146]. Later we found out that the $DO_3$ to $L1_2$ phase transitions in Fe-27Ga at around 450-470 °C leads to a noticeable difference in thermal expansion coefficients of different phases and to increase in local stresses at the $DO_3/L1_2$ boundaries [120, 147]. Temperature dependent changes of the atomic volume for the $DO_3 \rightarrow L1_2$ transition in Fe-27Ga are calculated by corresponding changes in unit cell parameters: the most pronounced difference between them is seen in the course of transition $DO_3 \rightarrow L1_2$ ($\Delta V/V \approx 1\%$).

If a cyclic stress, $\sigma=\sigma_0 \times \cos(\omega t)$, is applied to a sample, the internal friction resulting from this phase transition is described by eq. (1.11). Thus, it is not surprising that the first order $DO_3 \rightarrow L1_2$ transition is accompanied by a sharp transient IF peak at 450-500 °C in Fe-27Ga with $Q_{Tr}^{-1} \sim \dot{T}/\omega^n$ dependence [26]. Indeed the analysis of the transient IF peak in Fe-27Ga (Fig. 3.28) confirms this dependence. We have not recorded the $P_{Tr}$ effect at cooling as at cooling, this transition is smeared in a wide temperature range where both phases (A2 and $L1_2$) coexist.

Similar but smaller, nearly in one order of magnitude, transient effect is recorded in Fe-13Ga, Fe-19Ga (Fig. 3.28) and Fe-26.5Al alloys (Fig. 3.29) both at heating and cooling at around 550 °C. This peak temperature in Fe-26.5Al single crystal (SC) and polycrystals (PC) corresponds well to the $DO_3 \leftrightarrow B2$ second order transition according to the equilibrium diagram and to our in-situ neutron diffraction. The $DO_3$ and B2 ordered phases appear and disappear as a result of Al atoms ordering in $\alpha$-Fe-Al (A2) solid solution. In contrast with the $DO_3 \rightarrow L1_2$ first order transition in Fe-27Ga the $DO_3 \leftrightarrow B2$ transition in Fe-26.5Al is not accompanied by significant increase in internal stresses. The peak height is nearly inversely proportional to measuring frequency according to theory: it is not surprising that this peak was not reported in earlier papers [18, 27-29, 30, 37], in which only relatively high frequency tests (at least above 200 Hz) were used.

Several internal friction peaks can be distinguished at the TDIF curves in the temperature range of 350 to 600 °C: those of them which are thermally activated, i.e., their temperature increases with an increase in measuring frequency, are denoted as Z (Zener) peaks. Phase transition peak denoted as $P_{Tr}$ is non-thermally activated but that is related to phase transition: its temperature (~550 °C) does not depend on the measuring frequency; however, its height is reverse proportional to the measuring frequency. Upon instant cooling (after heating to 600 °C) with a cooling rate of 1 K/min, the Zener thermally activated internal friction peak can be distinguished again in both single and poly crystalline alloys. The transient $P_{Tr}$ peak is well recorded at cooling at about 550 °C for both SC and PC alloys as well. In this paper, we do not discuss effects that take place below 350 °C, as they were already reported: low temperature ordering in as-cast and as-quenched samples in [58] and anelastic effects in [135, 137].

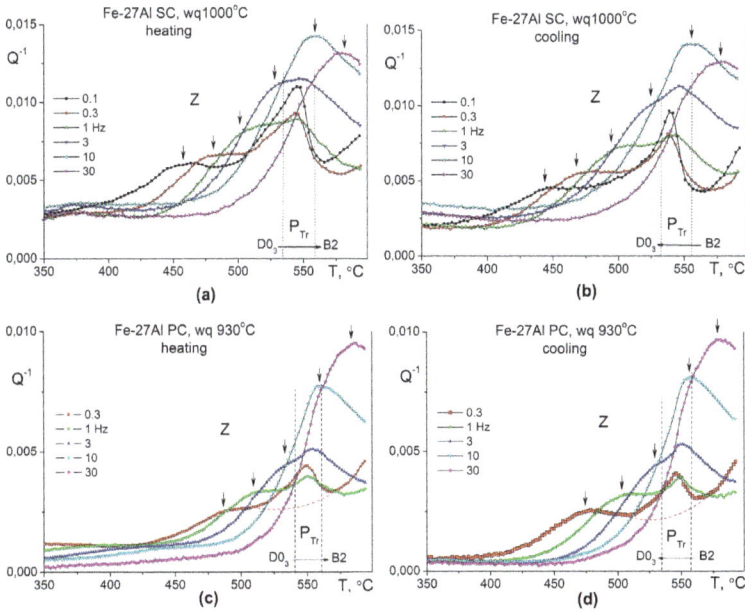

Fig. 3.29 TDIF curves for Fe-27Al single and polycrystal samples at heating (a and c) to 600 °C and subsequent cooling (b and d): f = 0.1, 0.3, 1, 3, 10, 30 Hz, $\varepsilon_0$ = 0.005%. DMA Q800.

We used a conventional Arrhenius treatment to evaluate the activation energy and pre-exponential time for recorded thermally activated processes (Table 3.2). It should be noted that the temperature control of relatively massive samples in the DMA equipment is not precise at least due to the temperature difference between inner and outer parts of the sample. Thus, the values presented in Table 3.2 should be considered as reasonable estimations that can be precise in future by using advanced equipment.

*Table 3.2 Parameters of the internal friction peaks in the single and poly Fe-27Al crystals.*

| Sample & heat treatment | $P_{Tr}$, °C | Zener relaxation | |
|---|---|---|---|
| | | H, eV | $\tau_0$, s |
| Fe+27Al PC, wq 930 °C | ~550 (heating) | 2.80 ± 0.16 | $2 \times 10^{-19}$ |
| | ~550 (cooling) | 2.27±0.45 | $2 \times 10^{-16}$ |
| Fe-27Al SC, wq 1000 °C | ~550 (heating) | 2.54 ± 0.17 | $4 \times 10^{-18}$ |
| | ~550 (cooling) | 2.21 ± 0.12 | $5 \times 10^{-16}$ |

The activation energy of the Z peak is observed in the single and poly Fe-27Al crystals at heating, cooling, and in subsequent heating-cooling tests varies from 2.2 to 2.8 eV and typical values for $\tau_0$ range from $10^{-17}$ to $10^{-19}$ s in agreement with the values of H from 2.5 to 3.2 eV and of $\tau_0$ from $10^{-18}$ to $10^{-20}$ s given in Ref. [138, 148] and H from 2.5 to 2.9 eV given in Ref. [125]. Even these activation parameters may correspond to both grain boundary relaxation, the presence of this peak in single crystal excludes grain boundary mechanism. The Zener peak (Z) in the Fe-27Al alloys overlaps with a transient frequency independent λ-type [6] IF peak, $P_{Tr}$, typical for second order transition, at 550 °C for both SC and PC samples at heating and cooling (Fig. 3.29).

Fig. 3.30a shows the scheme used to separate the Zener peak and a transitory effect. The main steps of this process are: 1) extrapolation of IF background between low- and high-temperature background (green), 2) fitting a thermally activated peak by the Debye equation using the activation energy from Table 2 (blue), and extrapolation of this peak until it meets background, and 3) subtraction from total experimental curves background and the Debye peak in order to outline the transient peak. Fig. 3.30b demonstrates that the temperature position of the Zener peaks measured at heating moves to higher temperatures with an increase in the measurement frequency and its height increases.

Transient and thermally activated peaks are close to each other, the temperature position of the $P_{Tr}$ remains practically independent on measuring frequency but its height decreases with increase in frequency. Fig. 3.30c shows Debye plot for the Zener relaxation at cooling and position for transient effect. The peak height of the transient peak $P_{Tr}$ vs. inverse measurement frequency is presented in Fig. 3.30c. The peak heights upon heating and cooling for the same frequencies are practically the same. The transient effect in single crystal is bigger for low frequencies $Q_{Tr}^{-1} = A/f$ (Fig. 3.30d).

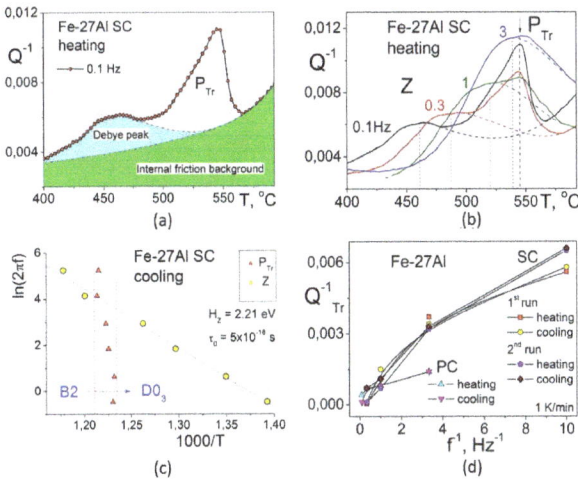

Fig. 3.30 Schematic illustration of the procedure used to separate a thermally activated peak and a transitory peak (a); change in the temperature position of the Z peaks at heating depending on the measurement frequency (b), Debye plot for Z relaxation at cooling and position of the $P_{Tr}$ effect on the same diagram (c), and height of transient peak vs. inverse measurement frequency (d).

Depending on the measuring frequency, Zener peak was measured in a different structural state of the sample: in the $D0_3$ state if the frequency is below 3 Hz or in the B2 state if the measuring frequency is above 3 Hz (Figs. 3.29 and 3.30). The Zener peak parameters depend on ordering-disordering of the Fe-Al alloys; and therefore, should be different in these phases. The calculations [57] indicate that Fe and Al atoms in the initial state and up to 500 °C are maximally ordered: the (1a) sites are completely filled with Fe atoms and 47 and 53% of the (1b) sites are filled with Fe and Al atoms, respectively.

Then, the degree of ordering decreases gradually and the ordering disappears stepwise at 720°C. Upon cooling, these processes repeat in the inverse order.

Indeed, the height of the Zener peak is different if it is measured using the frequency dependent tests for fixed temperatures either in the $D0_3$ or in the B2 temperature range (Fig. 3.31). The peak height increases with a step-by-step increase in temperature of the frequency dependent internal friction tests and corresponding decrease in the degree of ordering.

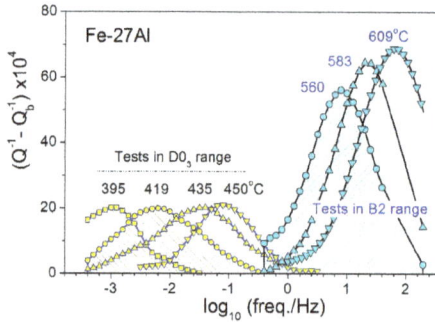

*Fig. 3.31 Frequency dependent internal friction of the Fe–27Al alloy measured at fixed temperatures during step by step heating at the temperature range of $D0_3$ and B2 phases. Forced torsion pendulum (A. Rivière).*

Using the Arrhenius plot for treatment of TDIF results without distinguishing between the points belonging to $D0_3$ and B2 phases introduces some mistake in the calculations. The relaxation strength of the Zener relaxation also depends on the degree of order, which decreases when the $D0_3 \rightarrow$ B2 transition takes place. This effect of order on the height of the Zener peak is clearly seen in the Fig. 3.11: when the peak is located in the B2 range, its height increases. Detailed observation of this effect was carried out using frequency dependent internal friction tests [141, 148]. Consequently, the isothermal mechanical spectroscopy frequency dependent tests [141, 148] demonstrate the difference in the activation energy of the Zener relaxation in $D0_3$ or B2 ordered states – Table 3.2.

The Zener peak parameters depend on ordering-disordering of the Fe-Al alloys, and therefore are different under iso- and anisothermal conditions of measurements. The isothermal IF technique gives the opportunity to study the Zener relaxation in different ranges of the Fe-Al phase diagram. From Table 3.2 it is easy to see that if the Zener peak is measured at fixed temperatures by varying frequency, the activation energy depends distinctly on ordering of the alloys.

*Table 3.3   Parameters of the Zener relaxation in Fe-21.7Al and Fe-25.9Al alloys and diffusion data for Fe-25.5Al [149, 151].*

| n | Alloy, at.% | Zener relaxation | | | | Activation energy of diffusion (Fe-25.5Al), eV [151] | | | Structure |
|---|---|---|---|---|---|---|---|---|---|
| | | $f,$ Hz | $T,$ K | $H,$ eV | $\tau_0,$ sec | Inter diffusion | Al tracer diffusion | Fe tracer diffusion | |
| I | Fe-21.7Al | $10^{-4}$-$10^2$ | < 730 | $2.82^{\pm0.06}$ | $9.1\times10^{-18}$ | | | | D0$_3$+A2 |
| | | | 730-773 | $2.48^{\pm0.06}$ | $4.3\times10^{-17}$ | | | | A2 |
| II | Fe-25.9Al | $10^{-4}$-$10^2$ | < 820 | $2.97^{\pm0.08}$ | $4.8\times10^{-19}$ | | | $2.89^{\pm0.05}$ | D0$_3$ |
| | | | > 820 | $2.44^{\pm0.02}$ | $8.0\times10^{-17}$ | $2.80^{\pm0.17}$ | $2.45^{\pm0.03}$ | $2.41^{\pm0.02}$ | B2 |
| | | | | | | $2.40^{\pm0.05}$ | | $2.26^{\pm0.08}$ | A2 |

The activation energies of the Zener relaxation in A2, B2 and D0$_3$ phases are in the same sequence as the activation energies of diffusion in these phases [149], i.e. $H_{A2} < H_{B2} <$ $H_{D03}$. The increase in the activation energy of the self-diffusion with increase in order parameter ($\eta$) is in an agreement with the Girifalco's theory [150]: $H(\eta) = H_{\eta=0}(1+ \alpha\eta^2)$, where $\alpha$ is a parameter. The activation energies of the Zener relaxation are lower than the activation energies of interdiffusion in B2 [151,152,153] and are practically the same as the activation energy of the Al tracer diffusion in B2 [151], i.e. the Al self diffusion in Fe-26Al, and slightly higher than the activation energy of the Fe tracer diffusion in the B2 and D0$_3$ [149]. Nevertheless, the fit between activation energy of the Zener relaxation and Al tracer diffusion in B2 range can be more or less incidental: first, the accuracy of the frequency dependent internal friction tests still can be improved in future, second, the reorientation of Al atom pairs involves also vacancies and Fe atom jumps. Most probable, that the activation energy of the Zener relaxation is below the activation energies for solvent and solute self diffusion as proposed in [34]. Anyway, it is notable that very limited number of diffusion data are available for Al self diffusion in the D0$_3$ range by "classical" diffusion methods due to low temperature of this phase existence while the internal friction technique provides an opportunity to study diffusion at relatively low temperatures, which also means - in relatively low-temperature phases.

The relaxation strength of the Zener relaxation depends on the degree of order, i.e. on the order parameter ($\eta$) which was varied in our experiments by varying temperature of the FDIF measurements. The theory of the Zener relaxation [34] proposes that for concentrated (in our case Fe-Al) alloys the relaxation magnitude ($\delta J$) is:

$$\delta J = \{f(\chi_o, C_{Al}) \times (C_{Al}^2(1-C_{Al})^2)/k_B T\} \times \beta a^2(dU_p/da_p), \tag{3.6}$$

where T is temperature, $k_B$ is Boltzmann's constant, the term $C_{Al}^2(1-C_{Al})^2$ exhibits the concentration dependence of $\delta J$: $C_{Al}$ is the atomic concentration of Al in iron; the function $f(\chi_o, C_{Al})$ reflects the effect of order of Al atoms on $\delta J$: $\chi_o$ is a parameter of short range order in absence of external stress, $f(\chi_o, C_{Al})=1$ for the disordered state and $f(\chi_o, C_{Al})=0$ for the complete ordered state (this means the Zener relaxation is impossible in 100% ordered alloy); $\beta$ is a dimensionless geometrical parameter, a is the interatomic spacing, $a_p$ – the same in the direction "p", $U_p$ is the ordering energy in the direction "p", p is a direction of applied stress.

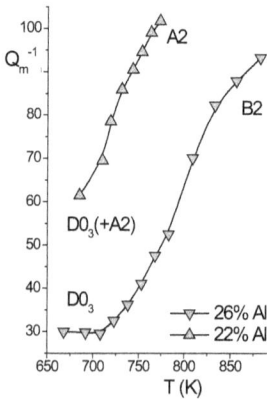

Fig. 3.32 Height of the Zener peak as a function of measuring temperature (A. Rivière).

It is clear from the eq. (3.6) that an increase in the Al concentration leads to an increase in the Zener peak height until it is possible to neglect the effect of ordering, i.e. until $f(\chi_o, C_{Al}) \approx 1$. Then the effect of ordering plays a more powerful role as compared with the effect of concentration, and the Zener peak height decreases. Such a behaviour for Fe-Al alloys was confirmed in many papers experimentally [122, 124, 126, 134, 138, 139]. Nevertheless, there is no well acceptable analytical solution of eq. (3.6) with respect to the effect of ordering as yet. The simplest linear approximation of the function $f(\chi_o, C_{Al})$ is $f(\chi_o, C_{Al}) = 1-\eta$. In case of a random distribution of Al atoms in iron (A2 range of the Fe-Al diagram) $\eta=0$, and $f(\chi_o, C_{Al})=1$.

In case of $\eta=1$, which in fact is impossible because of deviations from stochiometric composition of alloys and thermally activated disordering, $f(\chi_o, C_{Al}) =0$. In our case, the value of the order parameter is in the range $0 \leq \eta < 1$, $\eta \neq 1$ because of at least thermally activated disorder even in the $Fe_3Al$ composition. Quantitative analyses of the influence of the order parameter on the relaxation strength needs additional structural studies and together with a study of the influence of a "third" elements (Fe-Al-Me) is discussed in the next chapter. Qualitatively it is seen if the Zener peak height is compared in Fe-22Al and Fe-26Al as measured at the same temperature (Fig. 3.32): the peak in Fe-22Al is nearly two times higher as compared with the peak in Fe-26Al at ~470 °C (740 K). The effect of $C_{Al}$ should increase the peak height but this tendency is much weaker than compared with the effect of order: in Fe-22Al at the T=740 K value of $\eta \rightarrow 1$ (A2 range of diagram) and the function $f(\chi_o, C_{Al}) \rightarrow 1$, while T = 740 K in case of the Fe-26Al alloy corresponds to $DO_3$ order in Fe-Al phase diagram, i.e. $\eta$ and $f(\chi_o, C_{Al})$ are much smaller than one.

## References

[57]   Golovin, I.S.; Balagurov, A.M.; Bobrikov, I.A.; Cifre, J. Structure induced anelasticity in $Fe_3Me$ (Me = Al, Ga, Ge) alloys. J. All. Comp., 688 (2016) 310-319. https://doi.org/10.1016/j.jallcom.2016.06.277

[58]   Balagurov, A.M.; Bobrikov, I.A.; Mukhametuly, B.; Sumnikov, S.V.; Golovin, I.S. Coherent cluster atomic ordering in the Fe-27Al Intermetallic Compound. JETP Letters, 104, No. 8 (2016) 539-545. https://doi.org/10.1134/S0021364016200078

[59]   Golovin, I.S.; Palacheva, V.V.; Bazlov, A.I.; Cirfe, J.; Pons, J. Structure and anelasticity of $Fe_3Ga$ and $Fe_3(Ga,Al)$ type alloys. J. All. Comp., 644 (2015) 959-967. https://doi.org/10.1016/j.jallcom.2015.04.150

[60]   Dubov, L.; Shtotsky, Yu.; Akmalova, Yu.; Funticov, Yu.; Palacheva, V.; Bazlov, A.; Golovin, I.S. Ordering processes in Fe-Ga alloys studied by positron annihilation lifetime spectroscopy. Materials letters 171 (2016) 46-49. https://doi.org/10.1016/j.matlet.2016.02.051

[61]   Sykes, C.; Evans, H.: Transformation in iron-aluminium alloys. J. Iron Steel Inst., 131(1935) 225.

[62]   Quinn, C.J.; Grundy, P.J.; Mellors, N.J. The structural and magnetic properties of rapidly solidified Fe100-xGax alloys for 12.8<x<27.5. Journal of Magnetism and Magnetic Materials 361 (2014) 74–80. https://doi.org/10.1016/j.jmmm.2014.02.004

[63]   Golovin, I.S.; Jäger, S.; Mennerich, Chr.; Siemers, C.; Neuhäuser, H. Structure and anelasticity of $Fe_3Ge$ alloy. Intermetallics 15/12 (2007) 1548-1557. https://doi.org/10.1016/j.intermet.2007.06.004

[64] Yamamoto, H.; J. Phys. Soc. Jpn. 20 (1965) 2166-2169.
https://doi.org/10.1143/JPSJ.20.2166

[65] Drijver, J.W.; Sinnema, S.G.; van der F.Wonde J. Phys. F Metal. Phys., 6 (1976)
2165-2177. https://doi.org/10.1088/0305-4608/6/11/015

[66] Bradley, A.J.; Jay, A.H. "The formation of superstructure in alloys of iron and
aluminium" Proc. Roy. Soc. (London) A136 (1932) 210-232.
https://doi.org/10.1098/rspa.1932.0075

[67] Jones, F.W.; Sykes, C. "Atomic rearrangement process in the copper-gold alloy
Cu₃Au" Proc. Roy. Soc. (London) A166 (1932) 376-390.
https://doi.org/10.1098/rspa.1938.0099

[68] Wilson, A.J.C. "The reflexion of X-rays from the 'anti-phase nuclei' of AuCu₃"
Proc. Roy. Soc. A181 (1943) 360-368. https://doi.org/10.1098/rspa.1943.0013

[69] Scardi, P.; Leoni, M. Diffraction whole-pattern modeling study of anti-phase
domains in Cu₃Au. Acta Materialia 53 (2005) 5229-5239.
https://doi.org/10.1016/j.actamat.2005.08.002

[70] Warren, B.E. "X-ray diffraction" New York, 1969.

[71] Scardi, P. "Microstructural Properties: Lattice defects and Domain Size Effects" in
"Powder Diffraction. Theory and Practice" Ed-s R.F. Dinnebier, S.J.L. Billinge,
RSC Publishing, (2008) 376-413.

[72] Le Floc'h, D.; Loiseau, A.; Ricolleau, Ch.; Barreteau, C.; Caudron, R.; Ducastelle,
F.; Pénisson, J.M. Critical Behavior of Antiphase Boundaries in Fe₃Al close to the
D0₃ → B2 Phase Transition. Phys. Rev. Lett. 81 (1998) 2272-2275.
https://doi.org/10.1103/PhysRevLett.81.2272

[73] Stadler, L.-M.; Harder, R.; Robinson, I.K.; Rentenberger, Ch.; Karnthaler, H.-P.;
Sepiol, B.; Vogl, G. Coherent x-ray diffraction imaging of grown-in antiphase
boundaries in Fe₆₅Al₃₅. Phys. Rev. B 76 (2007) 014204.
https://doi.org/10.1103/PhysRevB.76.014204

[74] Shiiyama, K.; Ninomiya, H.; Eguchi, T. "Evolution of Antiphase Ordered Domain
Structure and Phase Separation Activated by Ordering" in "Research of Pattern
Formation" Ed. by R. Takaki, KTK Scientific Publisher, Tokyo (1994) 411-430.

[75] Golovin, I.S.; Pavlova, T.S.; Golovina, S.B.; Sinning, H.-R.; Golovin, S.A. Effect
of severe plastic deformation on internal friction of an Fe–26 at.% Al alloy and

titanium. Materials Science and Engineering A 442 (2006) 165–169. https://doi.org/10.1016/j.msea.2005.12.081

[76]    Bragg, W.L. The structure of a cold-worked metal Proc. Phys. Soc., 52 (1940) 105-109. https://doi.org/10.1088/0959-5309/52/1/315

[77]    Iveronova, V.I.; Minaev, A.I.; Silonov, V.M. X-ray diffuse scattering and temperature dependence of the Fe-Al alloys heat capacity. Physics of Metals and Metallography, 33 (1972) 978-983 (in Russian).

[78]    Warlimont, H.; Thomas, G. Two-Phase Microstructures of α-Fe–Al Alloys in the K-State. Metal. Sci. J. 4 (1970) 47-52. https://doi.org/10.1179/030634570790444158

[79]    Watanabe, D.; Morita, H.; Saito, H.; Ogawa, S. Transmission electron microscopic study of the K-state in iron-aluminium alloys. J. Phys. Soc. Jpn., 29 (1970) 722-729. https://doi.org/10.1143/JPSJ.29.722

[80]    Bhattacharyya, S.; Jinschek, J.R.; Khachaturyan, A.; Cao, H.; Li, J.F.; Viehland, D. Phys. Rev. B 77, 104107 (1-6) (2008) "Nanodispersed $DO_3$-phase nanostructures observed in magnetostrictive Fe–19%Ga Galfenol alloys".

[81]    Khachaturyan, A.G. The theory of structural transformation in solids. New York: Wiley; 1983.

[82]    Boisse, J.; Zapolsky, H.; Khachaturyan, A.G. Atomic-scale modeling of nanostructure formation in Fe-Ga alloys with giant magnetostriction: Cascade ordering and decomposition. Acta Materialia 59 (2011) 2656–2668. https://doi.org/10.1016/j.actamat.2011.01.002

[83]    Balagurov, A.M.; Golovin, I.S.; Bobrikov, I.A.; Palacheva, V.V.; Sumnikov, S.V.; Zlokazov, V.B. Comparative study of structural phase transitions in bulk and powdered Fe–27Ga alloy by real-time neutron thermodiffractometry. J. Appl. Cryst. 50 (2017) 198-210. https://doi.org/10.1107/S1600576716020045

[84]    Engdahl, G.: Handbook of Giant Magnetostrictive Materials, Academic, San Diego, CA 2000.

[85]    Etienne du Trémolet de Lacheisserie: Magnetostriction: Theory and Applications of Magnetoelasticity, CRC Press, Boston, MA, 1993

[86]    Atulasimha, J.; Flatau, A.B. A review of magnetostrictive iron–gallium alloys, Smart Mater. Struct. 20 (2011) 043001. https://doi.org/10.1088/0964-1726/20/4/043001

[87]  Clark, A.E. Ferromagnetic Materials, North-Holland Publ. Co., Amsterdam, 1980.

[88]  Coey, J.M.D. Magnetism and Magnetic Materials, Cambridge University Press, Cambridge, 2010. https://doi.org/10.1017/CBO9780511845000

[89]  Cullen, J.R.; Clark, A.E.; Wun-Fogle, M.; Restorff, J.B.; Lograsso, T.A. Magnetoelasticity of Fe-Ga and Fe-Al alloys, J Magn Magn Mater 226 (2001) 948–949. https://doi.org/10.1016/S0304-8853(00)00612-0

[90]  Clark, A.E.; Wun-Fogle, M.; Restorff, J.B.; Lograsso, T.A.; Cullen, J.R. Effect of quenching on the magnetostriction on $Fe_{1-x}Ga_x$ (0.13<x<0.21) IEEE Trans Magn. 37 (2001) 2678-2680. https://doi.org/10.1109/20.951272

[91]  Ueno, T.; Summers, E.; Wun-Fogle, M.; Higuchi, T. Micro-magnetostrictive vibrator using iron–gallium alloy, Sensors and Actuators A 148 (2008) 280–284. https://doi.org/10.1016/j.sna.2008.08.017

[92]  Kellogg, R.A.; Russell, A.M.; Lograsso, T.A.; Flatau, A.B.; Clark, A.E.; Wun-Fogle, M. Tensile properties of magnetostrictive iron–gallium alloys, Acta Materialia 52 (2004) 5043-5050. https://doi.org/10.1016/j.actamat.2004.07.007

[93]  Li, J.H.; Gao, X.X.; Zhu, J.; Bao, X.Q.; Xia, T.; Zhang, M.C. Ductility, texture and large magnetostriction of Fe–Ga-based sheets, Scripta Materialia 63 (2010) 246–249. https://doi.org/10.1016/j.scriptamat.2010.03.068

[94]  Guruswamy, S.; Srisukhumbowornchai, N.; Clark, A.E.; Restorff, J.B.; Wun-Fogle, M.; Strong, ductile, and low-field-magnetostrictive alloys based on Fe-Ga. Scripta mater., 43 (2000) 239-244. https://doi.org/10.1016/S1359-6462(00)00397-3

[95]  Kellogg, R.A.; PhD Thesis Engineering Mechanics, Development and modeling of iron-gallium alloys, Iowa State University, Ames, Iowa, 2003.

[96]  Clark, A.E.; Wun-Fogle, M.; Restorf, J.B.; Lograsso, T.A. Magnetostrictive properties of galfenol alloys under compressive stress. Mater. Trans. 43 (2002) 881-886. https://doi.org/10.2320/matertrans.43.881

[97]  Jayaraman, T.V.; Srisukhumbowornchai, N.; Guruswamy, S.; Free, M.L. Corrosion studies of single crystals of iron-gallium alloys in aqueous environments, Corros. Sci. 49 (2007) 4015-4027. https://doi.org/10.1016/j.corsci.2007.05.010

[98] Guruswamy, S.; Mungsantisuk, P.; Corson, R.; Srisukhumbowornchai, N. Rare-earth free Fe-Ga based magnetostrictive alloys for actuator and sensors, Trans. Indian Inst. Met. 57 (2004) 315-323.

[99] Andreev, A.V.; In: Buschow K.H.J., editor. Handbook of magnetic materials, vol. 8. Amsterdam: Elsevier; (1995) 59.

[100] Srisukhumbowornchai, N.; Guruswamy, S. Influence of ordering on the magnetostriction of Fe-27.5 at. % Ga alloys. J. Appl. Phys. 92 (2002)5371-5379. https://doi.org/10.1063/1.1508426

[101] Gou, J.; Liu, X.; Wu, K.; Wang, Yu.; Hu, S.; Zhao, H.; Xiao, A.; Ma, T.; Yan, M. Tailoring magnetostriction sign of ferromagnetic composite by increasing magnetic field strength. Applied Physics Letters, 109 (2016) 082404. https://doi.org/10.1063/1.4961668

[102] Ma, T.; Gou, J.; Hu, S.; Liu, X.; Wu, C.; Ren, S.; Zhao, H.; Xiao, A.; Jiang, C.; Ren, X.; Yan, M. Highly thermal-stable ferromagnetism by a natural composite. Nature communication 7:13937. https://doi.org/10.1038/ncomms13937

[103] Kawamiya, N.; Adachi, K.; Nakamura, Y. Magnetic properties and mössabauer investigations of Fe-Ga alloys. J. Phys. Soc. Jpn. 33 (1972) 1318–1327. https://doi.org/10.1143/JPSJ.33.1318

[104] Lograsso, T.A.; Summers, E.M. Detection and quantification of D0$_3$ chemical order in Fe-Ga alloys using high resolution X-ray diffraction. Materials Science & Engineering (A) 416 (2006) 240-245. https://doi.org/10.1016/j.msea.2005.10.035

[105] Lin, Y-C.; Lin, C-F. Effects of phase transformation on the microstructures and magnetostriction of Fe-Ga and Fe-Ga-Zn ferromagnetic shape memory alloys Journal of Applied Physics 117 (2016) 17A920-1.

[106] Li, M.; Li, J.; Bao, X.; Wang, J.; Zhao, Y.; Jiang, H.; Gao, X. Anomalous temperature dependence of Young's modulus in Fe$_{73}$Ga$_{27}$ alloys. Journal of Alloys and Compounds, 701 (2017) 768-773. https://doi.org/10.1016/j.jallcom.2017.01.175

[107] Palacheva, V.V.; Emdadi, A.; Emeis, F.; Bobrikov, I.A.; Balagurov, A.M.; Divinski, S.V.; Wilde, G.; Golovin, I.S. Phase transitions as a tool for tailoring magnetostriction in intrinsic Fe-Ga composites. Acta Materialia 130 (2017) 229-239. https://doi.org/10.1016/j.actamat.2017.03.049

[108] Ma, T.; Hu, S.; Bai, G.; Yan, M.; Lu, Y.; Li, H.; Peng, X.; Ren, X. Structural origin for the local strong anisotropy in melt-spun Fe-Ga-Tb: Tetragonal nanoparticles. Applied Physics Letters, 106 (2015) 112401. https://doi.org/10.1063/1.4915308

[109] Zhang, C.; Ma, T.; Sun, G. Tailoring volume magnetostriction of giant magnetostrictive materials by engineering magnetic domain morphology. Applied Physics Letters, 110 (2017) 062403. https://doi.org/10.1063/1.4975758

[110] Zhang, J.; Ma, T.; Yan, M. Magnetic force microscopy study of heat-treated $Fe_{81}Ga_{19}$ with different cooling rates, Physica B 405 (2010)3129-3134. https://doi.org/10.1016/j.physb.2010.04.027

[111] Emdadi, A.; Hossein Nedjad, S.; Badri Ghavifekr, H.; Effect of solidification texture on the magnetostrictive behavior of galfenol, Metallurgical and Materials Transactions A, 45 (2014) 906-910. https://doi.org/10.1007/s11661-013-2022-2

[112] Emdadi, A.; Cifre, J.; Dementeva, O.Yu.; Golovin, I.S. Effect of heat treatment on ordering and functional properties of the Fe–19Ga alloy, Journal of Alloys and Compounds 619 (2015) 58-65. https://doi.org/10.1016/j.jallcom.2014.08.231

[113] Emdadi, A.; Hossein Nedjad, S.; Badri Ghavifekr, H.; Kavanlouei, M.; Alijan Farzad Lahiji, F. Ramezankhani, V.; Microstructural dependence of magnetic and magnetostrictive properties in Fe-19Ga, Rare Metal, 2016 DOI: 10.1007/s12598-016-0800-x. https://doi.org/10.1007/s12598-016-0800-x

[114] Hubert A, Schafer R. Magnetic Domains: The Analysis of Magnetic Microstructures. 3rd ed. Berlin: Springer; 2000.

[115] Zhang, J.X.; Chen, L.Q. Phase-field microelasticity theory and micromagnetic simulations of domain structures in giant magnetostrictive materials, Acta Mater. 53 (2005) 2845–2855. https://doi.org/10.1016/j.actamat.2005.03.002

[116] Yasuda, H.Y.; Maruyama, T. Effects of Ga concentration, heat treatment and deformation temperature on pseudoelasticity of FeGa polycrystals, Materials Transactions, 54 (2013) 36-42.

[117] Golovin, I.S.; Balagurov, A.M.; Bobrikov, I.A.; Palacheva, V.V.; Cifre, J. Phase transition induced anelasticity in Fe-Ga alloys with 25 and 27%Ga, Journal of Alloys and Compounds, 675 (2016) 393-398.

[118] O'Handley, R.C.: Modern Magnetic Materials, 1st ed., Wiley, New York, NY, 2000.

[119] Jayaraman, T.V.; Corson, R.P.; Guruswamy, S. Ordering, magnetostriction, and elastic properties in Fe-27.5 at. % Ga alloy single crystals. J. Appl. Phys. 102 (2007) 053905.

[120] Golovin, I.S.; Balagurov, A.M.; Palacheva, V.V.; Bobrikov, I.A.; Zlokazov, V.B. In situ neutron diffraction study of bulk phase transitions in Fe-27Ga alloys. Materials Design, 98 (2016) 113-119. http://dx.doi.org/10.1016/j.matdes.2016.03.016

[121] Zhao, X.; Mellors, N.; Kilcoyne, S.; Lord, D. Neutron diffraction studies of magnetostrictive Fe–Ga alloy ribbons. Journal of Applied Physics, 103 (2008) 1-3,07B320.

[122] Fischbach, D.B. The Zener relaxation and a new magnetic relaxation effect in Fe-rich Fe-Al alloys Acta Met., 10 (1962) 319-326.

[123] Hren, J.A. The effect of atomic order and ferromagnetism on the elastic and anelastic properties of Fe-25 atom % Al. Phys. Stat. Sol. 3 (1963) 1603-1618.

[124] Tanaka K. Internal friction of iron-aluminium alloys containing carbon. J. Phys. Soc. Japan 30 (1971) 404-411.

[125] Golovin, I.S.; Blanter, M.S.; Pozdova, T.V.; Tanaka, K.; Magalas, L.B. Effect of substitutional ordering on the carbon Snoek relaxation in Fe-Al-C alloys. Physica Status Solidi (a) 168 (1998) 403-416.

[126] Golovin, I.S.; Pozdova, T.V.; Rokhmanov, N.Ya.; Mukherji, D. Relaxation Mechanisms in Fe-Al-C Alloys. Met. Mat. Trans.A 34 (2003) 255-266. doi:10.1016/j.intermet.2010.10.017

[127] Golovin, I.S.; Rivière, A. Mechanisms of anelasticity in Fe-13Ga alloy. Intermetallics, 19 (2011) 453-459.

[128] Golovin, I.S.; Cifre, J. Structural mechanisms of anelasticity in Fe–Ga-based alloys. Journal of Alloys and Compounds 584 (2014) 322-326. http://dx.doi.org/10.1016/j.jallcom.2013.09.077

[129] Snoek, J. Effect of small quantities of carbon and nitrogen on the elastic and plastic properties of iron. Physica 8 (1941) 711-733. and Physica 6 (1939) 591.

[130] Pascheto, W.; Johari, G.P. Annealing by Internal and Aging Friction of Interstitial C in a-Fe, As Measured by Internal Friction. Mat. and Met. Trans. 27A (1996) 2461.

[131] Blythe, H.J.; Kronmüller, H. A. Seeger, Phys. status sol. (a) 181 (2000) 233. https://doi.org/10.1002/1521-396X(200010)181:2%3C233::AID-PSSA233%3E3.0.CO;2-8

[132] Weller, M.; Anelastic relaxation of point defects in cubic crystals. J. de Physique IV, C.8, suppl. J. de Physiqie III 6 (1996) 63-69.

[133] Schaefer H.E., B. Damson, M. Weller, E. Arzt, E.P. George, Thermal Vacancies and High-Temperature Mechanical Properties of FeAl. Phys. Stat. Sol. (a) 160 (1997) 531-540.

[134] Hermann, W.; Ort, T.; Sockel, H.-G. Proc. 2. Int. Symp. Structural Intermetallics, 21-25 (1997) Champino PA, TMS Warrendale (Eds. M.V. Nathal, R. Dariola), 759-768.

[135] Golovin, I.S.; Golovina, S.B.; Strahl, A.; Neuhäuser, H.; Pavlova, T.S.; Golovin, S.A.; Schaller, R. Anelasticity of $Fe_3Al$. Scripta Materialia 50 (2004) 1187-1192. https://doi.org/10.1016/j.scriptamat.2003.12.020

[136] Blanter, M.S.; Golovin, I.S.; Sinning, H.-R. The mechanism of the anelastic X relaxation in the intermetallic compound $Fe_3Al$. Scripta Materiala 52 (2005) 57-62. https://doi.org/10.1016/j.scriptamat.2004.08.030

[137] Golovin, I.S.; Divinski, S.V.; Čížek, J.; Procházka, I.; Stein, F. Study of atoms diffusivity and related relaxation phenomena in $Fe_3Al$-(Ti,Nb)-C alloys. Acta Materialia 53/9 (2005) 2581-2594. https://doi.org/10.1016/j.actamat.2005.02.017

[138] Golovin, I.S.; Neuhäuser, H.; Rivière, A.; Strahl, A. Anelastisity of Fe-Al alloys, revisited. Intermetallics 12/2 (2004) 125-150. https://doi.org/10.1016/j.intermet.2003.10.003

[139] Shyne, J.C.; Sinnott, M.J. Trans AIME 218 (1960) 861-865.

[140] Nagy, A.; Harms, U.; Klose, F.; Neuhäuser, H. Mechanical spectroscopy of ordered ferromagnetic $Fe_3Al$ intermetallic compounds. Mater. Sci. Eng. A324 (2002) 68-72.

[141] Golovin, I.S.; Golovina, S.B. Effect of alloying α-Fe with aluminum, silicon, cobalt, and germanium on the Snoek Relaxation Parameters. The Physics of Metals and Metallography, 102, No. 6 (2006) 593-603. https://doi.org/10.1134/S0031918X06120064

[142] Golovin, I.S.; Rivière, A. Mechanical spectroscopy of the Zener relaxation in Fe-22Al and Fe-26Al alloys. Intermetallics 14 (2006) 570-577. https://doi.org/10.1016/j.intermet.2005.09.010

[143] Borah, M.C.; Leak, G.M. Metal Sci. 8 (1974) 315-316. https://doi.org/10.1179/msc.1974.8.1.315

[144] Golovin, I.S.; Neuhäuser, H.; Sinning, H.-R.; Siemers, C. Structure and anelasticity of ordered and disordered Fe-Ge alloys. Intermetallics 18 (2010) 913–921. https://doi.org/10.1016/j.intermet.2009.12.031

[145] Golovin, I.S.; Palacheva, V.V.; Bazlov, A.I.; Cifre, J.; Nollmann, N.; Divinski, S.V.; Wilde, G.; Diffusionless nature of $D0_3$ - $L1_2$ transition in $Fe_3Ga$ alloys. Journal of Alloys and Compounds 656 (2016) 897-902. http://dx.doi.org/10.1016/j.jallcom.2015.10.041

[146] Khachaturyan, A.G.; Viehland, D.D. Structurally Heterogeneous Model of Extrinsic Magnetostriction for Fe-Ga and Similar Magnetic Alloys. Metal Mater Trans A 38A (2007) 2308-2328.

[147] Golovin, I.S.; Balagurov, A.M.; Palacheva, V.V.; Emdadi, A.; Bobrikov, I.A.; Churyumov, A.Yu.; Cheverikin, V.V.; Pozdniakov, A.V.; Mikhaylovskaya, A.V.; Golovin, S.A. Influence of Tb on structure and properties of Fe-19%Ga and Fe-27%Ga alloys. Journal of Alloys and Compounds, 707C (2017) 51-56. http://dx.doi.org/10.1016/j.jallcom.2016.09.151

[148] Golovin, I.S.; Rivière, A. Internal friction in Fe-Al-Si alloys at elevated temperatures. Intermetallics 14 (2006) 1238-1244.

[149] Eggersmann, M.; Mehrer, H. Diffusion in intermetallic phases of the Fe-Al system. Phil. Mag. A, 80 (2000) 1219-1244.

[150] Girifalco, L.A. Vacancy concentration and diffusion in ordered – disordered alloys. J. Phys. Chem. Solids, 24 (1964) 323-333.

[151] Salamon, M.; Mehrer, H. Interdiffusion, Kirkendall effect, and Al self-diffusion in iron-aluminium alloys. Z. Metallkd, 96 (2005) 4-16.

[152] Helmut Mehrer. Diffusion in Solids: Fundamentals, Methods, Materials, Diffusion-Controlled Processes. Springer Berlin Heidelberg New York.

[153] Aloke Paul, Tomi Laurila, Vesa Vuorinen, Sergiy V. Divinski. Thermodynamics, Diffusion and the Kirkendall Effect in Solids. Springer Cham Heidelberg New York Dordrecht London, 2014.

Chapter 4

# Anelastic Relaxation Mechanisms in Iron and its Alloys

In this chapter several anelastic mechanisms in Fe-Al, Fe-Ga and Fe-Ge based alloys are considered. The relaxation mechanisms in Fe [16] and Fe-based alloys [11, 15, 154, 214] are associated with the diffusive motion under stress of point defects, the motion of dislocations or parts of them, and the motion of grain boundaries or other interfaces including the boundaries of the ferromagnetic domains. The hysteresis mechanisms are associated with dislocation motion and phase transformations.

Thermally activated point defect relaxation generally means an anelastic relaxation caused by a diffusive redistribution of point defects under the action of an applied stress ("diffusion under stress"). It is described by Debye equations (1.2). The main condition for relaxation process is that the symmetry of the local elastic distortions, caused by the defects in the crystal lattice, is lower than the symmetry of the lattice itself. The temperature of anelastic relaxation (i.e., of an internal friction peak at TDIF curves) is determined by the activation energy of diffusion of the point defect and by the frequency of vibrations. The relaxation strength is determined by the concentration of defects and by the strength of the individual, defect-induced distortions. Such a distortion field is also called an elastic dipole.

## Contents

**4.1  Thermally activated relaxation effects due to point defects**

**4.1.1  The Snoek relaxation**

The effect of anelastic relaxation in a steel tuning fork with a *b.c.c.* lattice, which later became classical, was for the first time experimentally described more than a century ago [155]. Its nature was related to the presence of interstitial atoms (carbon and nitrogen) in iron in 1939 [131]. Its full physical interpretation as an effect of directional diffusion of interstitial atoms under stress was given by Dr. Jacobus Louis Snoek in 1941 [17]. Later, a similar effect was revealed in a number of *b.c.c.* metals of group VB (V, Nb, Ta) and VIB (Cr, Mo, W) of the periodic table and was called the Snoek effect [6]. This phenomenon has been reviewed with respect to pure metals by Weller in [156].

The N and C atoms are located in the octahedral interstices of the *b.c.c.* metal. The two metal atoms nearest to the interstice (distance about $a/2$) move aside; the displacement of the other four atoms, located in a distance of $a/\sqrt{2}$, is about one order of magnitude smaller (Fig. 4.1). The resulting lattice distortions (elastic dipoles) are oriented mainly along one axis (*x, y, or z*) and have a tetragonal symmetry. Such an elastic dipole is described by the $\lambda$-tensor with two components the components $\lambda_1$ and $\lambda_2$ which are not equal to each other. The difference $|\lambda_2 - \lambda_1|$ determines the "elastic dipole strength".

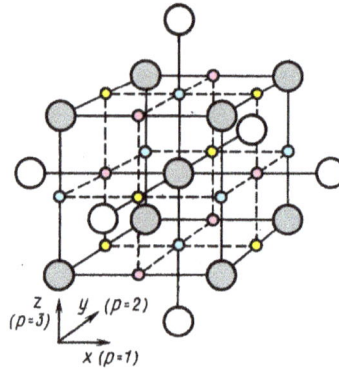

Fig. 4.1 Octahedral interstices in the b.c.c. crystal lattice: large circles are metal atoms; small circles are interstices of the sublattices $p = 1$, 2 and 3 (adopted from [6]).

The three types of interstices corresponding to the three lattice directions ($x$, $y$, $z$) form three sublattices (numbered $p = 1$, 2, 3). In the absence of external stresses, the dissolved IA are distributed uniformly among the interstices in all three sublattices: the related occupation probabilities $n_1$, $n_2$, and $n_3$ are equal to each other. By applying a tensile stress along one cubic cristal axis (e.g., "$x$" in Fig. 4.1), the arrangement of dissolved atoms in the octahedral interstices of the sublattice with the number $p = 1$ becomes energetically more favourable than those with $p = 2$ or 3. Therefore, the dissolved IA will diffuse from sublattices "2" and "3" into "1", and $n_1$ will be higher than $n_2$ and $n_3$. When the stress sign changes the reverse process sets in, and under the action of alternating periodic stresses this "diffusion under stress" of interstitial atoms (IA) causes periodic variations of the occupation numbers $n_1$ to $n_3$. This change in the distribution of IA among the sublattices of octahedral interstices causes an anelastic deformation of the crystal associated with a change in lattice spacings along the three main crystal axes.

The activation energy of the Snoek relaxation process is equal to the activation energy $H$ of the interstitial atoms diffusion: both processes have the same origin. The relaxation time $\tau$ of the Snoek relaxation process is associated with the diffusion of IA on the octahedral interstices, $D = D_0 \exp[-H/(RT)]$, where $D_0$ is the pre-exponential factor and $H$ is the activation energy of diffusion of interstitial atoms:

$$\tau = a^2/(36D), \tag{4.1}$$

where $a$ is the lattice parameter of the pure metal. Therefore, the activation energy of the Snoek relaxation is equal to the activation energy $H$ of the interstitial atoms diffusion. The Snoek peak temperature $T_m$ then follows from eq. (4.1) for $2\pi f \cdot \tau = 1$

$$T_m = H/\{R \ln[\pi a^2 f/(18 D_0)]\}. \tag{4.2}$$

$H$, $\tau_0$ and $T_m$, determined experimentally for $f = 1$ Hz, are listed in Table 4.1. Since the respective diffusion characteristics differ, the $T_m$ values are also different, but the nitrogen and carbon Snoek peaks overlap in α-Fe.

*Table 4.1  Snoek relaxation parameters [16].*

| System | $T_m$ [K] ($f$=1Hz) | $H$ [kJ/mol] | $\tau_0$ [$10^{-15}$ s] | $Q_m^{-1}$ |
|---|---|---|---|---|
| α – Fe-C | 314 | 83.7 | 1.89 | 0.215 |
| α – Fe-N | 300 | 78.8 | 2.38 | 0.200 |

(per one at.% of dissolved atoms, polycrystal)

The Snoek peak maximum $Q_m^{-1}$ in polycrystals depends on:

1) The atomic fraction of interstitial atoms in the solution ($C_0$),

2) The value of "elastic dipole strength" ($| \lambda_2 - \lambda_1 |$) produced by interstitial atoms (IA) located in octahedral interstices of the b.c.c. lattice of metals, and on the direction of the stress applied to the crystal lattice ($\Gamma$) according to equation (1.3c).

3) The peak maximum $Q_m^{-1}$ in single crystals depends on the direction of the stress applied to the crystal lattice (the orientation parameter $\Gamma$). This direction determines the change in the energy of a dissolved IA in the interstices of different sublattices if an external stress is applied. In a single crystal application of an external stress along the [111] direction deforms a crystal lattice equally along all three cubic axes and does not lead to relaxation.

For polycrystalline samples, averaging over all grain orientations gives for torsional vibrations

$$Q_m^{-1} = (0.4\ G/3) \cdot [cV \cdot (\lambda_2 - \lambda_1)^2 / (RT_m)], \tag{4.3a}$$

and for flexural vibrations

$$Q_m^{-1} = (0.4\ E/9) \cdot [cV \cdot (\lambda_2 - \lambda_1)^2 / (RT_m)], \tag{4.3b}$$

where $E$ and $G$ are Young's modulus and shear modulus respectively. The presence of a preferred crystallographic orientation of grains in a polycrystal, a texture, also leads to a dependence of $Q_m^{-1}$ on the type of oscillations and on the direction of applied stress. The height of the Snoek peak $Q_m^{-1}$ is proportional to $1/T_m$ (Eq.(4.3)), and therefore an increase in vibration frequency is associated with an increase of $T_m$ and decrease of $Q_m^{-1}$ for the same concentration of IA.

### 4.1.2 The Snoek-type relaxation

In $\alpha$-Fe carbon content up to 1 ppm can be detected in commercial steels [157]. Substitution atoms (SA) in a host lattice must influence the parameters of the original Snoek relaxation by SA-IA interaction, i.e. by a change in the activation energy of the relaxation, and by a change of the value $(\lambda_1 - \lambda_2)$, i.e. the relaxation strength. Some SA may not affect the position and height of the Snoek peak, others may reduce the peak height, lead to the appearance of additional peaks at higher temperature besides the height reduction, or suppress the Snoek peak and produce new peaks at higher temperatures. One can also find contradicting results in the literature.

Already in the early 1950s, first attempts were done to investigate the parameters of the Snoek peak by temperature dependence of internal friction not only in pure metals but also in b.c.c. alloys, first of all in iron-based alloys (Wert [158], Dijkstra and Sladek [159]). Finally, the Snoek relaxation can also be observed in many b.c.c. dilute alloys [11]. The theory of the influence of SA–IA interaction on the Snoek relaxation was given by Koiwa [18, 160] and has been proven in many experimental papers mainly for iron-based alloys.

The important question is if the *Snoek-type relaxation* can also be observed in b.c.c. alloys with significant concentration of substitutional atoms. The substitute atoms (SA) create energetically non-equivalent positions for interstitial atoms in the host b.c.c. lattice: the interaction between interstitial (i) and substitutional (s) atoms must influence the relaxation parameters. For this reason the Snoek-type relaxation in alloys is sometimes denoted in literature as the "i-s" peak. Nevertheless, this "i-s" relaxation in alloys can be explained in terms of the Snoek theory supplied with additional parameter of the SA–IA interatomic interaction in the crystal lattice which leads to a change in the activation energy of diffusion of dissolved IA near the relatively immobile SA. One can define this anelastic phenomena as the Snoek-type relaxation, i.e. the relaxation with the same mechanism originally proposed by Snoek.

Those trends which are known from the behaviour of heavy interstitial atoms (C, N) in low alloyed iron are inherited by Fe-based alloys and also in intermetallic compounds born from *b.c.c.* structure (e.g. the B2, D0$_3$ and L2$_1$ structures). Oxygen atoms do not contribute to the Snoek relaxation in iron.

Substitutional atoms in a host lattice must influence the parameters of the original Snoek relaxation via interstitial (i) and substitutional (s) atoms (or "i-s") interaction, i.e., by a change in the activation energy of the relaxation, and by a change of the value ($\lambda_1 - \lambda_2$), i.e., the relaxation strength. In spite of more than 50 years of study since the pioneering works of Wert on alloyed iron, the experimental situation about the influence of SA on the Snoek relaxation has remained not well systemised. Some SA may not affect the position and height of the Snoek peak, others may reduce the peak height, lead to the appearance of additional peaks at higher temperature besides the height reduction, or suppress the Snoek peak and produce new peaks at higher temperatures.

In the case of carbon Snoek relaxation in iron a clear effect of Al and Si substitute atoms has been noticed in the literature [150, 161, 162, 163, 164, 165, 166, 167, 168] in contrast with no-effect or very little influence of Mn, Ge, Ni. One can also find contradicting results in the literature for example with respect to Co and Cr in α-Fe. In some cases existing controversies can be explained by the fact that the authors studied alloys with different amounts of alloying elements. The nitrogen Snoek peak in iron is influenced by the presence of Cr [169], Mn [170, 171], Mo and V. The nitrogen Snoek peak plays a much less important role in Fe-Al-based alloys and intermetallics due to trapping of N into AlN. Examples of the carbon Snoek peak in Fe-Al, Fe-Co, Fe-Ge, Fe-Si with ~3% of substitute atoms are given in Fig. 4.2. Fitted Debye peaks are added in the insets.

The program employed [172] for the analysis of relaxation spectra is based on the equation for Debye internal friction peak as a function of temperature at a constant vibration frequency ($\omega = 2\pi f$) assuming an Arrhenius law for the frequency ($1/\tau$) of atomic jumps. In the case of a normal (Gaussian) distribution of the relaxation times $\tau$, the shape of the IF curve is determined by the broadening parameter $\beta$ (distribution of relaxation times):

$$Q^{-1}(T) = Q_{max}^{-1} \int_{-\infty}^{\infty} \frac{\exp(-z^2) \cdot \omega \cdot \tau \cdot \exp(\beta z) dz}{1 + [\omega \cdot \tau \cdot \exp(\beta z)]^2} =$$

$$Q_{max}^{-1} \cdot \omega \cdot \tau_0 \int_{-\infty}^{\infty} \frac{\exp(-z)^2 dz}{\exp(-(H/RT + \beta z)) + (\exp(H/RT + \beta z))}, \tag{4.4}$$

where $z = \ln\tau$, $H$ is the effective enthalpy of activation of the relaxation process, and $R$ is the universal gas constant. The program permits one to use two fitting parameters: the number of peaks that are introduced and the parameters of broadening of each peak. The enthalpy of the relaxation process can be taken from an independent source rather than be used as an adjustable parameter in the analysis of experimental data.

The fitting technique used for the analysis of frequency dependent IF spectra is similar: The program describes the relaxation Debye maximum of IF as a function of frequency at a constant temperature. The relaxation time $\tau_0$, the activation energy $H$, and the distribution of relaxation times $\beta$ can be fitted in the program. In case none of these parameters is known from the Arrhenius plot, they can be estimated from this fit for a given temperature of measurements, in case some of them are known from the Arrhenius plot they can be fixed in the program:

$$f(x,\beta) = \frac{Q^{-1}}{2Q_{max}^{-1}} = \frac{1}{\sqrt{\pi}} \int_{-\infty}^{\infty} \frac{\omega\tau \cdot \exp(\beta\xi)\exp(-\xi^2) \cdot d\xi}{1+(\omega\tau)^2 \exp(2\beta\xi)}, \qquad (4.5)$$

where $z = \ln(\tau/\tau_0)$ and $x = \ln(\omega\tau_0)$.

Results of the Snoek-type mechanism studies can be summarised with respect to alloying elements in *b.c.c.* iron in the following way: Every substitute (e.g., Al, B, Co, Cr, Cu, P, Si) reduces Snoek peak height even if the amount of solute C is the same, and this causes an increase in the proportionality constant between the Snoek peak height and carbon content. In the range where substitutes content is dilute, Co, Mn, Cr, Si, P and Al lead to decreases in Snoek peak height in this ascending order. Cu leads to additional damping component which is explained by C-Cu interaction. The solute carbon presence in the region where the lattice distortion around the substitute atoms is greater than the threshold value (the order of $10^{-3}$) cannot contribute to the Snoek peak and the volume of influence region increases as the difference in atomic size increases. The strain field generated by a substitutional atom due to the difference in atomic size is the reason for the reduction in Snoek peak height [168].

The Snoek peak in Fe-C-Ge and Fe-C-Co alloys is unimodal (Fig. 4.2a,b) at least up to a certain concentration of substitute atoms; its parameters are relatively close to those of the carbon Snoek peak in iron. Cobalt ($\leq 4\%$ [150]) increases the activation energy of the peak slightly, at higher concentration of Co in Fe the Fe-C-Co component of the Snoek peak can be observed [165]. The Snoek peak in Fe-C-Al alloys is formed by two components (Fig. 4.2c): a peak in which parameters correspond to those of the Snoek peak in pure iron (Fe-C-Fe) and a second peak whose parameters are determined by an additional interaction between C and Al atoms (Fe-C-Al). The Snoek peak in Fe–C–Si

(Fig. 4.2d) is at least broadened (as shown for Fe-3Si by dotted line); in some tests the Fe-C-Si relaxation component is clearly observed (Fig. 4.2d) [150, 163, 173]. The shape of the Snoek-type peak in Fe-Si alloys is sensitive not only to the heat treatment regime but also to inhomogeneous distribution of Si in Fe and can be slightly different from sample to sample: for that reason it is difficult to conclude if the peak at ~450 K (Fig. 4.2d) is a part of the Snoek peak or has another reason.

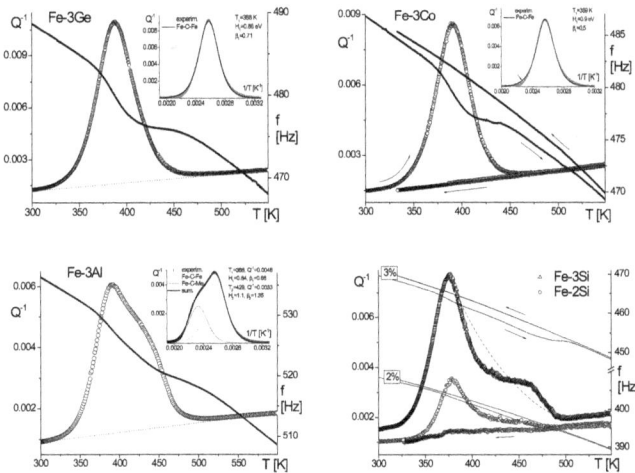

Fig. 4.2 Influence of Ge, Co, Si and Al (all SA ~3 at.%) on the carbon Snoek peak (vibrating reed, free-decay vibrations: S. Golovina. A. Strahl). Insets: fit of experimental data by Debye peaks.

The activation energy of the Fe-C-Al peak is higher than the basic peak by ~0.2 eV due to an additional (elastic) C-Al interaction in the solid solution. The Fe-C-Al peak is twofold as wide as the basic (Fe-C-Fe) peak, which reflects the existence of a set of different (in their energy state) positions for C atoms depending on their distance from an atom or atoms of Al. Based on the magnitude of the critical concentration at which the basic (Fe-C-Fe) peak vanishes and only the Fe-C-Al remains (~12%Al), i.e., at which the C atom always feels the Al atoms, the range of the C-Al interaction was estimated as equal to at least three coordination shells [150]. At a comparable concentration of Al and Si in the Fe-Al-Si alloys, Al exerts a greater effect on the profile of the temperature dependence of IF in the range of the Snoek relaxation, which is explained by its more efficient contribution to the long range elastic interaction with C atoms [150, 174].

In both binary (Fe-Al, Fe-Si) and ternary (Fe-Al-Si) alloys, an increase in the temperature of heating for quenching leads to an increase in the Fe-C-Fe component at the decrease in the Fe-C-Me component of the Snoek peak. Low temperature (<675 K) ageing of quenched in specimens leads to a decrease in both Fe-C-Fe and Fe-C-Me peaks, while ageing at higher temperatures changes TDIF and ADIF in several Fe-(1.5-4)Al-(2-4)Si alloys due to new carbon redistribution between solid solution and dislocations: The Snoek-type peak increases (C atoms go to solid solution), dislocation mobility and ADIF slope increases, and a new peak at ~600 K (450 Hz) due to motion of dislocations decorated with weak pinning points appears at higher temperatures as compared with the Snoek peak [150, 175].

Carbon forming elements (Cr, Mo, V, W) lead to a decrease in the Snoek-type peak height due to trapping of carbon to carbides, in a few papers the second Fe-C-Me peak was reported in these systems mainly for relatively "weak" carbide forming elements (Mn, Cr). Effect of other studied elements (e.g., As, Ni, Sn) is not well systemised.

Two additional components to the main nitrogen Snoek peak were noticed in the Fe-Mn-N system [138]: one (Mn-N-Mn component) below and one above (Fe-N-Mn) the Fe-N-Fe peak. Study of Fe–Cr–N alloys demonstrated clear appearance of the second component of the Snoek peak which can be called Fe-N-Cr for simplicity reasons. Within the framework of the simplified theory [18] where the influence of a substitutional atom is assumed to extend to the second neighbour shell, the observed relaxation spectra for Cr levels ranging from 0.05 to 0.2 at.% can be satisfactorily reproduced, and the binding energy of a Cr–N pair has been determined to be in the range between 0.16–0.18 eV [176]. Rare earth elements (As, Ce, La, Y) decrease and broaden the nitrogen Snoek peak, and a tendency to separate the peak into two peaks takes place [177]. According to the internal friction data, rare earth elements have a tendency to segregate. This short analysis of substitution atoms' influence on the Snoek relaxation in iron gives a key for interpretation of the contribution of interstitial atoms in Fe-based ordered alloys and even intermetallic compounds.

### 4.1.3 The Zener relaxation

The existence of solute next neighbour pairs or clusters results in a relaxation maximum, called „Zener peak", in a temperature range where the solute atoms are mobile and enable reorientation of the solute atom pair in the lattice under the action of the applied stress (Fig. 4.3). This applies for *f.c.c.*, *b.c.c.*, and *h.c.p.* alloys [6, 7, 11, 178] but not for pure metals including iron.

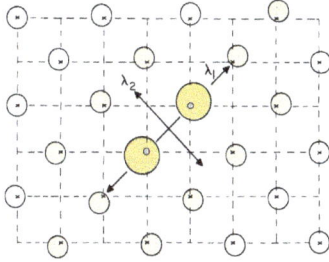

*Fig. 4.3 Schematic illustration of the atomic displacements about a pair of oversized solute atoms in the (100) plane of the fcc lattice. After C. Zener [3].*

As pairs of substitute atoms are involved, for dilute and random solid solutions the relaxation strength, which is proportional to the number of reorienting pairs, should be proportional to the solute concentration squared, $C^2$ (or rather, but in practice not distinguishable, $\propto C2(1-C)2$), with a Boltzmann factor including the binding energy $H_B$,

$$\Delta \sim C^2 \exp(H_B/kT) \quad \text{and} \quad \tau = \tau_0 \exp(H/kT). \tag{4.6}$$

As the atom movements during Zener relaxation (activation energy $H$, pre-exponential $\tau_0$) resemble those in ordinary diffusion, the rate of relaxation, or relaxation time, can be used to estimate the diffusion activation energies $H_D$ at temperatures far below those in ordinary bulk diffusion experiments, because in internal friction the site changes involve near neighbours only.

In the case of concentrated solid solutions the relaxation is accompanied by a change of the degree of short-range order. According to the theory of Le Claire and Lomer (1954) the Zener peak height is (see also eq. (3.4)):

$$Q_m^{-1} \propto [V_0\, f(\chi_0, C)\, C^2(1-C)^2 / (kT)] \sum_p (\lambda^{(p)})^2 , \tag{4.7}$$

where $V_0$ is the atomic volume; the coefficients $\lambda^{(p)} = (\partial\varepsilon/\partial\chi_p)_{\sigma,T}$ describe the crystal lattice distortion in the direction $p$, $\varepsilon$ is the deformation, $\chi_p$ is the number of bonds of one type in the direction $p$ ($\chi_p$ is the short-range order parameter), $\chi_0$ is the value of $\chi_p$ for zero stress $\sigma$. The function $f(\chi_0, C)$ varies between 0 for the totally ordered and 1 for the totally disordered alloy. Eq. (4.7) can approximately describe the Zener relaxation including the absence of the relaxation for ordered solid solutions.

## 4.2 Anelastic relaxation in binary Fe-Al alloys

### 4.2.1 Snoek-type relaxation

The Snoek-type relaxation in Fe-Al alloys was first discovered by Fishbach in 1962 [124], and reported later in many research papers [125, 126, 127, 128, 139, 140, 174, 179, 180, 181, 182].

*Influence of Al concentration.* The first systematical studies of the influence of Al content in iron on the Snoek-type relaxation were done by Jäniche et al. [179] and Tanaka [126]. Selected data from [126] are presented in the Fig. 4.4.a in normalised $((Q^{-1}-Q_b^{-1})/(Q_m^{-1}-Q_b^{-1}) = 1)$ form: original data are shown in the inset. The main result of C-Al interaction on the Snoek peak is that the peak shifts to a higher temperature, the activation energy of the peak and the peak width increase: the carbon atoms obtain new distribution in activation energies for diffusion jump 'under stress' [127]. The peak height decreases in accordance to the peak width: not the peak height but the area under the peak becomes proportional to the carbon content in Fe(Al). Two components of the Snoek peak are distinguished and computed [174, 181] (Fig. 4.4.b, data are normalised for better visualisation): the first peak corresponds to Fe-C-Fe positions, the second peak corresponds to Fe-C-Al positions for carbon atoms. The Fe-C-Fe peak decreases while the Fe-C-Al peak increases with each increase in Al concentration in iron.

*Influence of Al atom ordering.* For Fe-Al alloys with the Al concentration of more than 12%, the Fe-C-Fe peak is not distinguishable in the TDIF spectrum [150, 174]. On the other hand, the Al atoms ordering becomes possible if Al > 10-12%. The pronounced effect of ordering (annealing below 675K) and disordering (quenching from 1075-1275K) was recorded for Fe-22Al alloy [142]. Disordering results in the peak broadening due to an increase in different types of positions for C atoms in *b.c.c.* Fe(Al) solution, while the peak height decreases correspondingly. If ordering takes place, the number of positions for C atoms in Fe(Al) ordered solution decreases and the peak becomes more narrow but correspondingly higher. The effect of ordering depends on time and temperature of annealing [142]. The effect of ordering in as quenched and annealed alloys with a different Al content is shown in Fig. 4.5 [128]. The peak width in as quenched from disordered A2 range specimens increases significantly with increase in Al% in iron from ~10 to ~ 22%Al ($\beta_{max} > 3$), then decreases to $\beta = 1.5$-2.0 in the D0$_3$ and B2 ordered ranges. Similar effect in quenched and aged specimens is smaller because of short and long range ordering even for alloys with less than 22%Al, and $\beta \leq 1.5$ for alloys with Al > 22%.

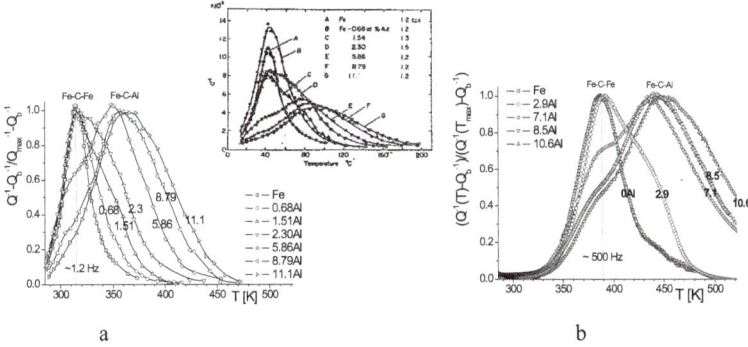

*Fig. 4.4  Influence of Al content (in the range between 0 and 12 at%) on "normalized" TDIF in bcc Fe-Al alloys: a) as measured at torsion at ~1.2 Hz by Tanaka [126], original data are presented in inset, and b) as measured at bending at ~500 Hz (A. Strahl et al. [174]). For better visualisation the peaks height is normalised to one: two components of the peak are seen in both cases. In all tests free-decay vibrations were used.*

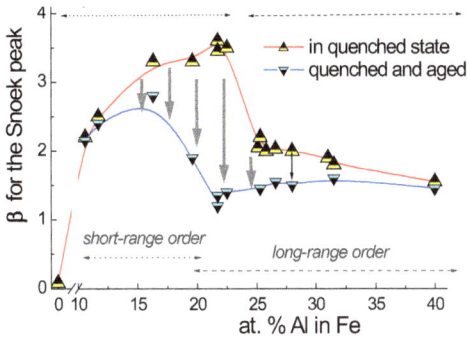

*Fig. 4.5  Influence of Al concentration and regime of heat treatment (quenched, i.e. partly disordered state, aged means the ordered state) on the width $\beta$ of the Snoek-type peak. Arrows indicate influence of ordering [128].*

***Influence of C content.*** The effect of carbon on the peak height was reported by several authors in 1970-1980 (e.g., [125, 126, 179, 180]) but questioned more recently in [135, 183], and reviewed in [140]. Practically linear increase in the peak height with increase in

C content in Fe-25/31Al alloys was reported [137]. The usage of single crystals of Fe-26Al confirms the typical for the Snoek relaxation orientation dependence of the $Q_m^{-1}$ on $\Gamma$ [184].

*Influence of other Me atoms.* The additions of carbide and non-carbide forming elements to Fe-(20-25)Al alloys acts differently on damping spectra. While non-carbide forming elements (Si, Ge, Co, Mn) produce like in $\alpha$-Fe little influence on the Snoek-type peak height, shape and temperature in Fe-Al, strong carbide formers (Nb, Ta, Ti, Zr) completely eliminate the Snoek-type peak confirming the carbon Snoek-type origin of this effect. Cr, being a less strong carbide-forming element, decreases peak, modifies peak shape and shifts it to a higher temperature.

*Contribution of vacancies.* The diffusivity of thermal vacancies in Fe-39Al (B2 phase of Fe-Al phase diagram) studied by positron lifetime spectroscopy [135, 183] are characterized by the effective enthalpies $H_F = 0.98$ eV for vacancy formation and $H_M = 1.7$ eV for vacancy migration. These values are close to the activation energy of the Snoek type and the so-called X relaxation (see next section). In case of quenching from high temperature these two types of relaxation (due to carbon and due to vacancies) can overlap in the same temperature range in Fe-Al alloys [128, 185]: increase in Al% and annealing temperature above 1000 °C before quenching increases contribution of vacancies. Nevertheless, the peak drastically decreases in the presence of carbide forming elements proving the dominating role of carbon (Fig. 4.6).

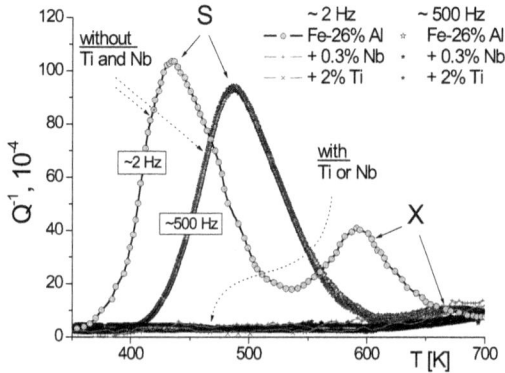

*Fig. 4.6 Influence of carbide forming elements (Nb and Ti) and measuring frequency (~2 Hz torsion and ~500 Hz bending) on the temperature dependent IF for Fe-26%Al alloy after water quenching from 1000 °C, free-decay vibrations [137].*

*Modelling.* Results presented in Figs. 4.4 and 4.5 were used in the simulations as a Snoek-type effect. They show that the Fe-C-Fe, i.e. "pure iron" component is not seen more in the Snoek-type peak if the Al concentration in iron is more than 10-12%. It means that if there are 10-12%Al in iron the Al atoms always affect C atom jumps in the Fe(Al) solid solution. From this the effective distance of the C-Al interatomic interaction can be estimated from $\sqrt{5} \cdot a / 2$ (*a* is lattice parameter) or as three coordination spheres as a minimum and $\sqrt{9} \cdot a / 2$ (six coordination spheres) as a maximum [174]. Monte Carlo simulations based on Khachaturian and Blanter approach using energies of the long-range strain-induced ('elastic') pair interatomic C-Al interaction supplemented by 'chemical' repulsion demonstrate [127] the strong influence of ordering reaction on carbon redistribution around Al atoms and correspondingly on parameters of the Snoek-type peak (Fig. 4.6): this effect is stronger in Fe-Al alloys with Al < 25% with the A2 to D0$_3$ transition [127] than in alloys with Al > 25at.% with the B2 to D0$_3$ transition [128]. The main factor which determines the effect of Al on the carbon Snoek-type peak is the long-range (up to five coordination shells [127]) 'elastic' interaction. The 'elastic' C-Al attraction significantly increases the peak temperature and the 'chemical' C-Al repulsion according to Lennard-Jones potential compensates this increase but not completely: the carbon Snoek peak temperature in Fe-Al is higher than in α-Fe.

Thus, the mechanism of the Snoek relaxation in metals may be extended to the Snoek-type mechanism in *b.c.c.* alloys, in particular in Fe-Al alloys. The Snoek-type mechanism means that the origin of this phenomena in alloys is the same with the Snoek relaxation in metals: an interstitial (in our case carbon) atom jumps under the stress in alloyed iron with *b.c.c.* stucture, and parameters of these jumps, i.e. parameters of the relaxation process are influenced by interstitial (C) – substitutional (Al) atoms interaction in solid solution. Foreign Me atoms and vacancies can modify parameters of the Snoek-type peak or contribute in the same range of temperatures independently.

### 4.2.2  Thermally activated X-relaxation

This relaxation effect with average activation energy about 1.7 eV has been repeatedly observed in Fe-Al alloys since 1996 by different authors [135, 183, 185, 186, 187]. The term "X peak" was introduced in our papers [128, 140] to call somehow this unknown relaxation effect. The temperature location and the peak height at TDIF for the X relaxation curve can be seen in Fig. 4.6 for $f \sim 2$ and 500 Hz. The X relaxation is practically not observed for alloys with Al < 25%, which is nevertheless possible is case of very quick quenching from high temperatures and using low frequency measurements even in Fe-12Al alloys.

**Proposed mechanisms.** An elegant interpretation of the X peak (e.g. in Fe-37.5Al alloy the peak parameters are: activation enthalpy $H = 1.7$ eV, $\tau_0 = 10^{-13}$ s and $Q_m^{-1} = 0.0012$) was suggested in terms of vacancy reorientation: as reorientation related effect of "pure" vacancies [183] or iron-cite-vacancy $V_{Fe}$-and-Fe atom complexes [135]. A strong argument favouring this approach is that the activation energy for vacancy migration and the activation energy deduced from X-peak are rather similar. A similar explanation but in terms of movement of Al atom by means of thermal vacancies was given more recently for the IF peak in Fe-29Al alloy with parameters: $H = 1.64$ eV, $\tau_0 = 1.2 \times 10^{-15}$ s and $Q_m^{-1} \sim 0.001$ in [188]. Contrary to the above vacancy-and-metal atom-related explanations it was noticed that a similar peak ($H = 1.6$ eV, $\tau_0 = 1.7 \times 10^{-14}$ s) in Fe-32Al depends on the carbon content and the peak was explained as the "second" Snoek-type peak (i.e. jumps of carbon atoms) in the presence of an additional phase, other than equilibrium D0$_3$ [186, 187, 189]. At the same time the "ordinary" Snoek-type peak in the D0$_3$ structure was reported at lower temperatures. The hypothesis of C-vacancy complex formation in Fe-Al-C-vac system is discussed in [190].

**Experimental results.** Several tests have been carried out to study the mechanism of the X relaxation peak in Fe-Al alloys (the X peak was also recorded in Fe$_3$Al intermetallic compound after quenching [142, 191]). The summary of these experiments shows that:

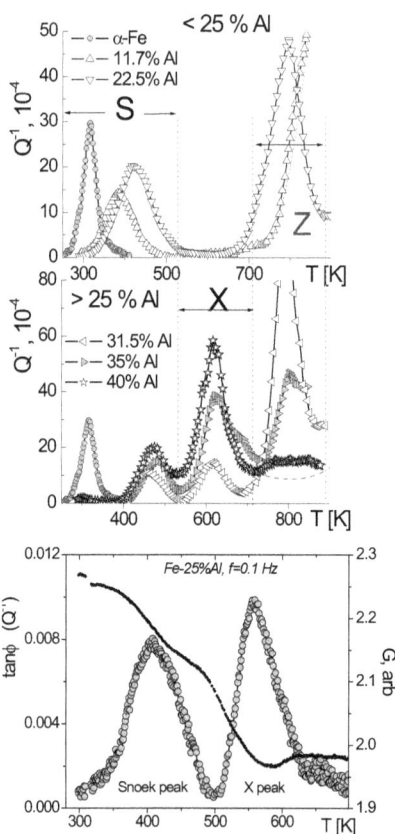

Fig. 4.7 a) Overview of the Snoek-type, X and Zener peaks in Fe-Al alloys with different Al% (~1 Hz: free-decay) and b) TDIF in Fe-25Al (0.1 Hz: forced vibrations), changes in relative shear modulus (right scale).

- The X-peak increases with increase in Al content at least between 25 and 40% (Fig. 4.7a) [142, 191]. In the range above 40%, the situation is more complicated; increase in annealing temperature before quenching (i.e. increase in concentration of thermal vacancies) increases the peak height. The difference of activation energies between the Snoek-type peak and the X peak remains about 0.5 eV in almost all tests [192];

- The X-peak height is very sensitive to the heating rate during TDIF measurements: the annealing of alloys decreases it drastically; moreover, the peak decreases at the same time with the measurements of the $Q^{-1}$ as a function of temperature. For this reason a decrease in resonance frequency, i.e. temperature of the peak, helps always to have a higher peak (Fig 4.7b): increase in the frequency from 0.1 to 500 Hz decreases the peak height by nearly one order of magnitude. This effect of heating rate on the peak height was studied by Han [185, 193].

- The X-peak dependence of the C content in as quenched Fe-25Al and Fe-31Al alloys was demonstrated mainly on a qualitative level [137] because of the peak instability with respect to heating during TDIF tests. In ternary Fe-Al-Me (Me = Cr, Ge, Mn, Nb, Si, Ta, Ti, Zr) alloys the peak was never recorded in the presence of strong carbide forming elements Ti, Nb and Zr, Ta; the same concerns also the Snoek-type peak. The peak is modified if Cr added in Fe-Al: peak shifts to higher temperature, i.e. higher activation enthalpy, and the peak height decreases. At cooling the X peak decreases more rapidly compared with the Snoek peak. The peak is only little changed if Si, Ge or Mn are added, some weak tendency to decrease in the peak temperature can be seen in the presence of Si and Ge in Fe-Al.

- The peak splits into two peaks if Al concentration in Fe-Al is above 30% (see e.g. Fig. 4.7a, curve for Fe-35Al), the height of each partial peak depends on quenching temperature and exhibits more complicated behaviour in alloys with more than 40%Al presenting probably both carbon-and-vacancy and triple vacancy complexes contribution. This later viewpoint was studied for alloys with >40%Al in [185, 193].

**Simulations.** Computer simulation of the X-relaxation in Fe₃Al was carried out using a model of carbon atom diffusion under the applied

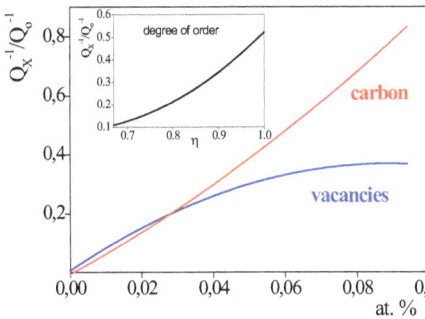

Fig. 4.8 Dependence of the X peak height (arb. units) on carbon concentration (1) at vacancy concentration 0.094% and on vacancy concentration (2) at carbon concentration 0.047 %. Inset: Dependence of additional peak high on a long-range order parameter (C=0.047 %, vacancies = 0.094%) (M. Blanter et al).

cyclic stress in an Fe$_3$Al-vacancy-C solid solution (see Appendix) and compared with experimental data. The following factors were taken into account in the modeling: concentration of C atoms, vacancies and degree of Al atom ordering (Fig. 4.8). The simulations show that carbon atom jumps near vacancies in the Fe$_3$Al intermetallic compound can lead to an anelastic relaxation with activation energy higher than that of the carbon Snoek-type relaxation. In terms of internal friction this type of anelasticity leads to the appearance of the X-peak. Thus, the X-peak is a complex effect in which both interstitial atoms and vacancies are involved [138, 193, 194].

From these experimental results and calculations it can be concluded that the peak is strongly affected by the presence of both carbon and vacancies in Fe(Al) solid solution. It can be suggested that complexes carbon-and-vacancy and vacancy-and-vacancy are responsible for the X-relaxation [192]. These two peak components are distinguished in [185, 193]: the 1$^{st}$ – smaller but more stable with respect to annealing - due to carbon-and-vacancy, the 2$^{nd}$ – due to vacancies complexes in Fe-Al-C-vac system. This second vacancy-related peak is very sensitive to heating, that is why it can be recorded only if a high heating rate (> 5 K/min) is used. At slower heating this peak decreases due to vacancy decrease during the measurements.

### 4.2.3  The Zener relaxation

Examples of the Zener internal friction peaks as measured as a function of temperature are presented in figures 4.6 and 4.7, and as a function of frequency – in figures 3.13 and 4.10.

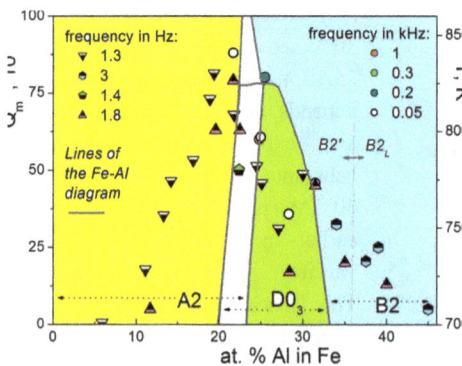

*Fig. 4.9   Zener peak height ($Q_m^{-1}$) vs. Al %. Experimental points from Refs. [124, 125, 135, 142, 174, 183, 186, 195, 196, 197].*

a) Temperature dependent internal friction (TDIF): overview. The Zener relaxation temperature in Fe-Al alloys as measured at frequency ~1 Hz is close to 790-805K, and it moves to a higher temperature if the frequency of measurements is higher [124, 125, 195]. This temperature range corresponds to the A2 disordered range of the Fe-Al phase diagram for alloys with Al <20-22%. This means that the relaxation strength of the r elaxation depends on Al% only because term f($\chi_o$,$C_{Al}$) ≈ 1 in eq. (4.7) in this range. Indeed, in this range Al < 20% the dependence $\Delta_Z \sim C_{Al}^2$ was observed in several papers [124, 125, 128, 141, 188, 195] as shown in Fig. 4.9. [196, 197]. Increase in Al concentration to 20% and higher leads to ordering ($f(\chi_o, C_{Al}) < 1$, in completely ordered state $f(\chi_o, C_{Al}) = 0$), and the peak decrease is clearly noticed for Al > 20%. The temperature for the $D0_3$-to-B2 order transition in binary Fe-Al alloys with Al concentration close to 25% is about 550 °C, i.e. close to the Zener peak position if measured at ~10 Hz. The disadvantage of the TDIF tests in the temperature range close to the order-disorder or order-order phase transformations is that the structural state of alloys changes during TDIF tests, which restricts applicability of the Arrhenius plot to determine activation diffusion of Al in iron. This effect will be considered below.

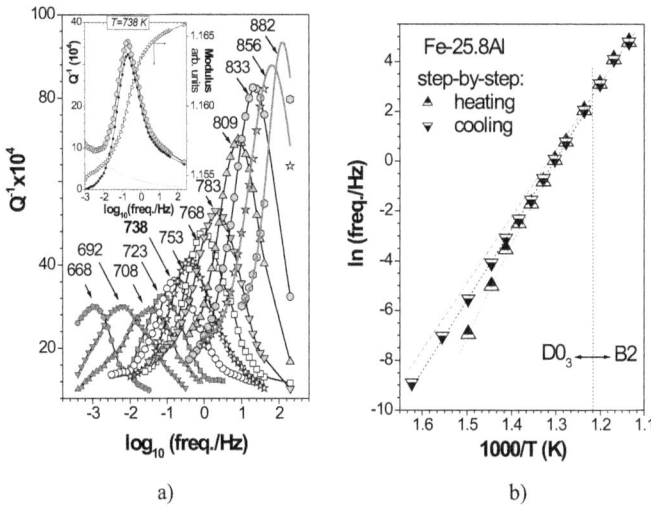

a)                                              b)

Fig. 4.10    Zener relaxation in the Fe-25.8Al alloy: (a) overview of Zener peaks measured at different temperatures between 668 and 882 K, forced torsion vibration; inset in left-upper corner: isothermal tests at 465°C: internal friction and relative modulus supplied with exponential background, and IF peak after background subtraction; (b) Arrhenius plot [144].

The TDIF tests demonstrate the Zener peak broadening in Fe-Al-Me (Me = Cr, Si) alloys. This effect comes from a contribution of Me atoms to the Zener effect [198, 199], i.e. to the short-range diffusion jumps. The Zener relaxation in Fe-Al-Si ternary alloys, as studied by TDIF, demonstrates a double-headed Zener peak caused by Al-Al and Si-Si pairs reorientation under stress Fe-Al-Si alloys [199]. The broadening of the Zener peak from high temperature side in Fe-Al-Cr alloys was observed by TDIF [198] and FDIF [200] tests.

**b) Frequency dependent internal friction (FDIF).** The isothermal FDIF spectroscopy allows one to measure a peak at a fixed temperature, i.e. at equilibrium conditions in different ranges of the phase diagram (Fig. 4.10a, inset). These tests demonstrate a difference in the Zener peak parameters in different phases, i.e. the difference in the relaxation strength (Fig. 4.10a) and activation energy (Fig. 4.10b) of the Zener relaxation in the $D0_3$ or B2 ordered and the A2 disordered states of the Fe-Al alloys [144, 150] and gives values of the activation energy in these phases very close to the diffusion experiments [151, 153].

The activation energies of the Zener relaxation in A2, B2 and $D0_3$ phases are in the same sequence as the activation energies of diffusion in these phases, i.e. $H_{A2} < H_{B2} < H_{D03}$ (Table 3.2). The increase in the activation energy of the Zener relaxation with increase in order parameter ($\eta$) is in agreement with the Girifalco's theory: $H(\eta) = H_{\eta=0} \times (1 + \alpha \eta^2)$, where $\alpha$ is a parameter. It is notable that the FDIF provides an opportunity to study diffusion at relatively low, untypical for diffusion experiments, temperatures, i.e. in low-temperature phases like the $D0_3$ phase in Fe-Al. The $f(\chi_0, C_{Al})$ function in eq. (3.4) can be analytically determined by *in situ* neutron diffraction study as discussed in the Chapter 3. The mechanism of the Zener relaxation in ternary Fe-Me1-Me2 alloys, e.g. in Fe-Al-Si and Fe-Al-Ga alloys remains to be interesting for further studies in order to find out contribution of Al-Al, Me-Me and Al-Me pairs in the relaxation process.

**c) Temperature dependent ordering and anelasticity in single and polycrystalline Fe₃Al-type intermetallic compound.** These results were partially considered in Chapter 3. To remind them in short:

- Upon instant heating, several phase transitions occur according to the neutron diffraction (Fig. 3.4): second order transitions between ordered phases (of B2 and $D0_3$ types) and "ordering B2 ↔ disordering A2" take place. Water quenching from 900°C prevents $D0_3$ ordering at room temperature but does not allow preserving the high temperature A2 structure, and the sample has the B2 structure at room temperature. In contrast, the $D0_3$ structure was recorded at room temperature in as- cast samples [57, 58]. Thus, at heating the transitions sequence is B2 → $D0_3$ → B2 → A2 for water quenched sample from 900°C or $D0_3$ → B2 → A2 for as-cast sample.

The B2 and $D0_3$ ordered phases appear and disappear as a result of Al atoms ordering in *b.c.c.* Fe-Al (A2) solid solution. First, the metastable B2 phase fixed at room temperature by water quenching transforms to equilinrium $D0_3$ phase at around 300-350°C. Then the equilibrium $D0_3$ phase transforms to equilibrium B2 phase at around 550°C. These processes are identified by the appearance and disappearance of D0₃ diffraction peaks of (111) and (311) and B2 peak of (100). Details of $D0_3 \rightarrow B2 \rightarrow A2$ transitions were recently described in our earlier papers [57, 58] including determination of the occupancy factors of sites by Fe and Al atoms. Ordering needs some time: that is why, the appearance of the D0₃ and B2 ordered phases depend on heating or cooling rate.

- Upon continuous cooling from 850°C, the $A2 \rightarrow B2 \rightarrow D0_3$ transitions take place.

For single crystal sample neutron diffraction was used mainly for testing its volume quality. Diffraction patterns measured for two directions are shown in Fig. 4.11. The profiles of diffraction lines are smooth; no indications of twins or irregular mosaicity were recognized confirming a high quality of this single crystal.

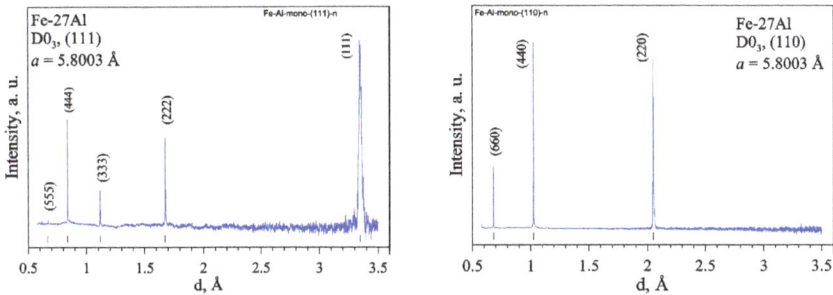

*Fig. 4.11 Fe-27Al single crystal diffraction patterns measured for (111) and (110) directions. Several orders of reflexions are seen, which calculated positions are shown as vertical bars.*

Fig. 4.12 shows the magnetization versus the temperature curve and the heat flow curve for the as-cast Fe-27 sample. The heating curve shows a decrease in the magnetization from ~130 emu/g at room temperature to ~30 emu/g at ~510 °C. Approximation of this curve to zero magnetization corresponds to ferromagnetic Curie temperature of the D0₃ phase (~525 °C) that is in a reasonable agreement with the magnetic phase diagram [37]. Further instant heating leads to a relative increase in total magnetization till ~550 °C. This confirms the structural change from D0₃ into the B2 phase [61]. The endothermic effect recorded in the heat flow curve (right scale at Fig. 4.12) at interval 510-560 °C agrees well with the temperature range recorded in the magnetization curve showing the D0₃→B2 transition temperature range. Upon further instant heating, the magnetization starts to decrease till zero at ~750 °C, which corresponds well to the structural change from the B2 to non-magnetic α-Fe (A2) phase. Upon cooling, the reverse transition of B2

→ $DO_3$ takes place practically at the same temperature range as at heating. The transition temperature of ~550 °C recorded using magnetometry and calorimetry techniques fit well with that registered by neutron diffraction.

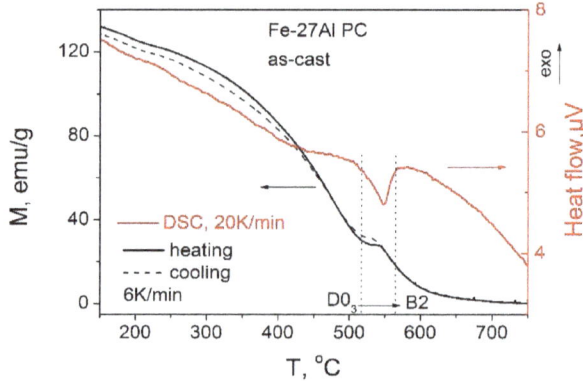

*Fig. 4.12  Magnetization versus the temperature curve (left scale, Chr. Mennerich) and the heat flow curve (right scale) for the as-cast Fe-27Al polycrystal sample.*

Temperature Dependent Internal Friction curves measured using five or six frequencies from 0.1 to 30 Hz and $\varepsilon_0 = 5\times10^{-5}$ at instant heating and cooling with a heating and cooling rate of 1 K/min are presented for Fe-27Al SC and PC alloys in Fig. 3.11. Several internal friction peaks can be distinguished at the TDIF curves in the temperature range of 350 to 600 °C: thermally activated Zener (Z) peaks and transition λ-shaped peak denoted as $P_{Tr}$ is non-thermally activated but that is related to phase transition at ~550 °C. Its height is reverse proportional to the measuring frequency in accordance with eq. (3.1). Upon instant cooling from 600 °C, the Zener peak can be distinguished again in both SC and PC alloys. Corresponding activation parameters for Zener relaxation are collected in Table 3.1. The transient $P_{Tr}$ peak is well recorded at cooling at about 550 °C for both SC and PC alloys as well.

The Zener peak overlaps with a transient frequency independent λ-type [6] internal friction peak, $P_{Tr}$, at about 550 °C for both SC and PC samples at heating and cooling (Fig 3.29). Figure 3.30 shows the scheme used to separate a thermally activated Zener relaxation and a transitory effect. Internal friction effect for second-order, often called lambda, λ, transition according to Landau's theory can be described by this equation: $Q_m^{-1}(T_C) = \chi^2 M/J_U \omega$.

### 4.2.4 High temperature relaxation: dislocations and grain boundaries

Grain boundary (GB) relaxation occurs in pure metals as well as in alloys. It may originate from GB sliding and disordering or reordering of atom groups in the GB, and movement of grain boundary dislocations. Grain boundary relaxation is one of the earliest examples of damping in polycrystals. Zener discussed the anelastic strain due to grain boundary sliding resulting from a shear stress and slip along the boundary of two adjacent crystals. The restoring force arises from the back stress built up at the triple junctions where the boundary ends. He found for polycrystals (grain size $d$) with random GB (width $\delta$) a relaxation time $\tau$ [21]:

$$\tau = \frac{\eta d}{G_U \delta} = \frac{\sigma d}{G_U v(0)} \quad , \tag{4.8}$$

where $\eta$ is an appropriate viscosity for grain boundary sliding (correlated with atomic self diffusion: $\eta = kT/Da$, $D = \frac{1}{2} v_j a^2$ is the diffusion constant, $v_j$ is the jump frequency, $a$ is the atomic distance), $G_U$ is unrelaxed shear modulus, $\sigma$ is acting stress, $v(0) = \sigma b/\eta$ is initial shear velocity.

A grain boundary sliding model with presence of particles at the boundaries has been proposed by Mosher and Raj [201]. They assume the rate of sliding to be controlled by diffusive accommodation across the particles involving either bulk ($D_V$) or boundary diffusion ($D_B$), and find

$$\tau = \frac{kT(1-\chi^2)}{\Omega E} \times \frac{p^4 d}{\lambda^2 \delta} \times \frac{1}{D_B} \tag{4.9}$$

(with $p$ = particle diameter, $\Omega$ = atomic volume), if the diffusion along the grain boundary (diffusion coefficient $D_B$) is dominant.

The corresponding damping peak is expected to occur at higher temperature than in pure metal without precipitates which retard the accommodation; however, the relaxation should have the same activation energy.

Finally, the models involving grain boundary dislocations should be mentioned (21,202), which also yield an Arrhenius-type expression for the relaxation time including the sum of the activation energies for jog formation and for self diffusion.

The variety of possible grain boundary microstructures, including faceting and the dependence of GB energies on GB orientation, implies a wide distribution of the relaxation times. Detailed account of the various structures of grain boundaries with different misorientations, applying the notions of incommensurate systems, and of the related internal friction (considered as due to a thermal excitation of the GB), has been provided by Darinskii et al.

The GB-related peak in Fe-Al is mentioned in several papers [125, 198, 203]. According to Hren [125] the GB peak in Fe-24.8Al (at ~700 °C, 950 Hz) is recorded only at ~50K higher temperature than the Zener relaxation for the same alloy. Wert [204] reported the GB peak in Fe-28.3%Al at higher temperature 750 °C (torsion frequency is not indicated in this paper but from the specimen size one can estimate it to be ~1 Hz). In both cases the increase in the grain size leads to a decrease in the IF peak height, and from this the GB origin of the peak was concluded. The internal friction peak at ~475 °C (~4 Hz) with $H \approx 1.8$ eV and $\tau_0 = 10^{-14}$ s in Fe-50%Al nanocrystalline alloy was also attributed by the authors to the GB peak [205].

Pavlova et al. [199] explained a decrease in the IF of Fe-Al-Si alloys in the temperature range from 870 to 980K with increase in annealing temperature of the specimens and corresponding growth of grain size by the GB relaxation. Lambri et al. [206] has repeated the measurements of the same Fe-Al-Si alloys to higher temperatures along with neutron diffraction studies, and he found that the IF peak at (~1000K) depends very much on the temperature of the $D0_3$ to B2 transition (Fig. 4.13). Contrary to a symmetrical shape of the peak in B2 phase, the high temperature peak is significantly suppressed in the $D0_3$ phase which can be explained by lower dislocation mobility in the $D0_3$ structure.

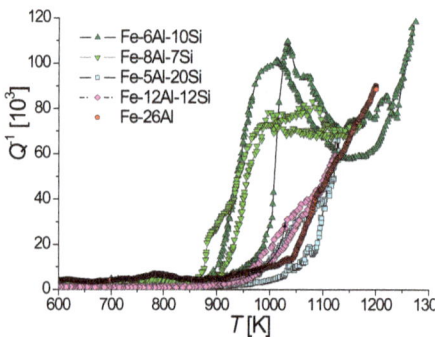

Similar conclusions were also done by Lambri et al [207, 208] with respect the influence of the order degree on the dislocations and grain boundary mobility in Fe-25Al-8Cr and Fe-25Al-25Cr alloys studied by means of mechanical spectroscopy and neutron diffraction studies. In ordered alloys independently of the type of order, $D0_3$ or B2, the mobility of dislocations and grain boundaries is markedly reduced, which is shown through small values of damping. In contrast, when the order decreases or disappears after "in situ" annealing both in the mechanical spectroscopy and neutron diffraction tests, the dislocation and

*Fig. 4.13 $Q^{-1}(T)$ curves in the range of elevated temperatures (f ~ 1Hz) for Fe-5Al-10Si, Fe-8Al-7Si, Fe-25Al, Fe-12Al-13Si, and Fe-5Al-20Si specimens after background subtraction (O. Lambri et al).*

grain boundary mobility increases which results in an increase in damping as compared with the ordered state. Similar interpretation for Fe-8Al-4Ge alloy [209] is doubtful as the $D0_3$ ordering is hardly possible in this composition especially at temperatures up to 670 °C. Even at room temperature our neutron diffraction study does not confirm any ordering neither in as cast nor in well annealed state.

High temperature IF peaks in binary Fe-38Al alloy were studied by San Juan et al. [206] using a forced torsion pendulum ($10^{-3}$ to 10 Hz). Two peaks have been observed at about 780 and 1100K (1 Hz), which are largely superposed in the intermediate temperature range. Both peaks were attributed to relaxation mechanisms. The low temperature peak was identified as the Zener relaxation of Al atoms. The activation energy of the high-temperature peak has been determined to be $H_{act}$ = 2.87±0.05 eV, a pre-exponential factor $\tau_0$ =$10^{-15}$ s, and a broadening of the peak of the Gaussian distribution with $\beta \approx 4$. These values were attributed to the dislocation glide relaxation mechanism. Authors concluded that the mechanisms operating in this temperature range should be related to those controlling creep in these materials.

The opportunity for the GB relaxation in this alloy was according to authors completely blocked by the presence of the $Y_2O_3$ particles, and other very small particles (less than 50 nm) of Y and Al oxides. The peak was attributed to the kink pair formation mechanism of the $\langle 100 \rangle$ dislocations on their $\{100\}$ glide planes. The measured $\tau_0$ (=$10^{-15}$ s), corresponds well to a dislocation motion controlled by the kink pair formation mechanism: in *b.c.c.* metals the value of $\tau_0$ for the so-called $\gamma$ peak due to kink pair formation on screw dislocations is between $10^{-13}$ and $5\times10^{-17}$ s [210], and in particular it is in between $4\times10^{-13}$ and $1.2\times10^{-15}$ s for pure iron [211]: (the activation energy of the $\gamma$ peak in iron is lower). Thus, this peak is associated with the intrinsic motion of such $\langle 100 \rangle$ dislocations in the B2 ordered phase which is in agreement with authors' conclusion about dislocation-related origin of this peak [207].

### 4.2.5 Low temperature relaxation: effect of deformation
A broad, sometimes plateau-like, internal friction peak in the temperature range of 200 to 300 K (activation enthalpy is roughly about 0.43 to 0.65 eV) was recorded in several Fe-(20-30) Al alloys at low temperatures (Fig. 4.14a) [137]. It was suggested to be caused by interaction between dislocations and point defects (vacancies, self interstitial atoms), and it is enhanced by deformation (D-peak). The D peak is observed in a temperature range where from plastic deformation studies it is known that the strain rate is controlled by single (partial) screw dislocations moving in the ordered lattice by nucleation and rapid sidewise motion of double kinks [212] with an activation enthalpy for double kink formation of $2H_k$ = 1.08 eV, i.e. roughly twice the D activation enthalpy. During

vibration with low amplitude, on dislocations which are in general inclined to the Peierls valleys, single kinks (generated at the anchoring points of dislocations) move in the microyield region according to the picture in [213], interacting with point defects. The large width of the D "peak" may be due to different kinds and configurations of obstacles for this sidewise kink motion and may be supposed to be composed of $\gamma$ peak [214] and Hasiguti peaks [214], consisting of reversible and irreversible components [215]. The inverse modulus ($E(T)$) effect accompanied the D peak in some tests [137] may also result from dragging of points defects according to [100], where "dragging" means in fact short-range diffusion in the dislocation core region.

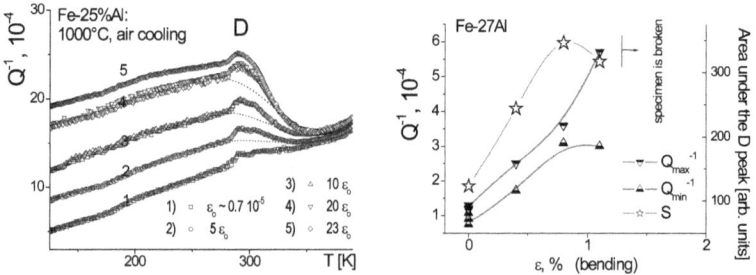

*Fig. 4.14. The D peak in Fe-25Al alloys: a) Influence of amplitude of measurements (curves 1 to 5) on TDIF in Fe-25Al alloy after air cooling from 1000 °C; b) Influence of cold-work deformation by bending on the D peak height (left scale) and area under the peak (right scale) for Fe-27Al. Before and between cold working the specimen was annealed at 500 °C, 1 h.*

A larger peak is observed after quenching than after slow cooling [137], quenching produces a high vacancy concentration (vacancies can help in dislocation kink formation and migration), and additional dislocations may have been also produced due to thermal stresses. The D peak divides IF background into two parts: below peak temperature of the low-temperature background is higher, and above the peak – background is lower. This amplitude dependent effect is bigger if higher amplitude of deformation is used (Fig. 4.14a). The decrease of the D peak itself with increase in Al concentration (in quenched alloys with 22 to 31%Al) indicates a contribution of vacancies whose concentration is known to increase [216]. For high Al concentrations (>30 at.% Al) the D peak seems to be suppressed. While the Snoek and X peak height clearly correlates with the C content, the D peak shows a complex behavior, possibly because the dislocation kink mobility is affected by carbon, vacancies and self interstitial atoms. The D peak is also observed in several ternary, e.g. in Fe-Al-Cr, alloys [217].

Fig. 4.14b shows an example of the D peaks in Fe-26.6at%Al after annealing at 500 °C and step by step bending deformation. The increase of $Q_{Dmax}^{-1}$ up to a strain of $\varepsilon \approx 1$ % reflects the increase of dislocation density in microyield, where long straight screw dislocations are produced by sidewise movement of the easily mobile edge components [213]. The observed dependence supports the idea that at least a part of the damping effect is due to kink nucleation and motion in dislocations.

As low ductility of Fe$_3$Al prevented a closer study of the influence of bending deformation, samples of the intermetallic compound Fe$_3$Al were subjected to high pressure torsion (HPT) deformation. Typically, the average grain size ($\bar{d}$) in as cast Fe$_3$Al is ~0.1 mm. Well-annealed (72 h at 300 °C) Fe$_3$Al is characterised by D0$_3$ atomic order with clearly visible domains and antiphase boundaries (Fig. 4.15a). After the HPT deformation (P = 3 GPa, $\gamma$ = 160) [218], $\bar{d}$ is comparable to the average size of the D0$_3$ domains: ultra fine grains with a mean size of ~100 nm (at 3 mm distance from the centre, Fig. 4.15b) are dominating in the outer parts of the HPT specimen. Further heating to 650 °C during the TDIF tests not only removes the IF peaks but also leads to recrystallization, grain growth, and atomic ordering (Fig. 4.15c).

The D-peak in Fe-25Al is shaped by HPT into a family of five IF peaks: 1, 2$_1$, 2$_2$, 2$_3$, 3 (Fig. 4.15d) which were separated by a numerical decomposition procedure (dashed lines). The D peaks family was also recorded in different Fe-Al-Me alloys (Me = Cr, Ge, Mn, Si etc.) [196, 219]. Comparison between the D peaks in Fe-26Al and Fe-26Al-5Cr [217] shows a difference in the magnitude of the partial peaks from the D-family (Fig. 4.15e). These single peaks are similar to the Hasiguti peaks [215] in cold worked or irradiated metals, i.e. they are caused by coupling of dislocations and different point defects [214].

A distinct group is formed by the "2$_n$" peaks, which are reduced by annealing more effectively than the other peaks (Fig. 4.15 c,d). These 2$_n$ peaks are apparently caused by relatively unstable configurations or associations of point defects. After heating to 230 °C with 1 K/min, neither a modulus defect nor the group of the 2$_n$ peaks are detected any more. Generally these peaks can be classified as Hasiguti peaks, which typically consist of a several Debye peaks caused by coupling of dislocations and point defects and their segregations. The interaction of the strain field of a dislocation with the strain field of point defects leads to relaxation effects if either the dislocation or the point defect move, or if the dislocation breaks away from the point defect with a characteristic thermally activated relaxation time. Better understanding of atomistic mechanism of the single peaks in $Fe_3Al$ should contribute to better understanding of both dislocation behaviour at low amplitudes of vibration in Fe-Al and instability at early stages of annealing of severely deformed alloys [217].

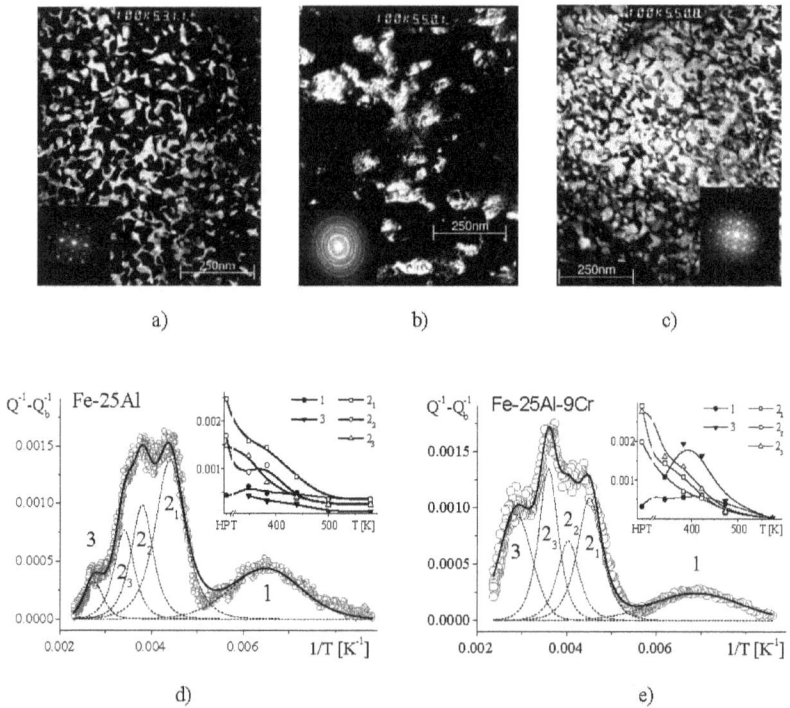

Fig. 4.15 Structure of Fe-25Al (dark field): a) antiphase $D0_3$ domains in annealed at 300 °C state ([110](111)) prior to deformation, b) size of grains after HPT deformation, and c) HTP specimen after heating to 650 °C: antiphase $D0_3$ domains. All images are taken in ~3 mm from the specimen center. Deconvolution of the D peak into several partial peaks for Fe-25Al (d) and Fe-25Al-9Cr (e): Insets - dependence of partial peaks from annealing temperature for Fe-25Al and Fe-25Al-9Cr alloys [214, 215].

### 4.2.6 Hysteretic effects

Amplitude dependent damping in iron and iron-based alloys consists of two contributions: magnetic - due to motion of magnetic domain walls (DW), and non-magnetic – mainly due to dislocation motion. Applying a stress to a ferromagnetic material causes a change of magnetisation due to the magneto-elastic coupling, which results in the so-called "ΔE effect", i.e. a reduction of Young's modulus below the value found in the magnetically saturated state, and also in a related dissipation of mechanical energy during loading and unloading or in case of vibration [16] the latter effect gives rise to a strong magnetomechanical damping with stress-dependent and stress-independent components. Four main mechanisms of magnetomechanical damping may be defined [220, 221]:

- macroeddy-current damping ($Q_a^{-1}$),
- microeddy-current damping ($Q_\mu^{-1}$),
- damping at magnetic transformations ($Q_{PhT}^{-1}$),
- magnetoelastic hysteresis damping ($Q_h^{-1}$).

Therefore, the total magnetomechanical damping $Q_M^{-1}$ in ferromagnets can be considered as a sum of all these components:

$$Q_M^{-1}(\varepsilon,f,T) = Q_h^{-1}(\varepsilon,f,T) + Q_a^{-1}(f,T) + Q_\mu^{-1}(f,T) + Q_{Ph.T}^{-1}.$$

It is notable that contrary to $Q_a^{-1}$ and $Q_\mu^{-1}$, the hysteretic contribution $Q_h^{-1}$ depends on the strain amplitude $\varepsilon$.

*Macroeddy-current* amplitude _in_dependent damping ($Q_a^{-1}$) is a result of the eddy currents induced temporarily in an electrically conductive sample as a response to a stress-induced change $dB/d\sigma$ in the total magnetic induction B.

*Microeddy-current* amplitude _in_dependent damping ($Q_\mu^{-1}$) is due to the microscopic eddy currents caused by local changes of the magnetisation arising during the stress-induced displacement of magnetic domain walls. Depending on the frequency range and on the type of motion assumed for the domain walls, different Debye-type and non Debye-type equations have been suggested for $Q_\mu^{-1}$.

*Damping at magnetic transformations,* $Q_{PhT}^{-1}$ (e.g. at Curie and Néel temperatures, spin-flip transitions etc.) is found in a narrow interval around Curie ($T_C$) or Néel ($T_N$) points. $Q_{PhT}^{-1}$ may result from the influence of an external stress on the processes of magnetic ordering. Therefor the relaxation strength of the peak decreases with increase in frequency of the tests. An example of IF spin-flip transition (~120K) and Néel temperature (~310K) in Chromium is given in Fig. 4.16 [222].

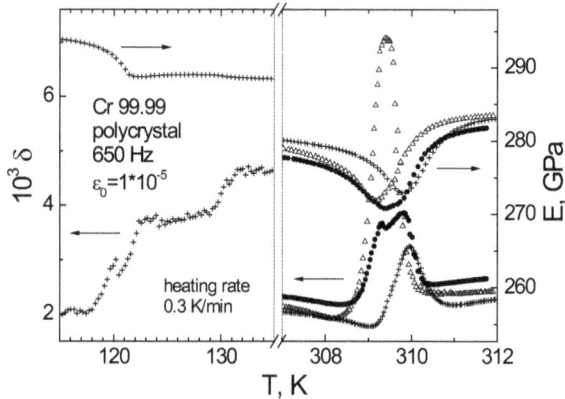

*Fig. 4.16. Effect of heat treatments on the behavior of the decrement and the Young's modulus near the Néel point and variation of the low-frequency acoustic properties in the vicinity of the spin-flip transition: • after quenching from 1000 °C, + after quenching from 1250 °C, Δ after quenching from 1000 °C and subsequent annealing at 460 °C for 2 hours.*

**Magnetoelastic hysteresis damping** $(Q_h^{-1})$ is the most important and powerful type of damping in ferromagnetic α-Fe and Fe-based *b.c.c.* alloys. It is due to the stress-induced motion of non-180° domain walls (while the 180° walls are stress-insensitive), including irreversible Barkhausen jumps beyond a critical stress $\sigma_{cr}$. At low applied shear stresses σ (or shear strains) (i.e. in the "Rayleigh region" of the magnetisation curve), the hysteretic energy dissipated is

$$\Delta W_h = \frac{4}{3}\frac{dG^{-1}}{d\sigma}\sigma^3, \text{ so that } Q_h^{-1} = \frac{\Delta W_h}{2\pi W}. \tag{4.10}$$

As the vibration energy $W$ varies with $\sigma^2$, this implies a linear amplitude dependence of internal friction. For higher stresses, however, the hysteretic losses $\Delta W_h$ no longer increase with σ3 but with a stress-dependent exponent $0 < n < 3$, and finally reach a saturation level; consequently, $Q_h^{-1}$ shows a *maximum* as a function of stress or strain amplitude. Relating the saturation value of $\Delta W_h$ to the magnitude of the "effective" internal stresses opposed to the movement of the domain walls. The position and height of the amplitude-dependent damping maximum of $Q_h^{-1}$ can be used to determine the level of these internal stresses, and found a value of 7 MPa for the case of high-purity iron if carefully recrystallized.

The temperature dependence of $Q_h^{-1}$, although always approaching zero at the Curie point, may vary from material to material. The frequency dependence of $Q_h^{-1}$ is not expected as long as the time for an irreversible domain jump, related to the relaxation time of the microeddy currents, can be neglected: thus, $Q_h^{-1}$ should decline at frequencies where the microeddy current peak appears [6].

To summarise, the different dependencies of magnetoelastic hysteresis damping on temperature, amplitude, frequency, and magnetic field are given schematically, in simplified form, in Fig. 4.17.

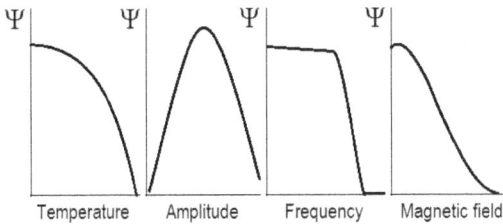

*Fig. 4.17 Schematic representation of the dependence of magnetomechanical damping (specific damping index) on temperature, strain amplitude, frequency, and magnetic field (adapted from [7]).*

Magneto-mechanical and non-magnetic dislocation-related contributions in Fe-Al can be separated using external saturated magnetic field in which magnetic contribution is completely suppressed. Both dislocations and magnetic domain walls contributions to amplitude dependent internal friction are also temperature dependent and practically frequency independent in Hz and kHz ranges. High damping capacity due to magnetomechanical hysteresis is determined by irreversible movement of domain walls (90°-type for Fe-Al alloys). The dissipated energy accompanying DW motion depends on the DW mobility. The mobility of DW is controlled by three factors:

a. Structure and size of magnetic domains and DW, i.e. magnetic parameters of alloys;

b. Structure of the crystalline lattice in which the DW moves;

c. Interaction between the DW and imperfections of the crystalline lattice.

There is only a limited number of papers considering all these three points together. Phenomenologically the energy loss $\Delta W$ due to magnetic domains for a vibration stress ($\sigma$) below some critical stress ($\sigma_c$) is given by the Rayleigh law:

$$\Delta W = D\sigma^3 \ ( \ \sigma < \sigma_c \ ; D \text{ is constant}). \tag{4.11.a}$$

For large stresses $\Delta W$ saturates and is given by:

$$\Delta W = k\lambda_S\sigma_c \ ( \ \sigma > \sigma_c \ ), \tag{4.11.b}$$

and the specific damping capacity $\Psi$ is defined as $\Psi = \Delta W/W$.

The magnetomechanical damping in ferromagnetic materials has its source in the stress-driven irreversible movement of the magnetic DW. The maximum damping at ADIF is proportional to $\lambda_S E/\sigma_i$ ($\lambda_S$ is the saturation magnetostriction, and $\sigma_i$ is the average internal stress opposing domain boundary motion, $k = 1$ is a constant characteristic of the shape of the hysteresis loop) [31]. For a Maxwell distribution of internal stresses the value of hysteretic damping ($Q_h^{-1}$) was described by Smith and Birchak as:

$$Q_h^{-1} = 0.34k\lambda_S E \ / \ \pi\sigma_i \text{ (at } \sigma \approx \sigma_{max}), \tag{4.12.a}$$

$$Q_h^{-1} = 4 \ k\lambda_S E \ / \ 3\pi\sigma_i^2 \text{ (at } \sigma < \sigma_{max}). \tag{4.12.b}$$

The dependence of $Q_h^{-1}$ on $\sigma_i$ for constant amplitude of external stresses $\sigma$: $Q_h^{-1} \sim 1/\sigma_i m$, where $m = 2$ ( for $\sigma \ll \sigma_i$ ) and $m = 1$ (for $\sigma = \sigma_{max}$).

The magneto-mechanical amplitude-dependent contribution to damping in Fe-Al alloys was studied in several research papers [51, 140, 223, 224 225, 226, 227]. In particular, it was shown that this contribution occurs at room and elevated temperatures, and that it is high enough to consider Fe-(10–12at.%)Al alloys as high damping materials with $\Psi$ up to 60% for free-decay vibrating-reed tests (Fig. 4.18). Here we would like to remind that absolute values for damping depends strongly on measuring technique and type of vibrations. Thus, one should consider several intrinsic and extrinsic factors which influence on damping capacity of Fe-Al alloys: the examples of intrinsic factors are heat treatment which influences on crystalline and magnetic structure of the alloys, grain boundaries contribution etc. External factors are amplitude and type of vibrations, external magnetic field and surface stress (e.g. due to polishing or clamping), temperature of tests. Damping capacity is the property which is strongly dependent on both intrinsic (structure) and extrinsic (test parameters) factors. This is the reason which makes comparison of damping capacity of materials measured by different techniques to be very difficult and often not reliable.

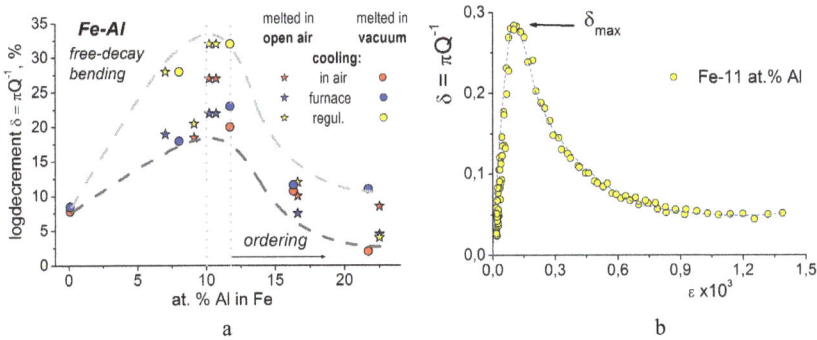

Fig. 4.18 a) Influence of Al content in Fe-Al binary alloys and heat treatment (cooling in air, in furnace, stepwise regulated cooling) on maximal damping (value of logarithmic decrement $\delta_{max}$ at ADIF curves), b) ADIF curve for Fe-11%Al with the peak due to magnetomechanical losses. Free-decay vibrations (I. Chudakov [224-226]).

### 4.2.7 Alloys with 10-12 at.%Al

The samples cut from a cold rolled Fe-5.4wt.%Al-0.05wt.%Ti (roughly Fe-11at.%Al) sheet were studied after annealing at different temperatures, annealing times, and cooling rates [228]. Fig. 4.19 shows an average grain size after annealing at different temperatures from 700 °C to 1050 °C and during different annealing time at 900 °C (inset). With an increase in the annealing temperature, the average grain size increases. Longer annealing times at 900°C also lead to larger grains.

Table 4.2 illustrates the dependence of the mechanical properties of the samples on the annealing temperatures and annealing times of the alloys. If the annealing temperature increases from 500 °C to 1000 °C, the hardness, the yield point, the tensile strength, and the elongation decrease, except for the temperature range of 600-700 °C, where the elongation increases sharply due to recovery, recrystallization and grain growth. As the annealing temperature increases, the grain growth is the main microstructural change in the samples, which leads to a decrease in the mechanical characteristics of this alloy. A sharp rise in elongation and a sharp hardness reduction in the temperature interval 600-700 °C is related to the recrystallization with a corresponding growth of new grains with low dislocation density.

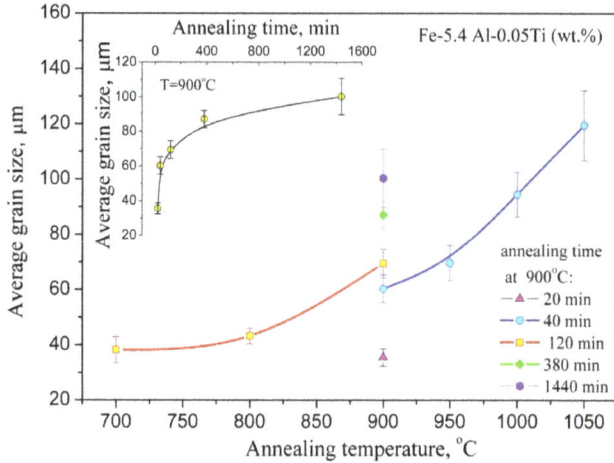

*Fig. 4.19 The grain size versus different annealing temperatures and times, inset shows the grain size as a function of annealing time at 900 °C (M. Nartey et al. [228]).*

*Effect of average grain size.* Annealing treatment can remarkably improve the damping capacity of ferromagnetic alloys [129, 229, 230, 231, 232, 233, 234]. Fig. 4.20a shows an amplitude dependent internal friction curves, ADIF, for bending forced vibrations (DMA Q800, single cantilever) for the Fe-5.4wt.%Al-0.05wt.%Ti sheet after different annealing regimes. Except for the cold rolled state, all the curves exhibit maximum at ADIF curves. This maximum characterizes the damping capacity of the alloy after a chosen heat treatment. Note, that damping capacity as measured using forced vibrations is about twice lower compared with free-decay tests (Fig. 4.18).

With an increasing annealing temperature the maximal damping (peak height at ADIF curves) increases and shifts to a lower amplitude of forced vibrations. This behavior is similar to the behavior of Fe-Cr based alloys [235]. Practically a linear correlation is found between the damping capacity and the grain size in the studied range of annealing regimes (Fig. 4.20b). The reason is related to the following factors:

(i) an annealing treatment lessens the crystal defects, such as dislocations, a crystal lattice strain, and residual stresses which are imposed on the material during a cold–rolling process,

(ii) a decrease in the grain boundary area due to the grain growth.

*Table 4.2   Mechanical properties of the Fe–5.4Al–0.05Ti alloy at different annealing states.*

| Annealing temperature (°C), time | HV | Yield Strength $\sigma_Y$ (MPa) | Tensile Strength $\sigma_U$ (MPa) | Elongation (%) |
|---|---|---|---|---|
| Cold rolled | 339±12.5 | 793 | 807 | 5.6 |
| 500, 2h | 313±7.9 | 679 | 715 | 4.6 |
| 600, 2h | 279.6±11.4 | 467 | 548 | 7.4 |
| 700, 2h | 201.1±6.1 | 324 | 454 | 40.0 |
| 800, 2h | 192±4.1 | 304 | 433 | 37.0 |
| 900, 40min | 189.6±7.4 | – | – | – |
| 900, 2h | 180.1±8.3 | 296 | 415 | 33.2 |
| 1000, 40min | 187.7±7.6 | 297 | 403 | 27.5 |

Crystal defects, internal stresses, and grain boundaries decrease the mobility of domain walls by pinning magnetic domain wall motions during a vibration stress; and therefore, they reduce the damping capacity of the materials [235, 236, 237, 238, 239]. The same factors decrease yield and tensile strength of the studied alloy (Fig. 4.21) exhibiting inverse dependence between damping and strength of the alloy. The highest damping was achieved in these tests after 40 min annealing at 1050 °C and corresponds to the average grain size of 120 μm. This is in agreement with the results described in paper [229], where a increase in the damping capacity, measured at free-decay conditions at frequency of 120 Hz, with an increase in grain size was reported below critical grain size of about 200 μm.

The energy loss occurs if the vibration stress exceeds the local stress barrier opposing domain wall motions [240]. With an increase in the annealing temperature, the maximum strain amplitude $\varepsilon_{max}$ shifts to lower values (Fig. 4.20). It reflects that high temperature annealing followed by air cooling lowers the amount of internal local stresses: a lower stress requires overcoming the barrier and mo ves magnetic domains. A similar behavior for $\varepsilon_{max}$ was reported for Fe-16Cr-4Mo ferritic steel [229].

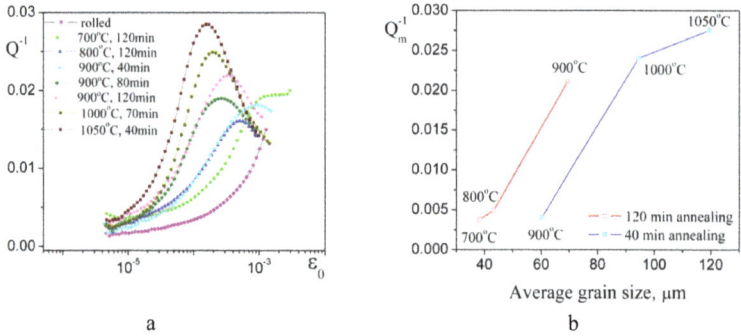

a                                                                                      b

Fig. 4.20 a) The annealing effect on the damping capacity for the annealed and air cooled samples; b) Damping capacity against the grain size (M. Nartey). DMA, forced vibrations.

To better clarify the contribution of grain boundaries in the damping capacity, we also measured the damping capacity of a polycrystalline and a single crystalline Fe-Al alloy cooled in air (see next section). The damping capacity of the single crystalline sample was nearly five times higher than that of the polycrystalline sample. It reflects a strong negative role of grain boundaries on the damping capacity due to pinning magnetic domain wall motions in response to the vibration stress.

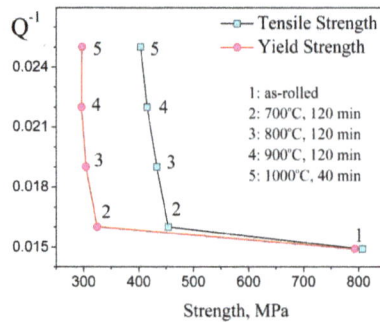

Fig. 4.21 Damping capacity against strength of the Fe-5.4Al-0.05Ti alloy after different heat treatments. DMA Q800, forced vibrations.

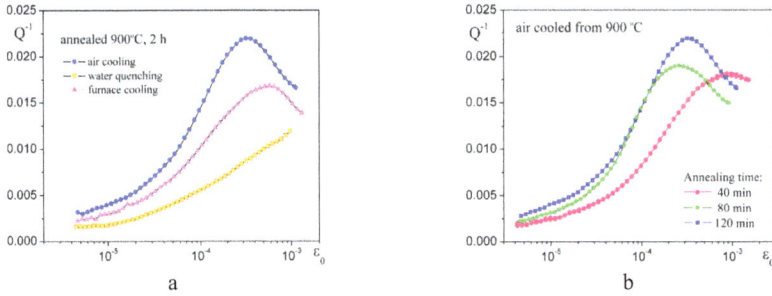

a        b

*Fig. 4.22 (a) Effect of cooling rate on damping capacity. (b). Effect of annealing time on damping capacity of the air cooled sample. DMA Q800, forced vibrations.*

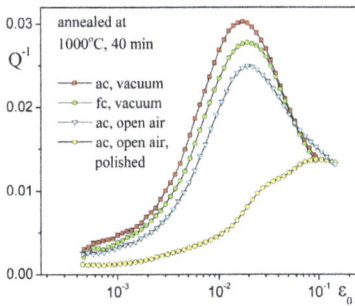

*Fig. 4.23 The effect of cleaning of the samples surface and vacuum treatment on the damping capacity. DMA Q800, forced vibrations.*

*Effect of cooling rate and annealing time.* Cooling rate is known to play an important role in achieving desirable damping capacity [140, 241, 242]. Fig. 4.22a illustrates a difference between the damping capacity of the samples annealed for 120 min at 900 °C with an average grain size of 70 μm and cooled in three different regimes: in water (water quenching, $wq$), open air (air cooling, $ac$), and in a furnace (furnace cooling, $fc$). Water quenching completely suppresses a magnetomechanical damping peak, whereas air cooling shows a peak slightly higher than 0.02. Furnace cooling exhibits a smaller magnetomechanical damping peak. The reason why the damping capacity depends on the cooling rate is discussed below. Fig. 4.22b shows that the variation of the annealing time for a fixed cooling regime also contributes to the damping optimization. A peak emergence due to the annealing time effect can be mainly attributed to recrystallization and grain growth. The average grain size was found to be 60.4±5.0 and 69.4±5.1 μm respectively for 40 and 120 min annealing at 900 °C (Fig 4.19) but since the damping capacity is greatly dependent on the cooling rate, it is difficult to distinguish between these two factors: cooling in air is not well defined and, consequently, well reproducible regime of heat treatment.

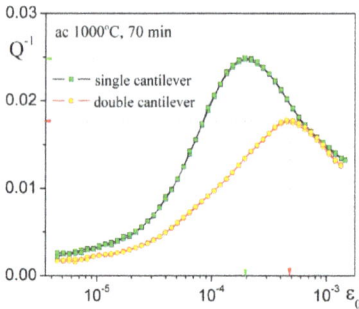

*Fig. 4.24. The influence of a measuring mode on the damping capacity. DMA Q800, forced vibrations.*

A magnetomechanical damping behavior has its origin in an irreversible stress-induced movement of magnetic domain walls. Therefore, the damping capacity is dependent on magnetic domain structures. The domains inside each grain are more regular and finer in the air cooled sample as compared to the *wq* and *fc* samples. The finer the domains are the more domains exist in the same grain and the domain wall area per unit volume in a grain is larger in the ac sample, making more vibration energy exhausted and resulting in a higher internal friction value [228]. A similar phenomenon was observed in the former research into Fe-Cr-Mo alloys [243]. Internal stresses do not only create the domain structures disorder but they also pin the domain wall and subsequently deteriorate the damping capacity. In the *fc* sample, the domains are not uniformed and somewhere irregular maze like domains with a low degree of alignment can be seen. This type of domains can be related to a short range ordering of Al atoms in α-Fe lattice as reported in many papers and denoted at the phase diagram as K1 state [37]. Two steps cooling regime: slower cooling to 550-400 °C to avoid thermal stresses and to allow only very weak short range ordering followed by faster cooling at lower temperatures can be used to increase damping capacity of Fe-(12-18)%Al alloys. The domain morphology strongly relies on the cooling rate, affecting the magnetomechanical damping capacity of an alloy.

*Effect of mechanical cleaning of the sample surface and treatment in vacuum.* For the samples annealed in open air, the samples surface is oxidized. Cleaning of the surface by grinding and subsequent polishing drastically decreases the damping capacity (Fig. 4.23) of the samples due to the local stresses introduced into the sample mechanical treatment. The internal stresses impede magnetic domain wall displacements, resulting in a significant lowering of the samples damping capacity. Consequenctly the samples were put inside quartz ampules and air was pumped out, then they were treated similarly to those samples annealed in open air.

For the samples annealed in vacuum, air cooling proved to be the best cooling medium for the rolled sheet in the optimization of the damping capacity. Cooling in air naked samples and the samples inside the quertz ampules are characterised by different cooling rates. It is very likely from the obtained results that the annealing for longer times in vacuum at 1000 °C will further enhance the damping capacity in comparison with the other treatments. Nevertheless, this conclusion is not supported by the results of the paper [137], in which a decrease in damping is reported if a grain size becomes larger than 200 μm. In order to check our hypothesis, we have compared damping in two Fe-Al alloys (see next section) – a poly and single crystal. Damping in a single crystal was about five times higher than that of a polycrystal. This confirms that an increase in the grain size enchances the magnetomechanical damping. As can be seen in Fig. 4.23, air cooling in both cooling media enhanced the damping but the sample treated in vacuum has higher damping capacity.

*Effect of testing method and static stress.* The dependence of internal friction at amplitude 0.02%, which is close to the amplitude of the peak maximum at ADIF curves, on frequency of forced vibrations was studied in the range between 0.01 and 30 Hz. The damping capacity is rather stable in this range: a weak decline in the damping capacity with an increasing frequency ($Q_{max}^{-1}$ from 0.0247 at 0.01 Hz to 0.0240 at 20 Hz) supplied with a simultaneous growth of the elastic modulus ($\Delta E = 0.4$ GPa) is a result of a decrease in relaxation time with an increase in frequency.

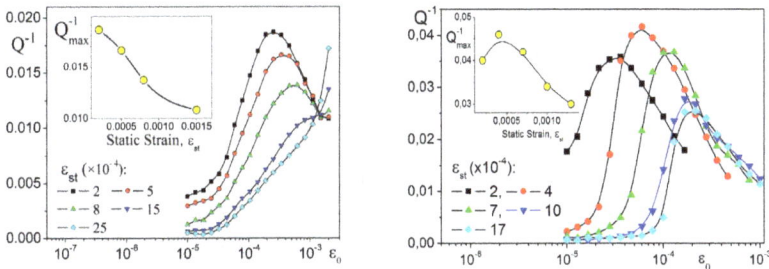

*Fig 4.25. The effect of the static stress ($\varepsilon_{st}$) on the damping capacity: (a) cantilever, (b) three point bending. Inset shows the maximal damping capacity as a function of the static stress. DMA Metravib, forced vibrations (R. Barbin).*

The setup configuration plays a vital role in damping measurements. Absolute damping values depend on test conditions – a type of vibrations and clamping details. The samples were tested at bending using forced vibrations in single and double cantilever configuration, in three point bending and free-decay in free-clamped vibrating reed setups. Experimental data for a single cantilever and a double cantilever are presented in Fig. 4.24. It follows from the results that a single cantilever produces higher damping at relatively lower amplitudes compared with the double cantilever configuration. This difference comes from a different number of sample clamps, which produce compressive stresses and prevent the free movement of magnetic domain walls, leading to a decrease in the damping capacity. For the double cantilever mode, the higher stress concentration exists on the sample and it requires a greater stress to overcome the barrier and move magnetic domains. As a result, the maximum strain amplitude $\varepsilon_{max}$ and the maximum damping capacity $Q^{-1}$ respectively shift to the higher and the lower values as compared to the single cantilever mode. Three point bending tests and tests in free-clamped vibrating reed give higher values of maximal damping: 0.04 and 0.07, correspondingly, which can be explained from the same viewpoint, i.e., by an decrease of clamping influence of a static stress in the sample.

The internal stress caused by the defects such as dislocations, alloying atoms, grain boundaries, secondary phase particles or residual stresses provide a barrier to the movement of magnetic domain walls and a decrease of the damping capacity [244, 245, 246]. The damping capacity is also sensitive to the applied static stress. Characterization of the damping behavior of Hidamets under static preloads is of great importance for determining their true damping effectiveness in practical application environments. Figure 4.25 shows the variation of the damping capacity $Q^{-1}$ with a strain amplitude $\varepsilon$ for the Fe–5.4Al–0.05Ti alloy measured under different pre-loaded static strains ($\varepsilon_{st}$) as measured by two techniques: dual cantilever (a) and three point bending (b). An increase in a static strain from 2.0 to $25 \times 10^{-4}$ lowers the damping under a peak of magnetomechanical damping in both cases. Nevertheless, for cantilever tests the damping capacity decreases monotonously, while for three points bending in passes through a peak. Similar results were also obtained for Fe-Ga-Al alloys (see Chapter 5). The reason is the fact that with applying the static stress, the magnetic domain walls have already driven to move prior to damping test to their new and stable positions: their mobility under a cyclic stress becomes significantly lower and the damping capacity decreases. Inset in Fig. 4.25 illustrates the $Q_{max}^{-1}$ as a function of the static strain $\varepsilon_{st}$. The damping capacity of the alloy continuously decreases with an increase in the static strain in the range $(2-15) \times 10^{-4}$. These results are in agreement with the results for Fe-15Cr, Fe-6Al [247, 248] and Fe-Ga-Al [34] ferromagnetic damping alloys. The maximum strain amplitude $\varepsilon_{max}$ and the maximum damping capacity $Q^{-1}$ respectively shift to higher and lower values with an increase of the static stress. It suggests that (i) a higher stress concentration exists on the sample and (ii) a greater vibration stress requires overcoming the local stress barrier opposing magnetic domain wall motions.

Thus, annealing of the cold rolled sheet at higher temperatures or longer times leads to an increase in the average grain size, which enhances the damping capacity. At least one reason for this effect is the fact that with an increase in the average grain size, the grain boundaries area decreases which act as a barrier for magnetic domain wall motions. This allows optimising the ratio between mechanical and damping properties of the alloy. The study of the magnetic domain morphology shows a much more regular and finer domain structure for the air cooled sample reflecting the lower amount of internal stresses and larger domain-wall area per unit volume, which makes more vibration energy exhausted and results in a higher internal friction value. The presence of irregular domains with a low degree of alignment in the furnace cooled sample may be ascribed to the formation of ordered structures. Local internal stresses distributed in the samples during water quenching and mechanical treatment (cleaning of the sample surface) lower damping capacity due to pinning effect of internal stresses against magnetic domain wall displacements. Applying static strain both by clamps or preloading lowers the damping under the peak of magneto-mechanical damping due to a decrease in magnetic domains mobility under a cyclic stress.

### 4.2.8 Amplitude dependent damping in a single crystal Fe$_3$Al sample

*Effect of magnetic field.* After furnace cooling from 1000 °C, the ADIF curve pronounced the IF peak due to magnetomechanical damping (Fig. 4.26) [249]. The magnetomechanical mechanism of this peak is further proven by applying an external magnetic field. In a saturated magnetic field (MF), magnetic domains inside the sample were fixed and they did not contribute to the energy dissipation during the vibrations in the elastic range of cyclic loading. The construction of a DMA Q800 apparatus did not allow placing the specimen inside the electrical coil in order to create the saturated magnetic field as often used in torsional pendula. Thus, we attached two NdFeB magnets ($B_r$ = 1.2-1.3 kGs, $H_{ci}$ = 12-15 kOe, $(B_H)_{max}$ = 30-35 MGOe) directly to the sample. This method does not allow creating a saturation field in the sample and to fix all magnetic domains in the magnetic field of attached magnets. However, this simple experiment demonstrates that total damping becomes significantly lower compared to the measurements without magnets. Similar influence of external magnetic field on damping at torsion was demonstrated for the Fe-13Ga and Fe-18(Ga+Al) alloys (next subchapter). The magnetomechanical damping

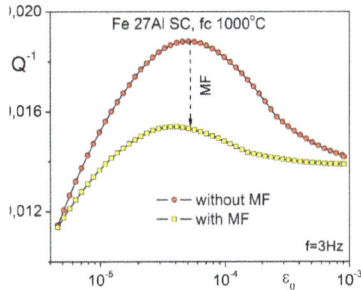

*Fig. 4.26 Effect of magnetic field on the damping capacity of the Fe-27Al single crystal sample furnace cooled from 1000 °C. DMA Q800, forced vibrations (M. Zadorozhnyy).*

at the ADIF curves decreases with an increase in the test temperature and reaches zero at Curie point ($T_C$) [248].

*Effect of heat treatment.* Fig. 4.27 illustrates a huge difference between the damping capacity of the SC samples annealed for 40 min at 1000 °C and cooled in four different regimes: in water (water quenching, wq), in open air (air cooling, ac), and in a furnace (furnace cooling, fc). After annealing at 1000 °C for 40 min, the sample was moved to another furnace with a temperature of 200 °C and annealed at this temperature for 15 min followed by cooling in open air ($ac_{200}$). Clearly, water quenching completely suppresses the damping peak and leads to formation B2 structure according to neutron diffraction results, whereas air-cooling and furnace cooling (with $D0_3$ structure) show a pronounced magneto-mechanical peak slightly less than 0.02 and 0.025, respectively.

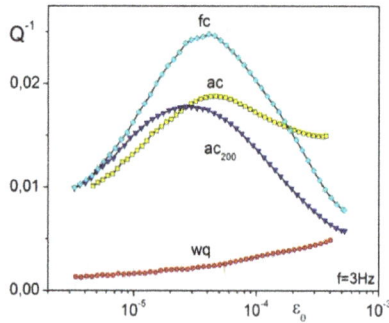

*Fig. 4.27 Effect of the cooling rate on the damping capacity of the single Fe-27Al crystal after annealing at 1000 °C for 40 min followed by water quenching (wq), air-cooling (ac), and furnace cooling (fc). The sample labelled as $ac_{200}$ was first annealed at 1000 °C for 40, then it was moved to another furnace with the temperature of 200 °C and additionally annealed at this temperature for 15 min followed by cooling in air. DMA Q800.*

The suppression of the damping peak after water quenching is due to huge internal stresses introduced to the sample. Internal stresses prevent magnetic domain wall motions and subsequently deteriorate the damping capacity. A similar tendency was discussed above for the Fe-5.4wt%Al alloy quenched in water after annealing at 900 °C for 2h. By decreasing level of internal stresses applied to the sample during air cooling or furnace cooling, the damping capacity increases correspondingly with a decrease in the cooling rate. The $ac_{200}$ sample exhibits almost the same value of $Q^{-1}$ compared to the air cooled sample, however, its maximum strain amplitude, $\varepsilon_{max}$, as estimated from the IF peak position, shifts to the lower values. It implies that a higher level of internal stresses are introduced to the sample cooled in air from annealing temperature of 1000 °C compared to the sample first cooled down inside the furnace to 200 °C followed by air-cooling.

*Effect of external stresses via sample clamping.* When mounting samples a single cantilever, an accurate, reproducible torque must be applied to mount screws on the sample clamp using a torque wrench. Fig. 4.28 exhibits the effect of different torque wrench forces on ADIF of the Fe-27Al single crystal sample cooled in open air after annealing at 1000 °C for 40 min, and inset shows the peak height $Q_m^{-1}$ vs. the torque wrench force level F.

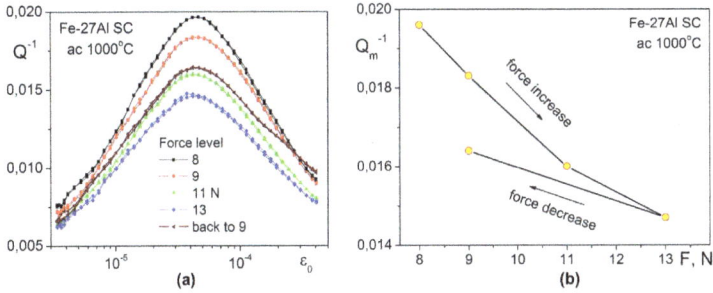

*Fig. 4.28 The effect of different torque wrench forces on ADIF of the Fe-27Al single crystal cooled in open air after annealing at 1000°C for 40 min (the force level first increased from 8 to 13 N, and then decreased back to the standard force level of 9 N). The inset shows the peak height $Q^{-1}_m$ vs. the force level F. DMA Q800, forced vibrations.*

We first increased the force level from 8 to 13 N, and then decreased the force back to the standard force level of 9 N. With an increase in the force level from 8 to 13 N, the $Q^{-1}$ reduces from 0.020 to 0.015 monotonously. After reducing the screwing force from 13 back to 9 N, the obtained values of IF are not completely recovered and do not coincide with the those measured initially at the same force (9 N). By increasing the tightening force, we produce new dislocations and induce micro-deformation to the sample. Therefore, the state of the sample is altered and the second measurement at the screwing force of 9 N after decreasing the force level from 13 N does not coincide with that measured initially at the force level of 9 N. The dislocations reduce the mobility of the domain walls by pinning the domain wall motion; and therefore, they decrease the damping capacity. The results demonstrate the high sensitivity of amplitude dependent damping of the Fe-27Al SC samples to the tightening force.

*Damping capacity of single crystal vs. polycrystal.* Grain boundaries have a significant influence on magneto-mechanical damping since they oppose the MDW motions during the vibration stress. Fig. 4.29 shows the ADIF plots for the Fe-27Al single and polycrystalline samples (measuring frequency is 3 Hz, test temperature is $30^{\pm0.5}$ °C). The absolute value of maximum damping $Q_m^{-1}$ in the SC sample is considerably higher than that of in the PC sample (400% or ~5 times). The maximum strain amplitude $\varepsilon_{max}$, as estimated from the IF peak position, in the SC sample is 1.7 times less than that of in the PC sample.

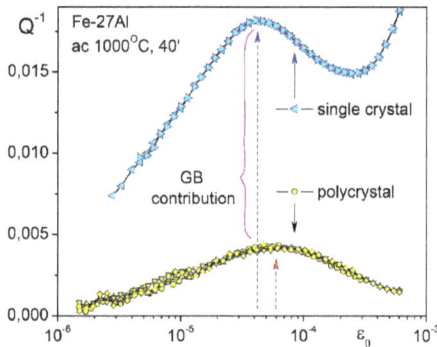

*Fig. 4.29 The ADIF plots for the Fe-27Al single- and polycrystalline samples (measuring frequency is 3 Hz, test temperature is 30 ± 0.5 °C) after annealed at 1000 °C for 40 min followed by air-cooling. DMA Q800, forced vibrations.*

This result suggests that a higher internal stress barrier exists on the polycrystalline sample; and therefore, a higher stress is required to overcome the local stress barrier opposing the MDW motions. The internal stress, $\sigma_i$, opposing the domain wall motion can be calculated using the Smith-Birchak model [31] for a maxwell distribution of the internal stresses from the $Q_m^{-1}(\varepsilon)$ position. As a first approximation, it is possible to use the following relation [250]:

$$\sigma_{max} \sim 0.7256\, \sigma_i$$

where $\sigma_{max}$ can be taken from the strain position of $Q_{max}^{-1}(\varepsilon)$ as $E \times \varepsilon_m$.

The obtained values of the internal stresses are 7 and 10 MPa for single and polycrystals, correspondingly. These value are in good agreement with those reported for the Fe-16Cr-4Mo alloys cooled in a furnace from 1000 °C to 600 °C followed by air cooling [235]. The higher value of the internal stress $\sigma_i$ for the polycrystalline sample corresponds to the contribution of the grain boundaries in suppression of magnetomechanical damping. The grain boundaries in the polycrystalline sample decrease the mobility of the domain walls by pinning the domain walls motions during the vibration stress.

Smith and Birchak model states that magnetomechanical damping in ferromagnetic materials originates from stress-driven irreversible magnetic domain walls displacement [31], and the maximum damping resulting from magneto-mechanical effect ($Q_{max}^{-1}$) is proportional to $\lambda_S E/\sigma_i$, where $\lambda_S$, E, and $\sigma_i$ correspond to the saturation magnetostriction, Young's modulus, and the average internal stress against domain boundary motion, respectively. Fig. 4.30 displays the magnetostriction ($\lambda$) versus the external applied magnetic field (H) for as-grown Fe-27Al single crystal and as-cast polycrystalline samples.

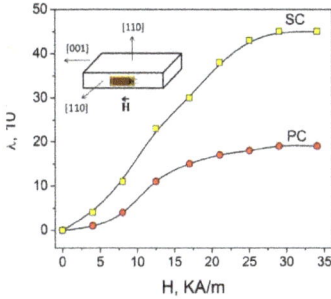

Fig. 4.30. Magnetostriction ($\lambda$) versus external applied magnetic field (H) for as-grown and the Fe-27Al single crystal samples and as-cast PC sample (A. Emdadi). The inset shows the scheme of the tests.

The magnetostrictive strains were measured in the direction of an applied magnetic field up to saturated value of 35 KA/m by the standard strain gauge method. According to the scheme shown as an inset in Fig. 4.30, the magnetostriction strains were measured along [001] direction ($\lambda_{001}$) of the single crystal sample, which was parallel to the H direction. The measured value of $\lambda_{001}=45\times10^{-6}$ for the single crystal sample is in good agreement with that of $\lambda_{001} = 43\times10^{-6}$ reported in [251]. The saturation magnetostriction of as-grown single crystal sample is 2.3 times larger than that of as-cast polycrystalline sample. An increase in the slope of the $\lambda$(H) curves in the single crystal alloy corresponds to an increase in the magnetic permeability, which is particularly desirable for a practical sensor or actuator applications.

Thus the following anelastic phenomena (relaxation peaks) are recorded in Fe-Al alloys and contribute to temperature dependent properties of Fe-Al alloys:

1. The Snoek-type relaxation (carbon atom jumps in Fe-Al solid solution: $H = 0.9$-$1.25$ eV),

2. The Zener relaxation (substitution atom (Al) pairs reorientation: 2.3-2.9 eV),

3. The vacancy-and-carbon related relaxation (the "X" peak: 1.5-1.8 eV),

4. The vacancy complexes reorientation phenomena (1.6-1.8 eV),

5. The grain boundary peak, also interpreted as dislocation-related peak (~ 3 eV),

6. The deformation and dislocation related ("D") relaxation at subzero temperatures (0.43-0.65 eV).

At present the anelastic effects related to the point defects (the Snoek-type and Zener peaks) diffusional reorientation are studied both on qualitative and quantitative levels. The "map" of activation energies for three peaks: the Snoek-type (S), Zener (Z), and the X peak is given in Fig. 4.31 as a function of Al content in iron. More complex anelastic effects due to complexes of point defects (the Hasiguti and X peaks) are studied mainly only on a qualitative level, and their better understanding is needed prior to their possible applications for structural studies. Nevertheless they can give some information about vacancy contribution to anelastic properties or about the beginning of recovery processes of severely deformed alloys (the "D" peak family) at least on the qualitative level.

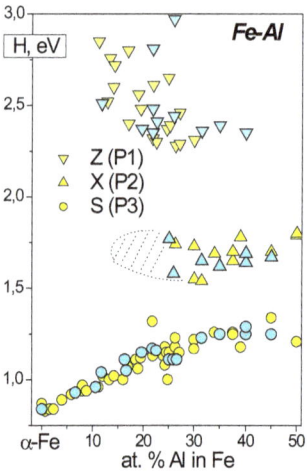

Fig. 4.31 The Snoek-type (S or P1), X (or P2), and Zener (Z or P3) peak activation energies as a function of Al concentration in iron. Nose for the X peak from ≈12%Al means that X peak has been recorded for these alloys but there is no sufficient data to evaluate activation energy. Yellow points – from literature, blue points - authors results [140].

There is a reasonable understanding of magnetic contribution to mechanical damping in Fe-Al alloys which allows one to use some of them for high damping applications.

7. Powerful amplitude dependent magnetomechanical damping in Fe-(10-12)Al alloys which is dependent on structure, i.e. heat treatment regimes of the alloys.

8. Significant increase in damping capacity in single crystal of Fe$_3$Al type compared with similar polycrystalline sample.

Internal stresses introduced to the sample during water quenching completely suppress the amplitude dependent internal friction peak; whereas, air-cooling and furnace-cooling show a pronounced magneto-mechanical damping peak. By increasing the measurement temperature 30 °C to 400 °C and applying the magnetic field, the damping capacity is considerably decreased. The same effect is produced by applying a larger tightening force to the sample in the clamps or by sample granding - the damping capacity decreases due to an increase in internal stresses, growth of dislocation density; and consequently, pinning of the magnetic domain wall motions during forced vibrations.

The study of the amplitude dependent internal friction revealed that the absolute value of maximum damping $Q_m^{-1}$ in the single crystal sample is about five times higher than that of in the polycrystal sample at room temperature. This result correlates with the opposing effect of grain boundaries presented in the polycrystal sample on the magnetic domain wall motions. Last but not least, absolute values of damping capacity depends on testing method.

### 4.3  Anelastic relaxation in binary Fe-Ga alloys

Contrary to the Fe-Al and Fe-Si phase diagrams which are well known about seventy years, the Fe-Ga diagram becomes in the focus of systematic studies only in the end of the last century and it is still under discussions. Our studies of $Fe_3Ga$-type alloys (with 25-29 at.% Ga) by *in situ* neutron diffraction contribute to better understanding of this system, they are discussed in the third chapter. Contrary to Fe-Al and Fe-Si systems, very few results on anelasticity in Fe-Ga alloys can be found in the literature.

Fe-Ga alloys (Galfenols) have recently become the focus of attention due to their giant magnetostriction in low saturation magnetic fields. They are promising rare earth free materials suitable for applications in sensors, actuators, energy-harvesters and spintronic devices. Galfenols have very low hysteresis, high tensile strength (~500 MPa) and their magnetomechanical properties have limited dependence in the range of climatic temperatures.

The magnetostriction ($\lambda$) of single Fe-Ga crystals approaches 400 ppm along the <100> direction [34]. Galfenols are used for magnetostrictive actuators and sensor devices. According to the Smith and Birchak's theory [31], the maximal value of damping is proportional to magnetostriction of ferromagnetic materials: $Q_{max}^{-1} \sim \lambda$. Thus, Fe-Ga alloys are also candidates for damping applications.

In some publications [252], it is believed that an increase in magnetostriction in Fe-Ga alloys is due to the preferential (110) *Ga-Ga* pairing in the disordered body centered cubic (*b.c.c.*) structure. On the other hand, Fe-Ga alloys are known for their ordering of Ga atoms in *b.c.c.* and fcc iron. Ordering decreases mobility of magnetic domain walls and dislocations, and leads to low ductility and reduce the damping capacity.

The formation of the equilibrium fcc-derivative $L1_2$ ordered phase below 630 °C according to the *equilibrium* Fe-Ga diagram is rather slow (see Chapter 3), and in most cases phase transitions develop in accordance with the *metastable* diagram. Starting from the paper by Ikeda et al [43], most scientists have accepted that the equilibrium state in Fe-Ga alloys according to the phase diagram is hardly achievable due to sluggish diffusion. Thus, in most cases a metastable phase diagram with two body-centered cubic b.c.c.- derivative ordered phases, namely, the B2 and $D0_3$ phases, describes the structure of Fe-Ga alloys far better. Atomic arrangement from A2 disordered phase to B2, $D0_3$ or $L1_2$ ordered phases affects both the value and the sign of magnetostriction coefficients.

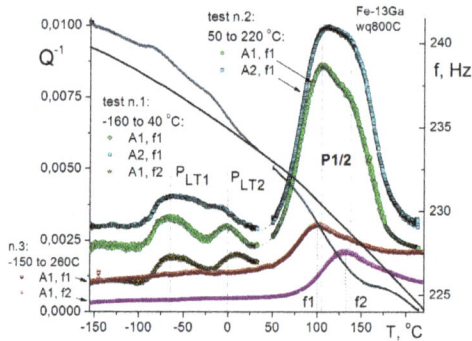

*Fig. 4.32 Temperature dependence IF (left scale) and resonance frequency (right scale) for Fe-13Ga alloy water quenched from 800 °C – free decay bending. Tests are carried out using two different amplitudes (A1 ($\varepsilon_0 = 10^{-5}$) and A2 ($5\times10^{-5}$) and two frequencies f1 – resonance frequency is shown on the right scale and f2 – the first obertone which is close to 2 kHz at -150 °C). Three subsequent temperature runs are presented in this figure: n.1 from -160 to 40 °C, n2 from 50 to 220 °C, and n.3 from -160 to 250 °C.*

## 4.3.1 Snoek-type relaxation

Temperature dependent internal friction, TDIF, was studied using free-decay vibrations in kHz (bending) ranges (Fig. 4.32) of resonance frequencies. Three groups of IF peaks have been recorded:

- Two "low temperature" IF peaks below 40 °C (denoted as $P_{LT1}$ and $P_{LT2}$),
- A broad peak between 50-250 °C composed of at least two peaks denoted consequently as P1/2 and P3 peaks,

TDIF tests were carried out using two different amplitudes: Amp1 and Amp2 which roughly correspond to $\varepsilon_0 = 1\times10^{-5}$ and $5\times10^{-5}$, respectively; and at two frequencies f1 – resonance frequency of free-clamped specimen (which is presented on the right scale in

Fig. 4.32) and f2 – the first obertone which is close to 2 kHz at -150 °C. Three subsequent temperature runs are shown: n.1 from -160 to 40 °C, n.2 from 50 to 220 °C, and n.3 from -160 to 250 °C.

It is easy to notice that:

- Increase in amplitude from $(1-5)\times10^{-5}$ increases damping of the specimen,
- increase in frequency shifts the $P_{LT2}$ (run n.1) and the P1/2 (run. 3) peaks to higher temperature,
- heating to 220 °C drastically decreases the P3 peak height.

Having limited experimental results on the $P_{LT}$ peaks we cannot provide further analyses of these effects. The "low temperature" $P_{LT2}$ peak is thermally activated with rough estimation of activation energy to be below 1 eV. These "low temperature" peaks are similar to the peaks observed in Fe-12Ge and -19%Ge alloys (next section). Increase in amplitude of vibrations leads to less pronounced separation between $P_{LT1}$ and $P_{LT2}$ peaks. We suppose that these peaks correspond to Hasiguti-type relaxations of dislocations with neighbouring self lattice defects. The β and γ relaxations in *b.c.c.* structures, i.e. kink pair formation in $a_0\{111\}/2$ screw dislocations on {110} and {112} slip planes, respectively, cannot be excluded completely as the reason for these peaks. But this hypothesis does not explain an increase in the peak height with deformation.

Increase in damping with the rise in amplitude of vibrations might have different reasons but in case of the Fe-Ga alloys the main source of damping is magneto-mechanical damping. This is proved by applying axial magnetic field which drastically decreases the damping level. In order to minimize magneto-elastic contribution to damping, we have used an amplitude of vibrations lower than that corresponding to maximum of magnetomechanical damping ($\varepsilon_0 \approx 10^{-4}$).

Tests at forced bending vibrations with the frequencies between 0.5 and 32 Hz confirm a complicated nature of the peak family above room temperature [129]. The $P_{1/2}$ peak temperature clearly depends on vibrating frequency exhibiting the activation energy between H = 1.16 eV for air cooled specimen and 1.08 eV for water quenched specimen, and pre-exponential relaxation time $\tau_0 \approx 10^{-16}$ s.

In contrast, the P3 peak temperature is independent on vibrating frequency and it is accompanied by Young's modulus plateau for furnace cooled specimen. In the case of water quenched specimen Young's modulus has got clear inverse behaviour in the temperature range of 150-350 °C both at free-decay and forced vibrations. The temperature of this inverse modulus behaviour decreases if we use the quenched specimen instead of the annealed one. Such modulus behaviour may take place in case of recrystallization (also known as Köster effect), decrease in vacancy concentration or ordering [154], or ferro- to paramagnetic transition. Similar inverse modulus effect at around 300 °C in Fe-17Ga alloy was explained by ferro- to paramagnetic transition in paper [36]. In Fe-13Ga composition with the Curie point around 720 °C, this second order transition can hardly be the reason for inverse modulus effect. The P3 peak takes place at the beginning of this modulus increase: the P3 and P1/2 peaks are better separated if low frequencies are used for the test, the peaks are merged at higher frequencies. The temperature range of inverse modulus behaviour corresponds to disorder – order ($DO_3$) transition. $DO_3$ order was found in Fe-14%Ga composition after slow-cooling [253].

The study of the TDIF in Fe-13Ga alloy within ten subsequent 'heating - cooling' runs with step by step increasing in highest annealing temperature from 220 to 430 °C demonstrates two experimental facts [129]. The P3 peak is very sensitive with respect to heating and completely disappears after heating to 400 °C. The P1/2 peak also decreases with annealing but does not completely disappear. Fit to Debye equation shows that the P1/2 peak consists of a big P1 peak and a small P2 peak at high temperature shoulder of the P1 peak. The total P1/2 peak maximum is shifted to higher temperature with increase in frequency. This allows to calculate the following activation parameters for the process responsible for mainly the P1 peak: the activation energy $H_1 = 0.92$ eV, and $\tau_{01} = 2 \times 10^{-16}$ s.

Similar IF peak with the activation parameters $H_1 \approx 0.87$ eV and $\tau_0 \approx 10^{-16}$ s is recorded in Fe-3Ga alloy. These P1 peaks in both Fe-3Ga and Fe-13Ga alloys are the carbon Snoek-type peaks and they correspond to carbon atom jumps in iron *b.c.c.* lattice (Fe-C-Fe). The peak (P1) is broadened from high temperature side due to carbon – gallium atomic interaction (Fe-C-Ga) that originates from the difference in atomic radii (in the atomic, Goldschmidt, radii the relative distortion ($\Delta r/r_{Fe}$) is Ga/Fe = 5.5%). The slight broadening of the peak from the high-temperature side suggests the appearance of a second Snoek-type peak (P2) at higher temperature due to interstitial (C) – substitutional (Ga) atoms interaction. The broadening of the Snoek-type peak due to appearance of the second peak is typical for several Fe-C-Me (where Me is a metal with a weak tendency to form carbide) [11], and has been reported for the Fe-Al, Fe-Ge and Fe-Si systems [154, 254]. It has been shown that the two Snoek-type peaks result from carbon jumps within the Fe-C-Fe and Fe-C-Me (Me = Al, Ge, Si) surroundings, respectively, indicating a long-range elastic interaction between the interstitial C and substitutional solutes in iron. No clear dependence of the P3 peak temperature on frequency was recorded in our experiments while its position depends on heat treatment of the specimen: quenching decreases peak temperature and corresponding modulus effect.

**Amplitude dependent internal friction** curves for free-decay torsion and forced bending vibrations for the Fe-13Ga sample (Fig. 4.33a) exhibit a peak at ADIF curve (curves 1 and 2, correspondingly). If external magnetic field is applied, which is technically possible in torsional inverted pendulum, this peak is nearly suppressed (curve 3). In a magnetic field the magnetic domain configuration is fixed. A little remaining internal friction peak results from the fact that the magnetic field of 150 Oe is not enough to suppress motion of magnetic structure in applied stress field and therefore to completely suppress magnetic losses. The difference between curves 1 and 3 is magnetoelastic damping (curve 4), while curve 5 represents roughly the contribution of non-magnetic sources to damping. The vertical dash line shows the value of maximal deformation $\varepsilon_0 = 5 \times 10^{-5}$ used in all our temperature dependent internal friction tests with forced bending vibrations.

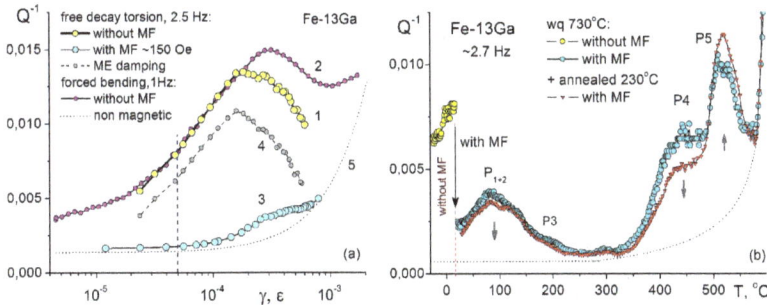

Fig. 4.33 Fe-13Ga alloy ADIF and TDIF tests: a) ADIF curves (water quenched from 900 °C) 1 and 2 at torsion and bending, 3 – torsion in magnetic field of 150 Oe, 4 – magnetomechanical damping (curve 1 – curve 3) torsion, 5 – estimation of non-magnetic IF; b) TDIF without (below 20 °C) and with (above 20 °C) magnetic field, grey arrows show influence of annealing on the peaks, dotted line – non-magnetic IF background (T. Pavlova).

Consequently the total damping at room temperature at the chosen amplitude of vibrations consists roughly of 2/3 parts of magnetoelastic and 1/3 part of non-magnetic (dislocations, point defects) contributions. Temperature dependent internal friction measured using torsion pendulum is presented in Fig. 4.33b: measurements below ~20 °C were carried out without magnetic field, above 20 °C - in a magnetic field of 150 Oe. A clear jump-like decrease in internal friction level takes place if a magnetic field is applied. Several IF peaks, the nature of which is discussed below, and effect of annealing (heating to 230 °C) can be seen at the TDIF curve.

To study these effects we used DMA Q800 setup and forced bending vibrations. Low temperature anelastic effects (0-300°C) in studied alloys (Fig. 4.34) are the result of several different mechanisms. Those are the thermally activated Snoek-type relaxation

[255], composed of the P1 and P2 peaks due to carbon atom jumps in the Fe–C–Fe and Fe–C–Ga surroundings (total loss peak is denoted as the $P_{1+2}$ peak), and the frequency independent P3 peak supplied by increase in elastic modulus caused by structural transformation - ordering of substitute atoms in $\alpha$-Fe based solid solution. Temperature positions of these effects depend on several factors, namely: measuring frequency, amount and type of substitute atoms, quenching temperature and rate and, consequently, might be either well separated from each other or merged.

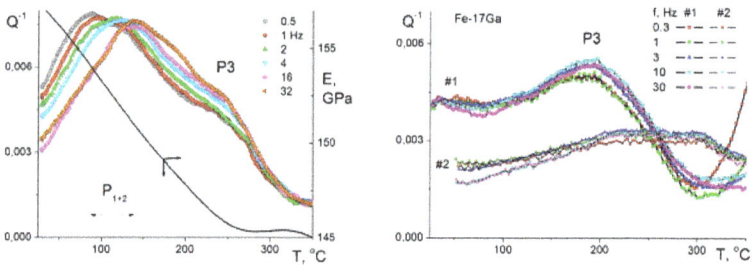

Fig. 4.34   Thermally activated $P_{1+2}$ peak and transient P3 effects in Fe-13Ga (a). In contrast with Fe-13Ga, the Fe-17Ga alloy (b), which does not contain carbon, shows only the P3 effect in the first heating test #1 and no peaks in the second heating test #2.

Annealing has a complex effect: the thermally activated Snoek-type relaxation decreases as a result of decrease of carbon content in solid solution. The higher annealing temperature is, the lower the Snoek-type IF peak is. It was suggested in [36], that the reason for decrease of the frequency independent effect (P3 peak in this paper) is the changing of magnetic structure of the sample at heating; and demagnetization of the sample restores its damping capacity. Thus, no positive effect of demagnetization is recorded in our test [130]. In contrast, nearly the initial high level of damping was restored after second quenching. Thus, (i) annealing decreases damping, (ii) demagnetization does not, but new quenching restores damping capacity, (iii) it is ordering/disordering that plays the dominant role in formation of damping level of the samples. In all tests at least a part of this complex effect is caused by thermally activated process. Consequently, we suggested that the P3 effect is connected not with demagnetization but with ordering of the Fe-Ga alloys [129].

### 4.3.2 Zener relaxation

As cast and water quenched Fe-Ga samples with Ga content less than 19% belong to a single phase (A2) range of both equilibrium and metastable phase diagram. Many papers report using XRD analysis a weak $D0_3$ or $D0_3$ – modified ordering in Fe-(18-19)%Ga alloys after furnace cooling. Our in situ neutron diffraction studies confirms this conclusion and, moreover, they allow to follow disordering reaction from D03 to A2 at instant heating above 400 °C and ordering reaction from A2 to D03 at instant cooling below 350 °C. Due to relatively low intensity of superstructure (111) и (311) reflections from the $D0_3$ phase, it is difficult to conclude reliably about 'cluster' character of the ordering similar to that discussed by us for Fe-Al alloys. In the Fe-Ga alloys with > 20% Ga, the $D0_3$ ordering of the A2 phase during furnace cooling or low temperature annealing (100-300 °C) occurs. The fcc-derivative $L1_2$ ordered phase appears at elevated temperatures in the as cast alloys with 25-27% Ga (at a heating rate of 2 K/min above 450 °C): the kinetic of $D0_3$ ordering and the growth of the $L1_2$ phase and its influence on the magnetic properties in the $Fe_3Ga$-type were reported in chapter 3.

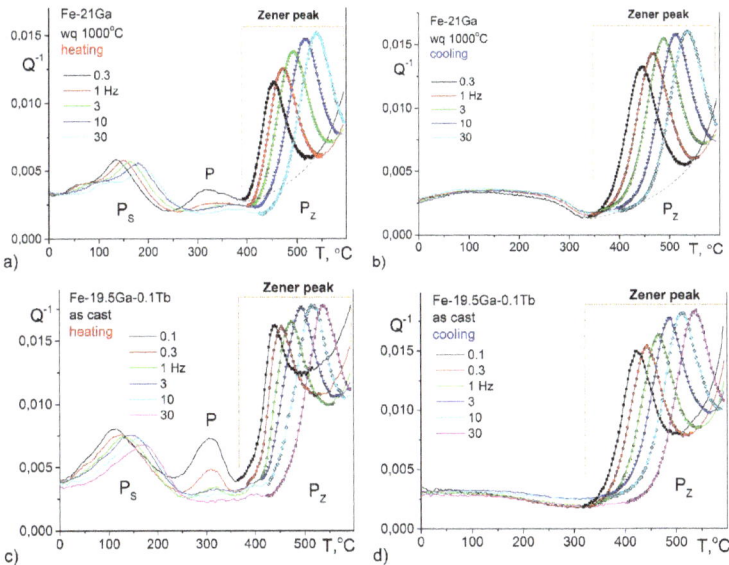

*Fig. 4.35  The TDIF tests of Fe-21Ga (a, b) and Fe-19Ga-0.15Tb (c, d) alloys at heating (a, c) and cooling (b, d). Frequencies and temperature range of the Zener relaxation (experimental points) are underlined. DMA Q800 (J. Cifre).*

*Concentration dependence of Zener relaxation in binary Fe-Ga alloys.* Fig. 4.35 shows typical TDIF curves measured at different frequencies at heating for the Fe-Ga alloys. There are several IF peaks both of thermally activated and non-thermally activated nature in these spectra denoted as $P_S$, P and $P_Z$ in order of appearance with an increase in temperature. The $P_S$ and P peaks are recorded only at heating, whereas the $P_Z$ peak is recorded both at heating and cooling. Zener relaxation ($P_Z$) measured at different frequencies is underlined by experimental points both at heating and cooling.

*Table 4.3 Activation parameters for the Zener relaxation in the Fe-Ga binary alloys.*

| # | Alloy | %Ga | Regime | Heat treatment | H, eV | $\tau_0$, s |
|---|-------|-----|--------|----------------|-------|-------------|
| 1 | Fe-8Ga | 7.7 | C | Wq 1000 °C, 40min | 3.17±0.09 | $2\times10^{-20}$ |
|   |        |     | 2H |                   | 3.18±0.11 | $2\times10^{-21}$ |
|   |        |     | 2C |                   | 3.07±0.06 | $1\times10^{-20}$ |
| 2 | Fe-13Ga | 13.0 | H | Wq 1000 °C, 40min | 2.70 | $2\times10^{-16}$ |
|   |         |      | H |                   | 2.53±0.05 | $8\times10^{-18}$ |
|   |         |      | C |                   | (2.48±0.07) | $(1\times10^{-17})$ |
|   |         |      | "H" | *FDIF test with step by step heating and cooling* | *2.9±0.1* | *$3\times10^{-20}$* |
|   |         |      | "C" |                   | *2.7±0.1* | *$4\times10^{-19}$* |
| 3 | Fe-17Ga | 17.0 | H | Wq 1000 °C, 40min | 2.86 | $3\times10^{-20}$ |
|   |         |      | C |                   | 2.89±0.10 | $1\times10^{-20}$ |
| 4 | Fe-18Ga | 18.2 | H | Wq 1000 °C, 40min | 2.87±0.08 | $6\times10^{-21}$ |
|   |         |      | C |                   | 2.71±0.09 | $6\times10^{-20}$ |
| 5 | Fe-19Ga-Tb | 19.5/0.1 | H | As cast | 2.79±0.09 | $3\times10^{-20}$ |
|   |         |      | C |                   | 2.89±0.07 | $7\times10^{-21}$ |
| 6 | Fe-21Ga | 20.3 | H | Wq 1000 °C, 40min | 2.84±0.03 | $8\times10^{-21}$ |
|   |         |      | C |                   | 2.69±0.09 | $6\times10^{-20}$ |
| 7 | Fe-23Ga | 22.4 | H | Wq 1000 °C, 40min | 2.62±0.14 | $6\times10^{-20}$ |
|   |         |      | C |                   | 2.70±0.13 | $3\times10^{-20}$ |
| 8 | Fe-24Ga | 24.3 | H | Wq 1000 °C, 40min | 2.17±0.21 | $7\times10^{-17}$ |
|   |         |      | C |                   | 2.54±0.49 | $1\times10^{-19}$ |
| 9 | Fe-26.7Ga | 26.7 | H | As cast | 1.64±0.12 | $2\times10^{-14}$ |
| 10 | Fe-27Ga | 27.6 | H | Wq 1000 °C, 40min | 1.65±0.06 | $1\times10^{-14}$ |
|   |         |      | H | As cast | 1.83±0.06 | $5\times10^{-16}$ |

Wq – water quenched; H – heating; C – cooling; 2H and 2C – 2$^{nd}$ heating and 2$^{nd}$ cooling

The values of the activation parameters in the Fe-Ga binary alloys listed in Table 4.3 give us reasonable estimations rather than precise values of the activation energy and the frequency factor for the Zener relaxation. The reasons can be as follows:

- Due to precision for determination of the peak temperatures at DMA setup which works in open air with a sample much larger (typically $30 \times 4 \times 1$ mm$^3$) as compared with the thermocouple. Therefore, with some temperature distribution along length and cross section giving some uncertainties to Arrhenius plot,

- Due to different structural processes (ordering/disordering, precipitation of phases) which take place at heating of some Fe-Ga compositions. This influence is partly discussed in the next sections.

- Further reasons for different experimental mistakes can be found in Zener's lecture by L. Magalas "Logarithmic decrement and elastic modulus: High resolution characterization of dissipation of mechanical energy in solids" to be published soon in Materials Research (https://en.wikipedia.org/wiki/Zener_Prize).

The peak temperature for a fixed frequency decreases slightly with an increase in the Ga content when TDIF is measured at cooling from 600 °C. This effect is less pronounced at heating as it interferes with structural transitions which mask the true position of the Zener peak.

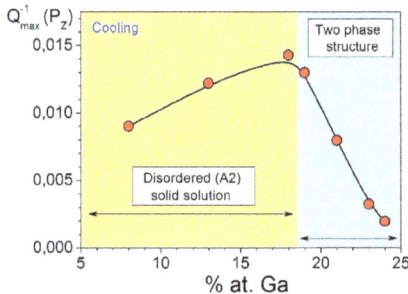

*Fig. 4.36 Zener peak height as a function of Ga content in Fe-Ga alloys as measured at cooling from 600 °C.*

The relaxation strength ($\Delta_Z = 2Q_m^{-1}$) depends on the Ga concentration in the binary Fe-Ga alloys not monotonously (Fig. 4.36): an increase in the Zener peak height, $Q_m^{-1}$, with an increase in the Ga content up to 19 % is followed by a pronounced decrease for the alloys with > 20 %Ga. In contrast with alloys with 19.5-21%Ga (Fig. 4.35), the peak height for several 13-19 %Ga alloys depends on measuring frequency non-monotonously (Fig. 3.7): this unusual effect results from overlapping of a frequency-dependent thermally activated Zener peak with a frequency-independent transient effect due to the structural transition. This effect is discussed in the next section.

The relaxation strength for "true" Zener relaxation, i.e. internal friction contribution apart of the additional frequency-independent transient contribution previously mentioned depends on the concentration of substitute atoms ($C_{Ga}$) and the degree of order, i.e. on the order parameter ($\eta$). Order parameter was varied in our experiments by varying the temperature of the frequency dependent IF measurements. The theory of the Zener relaxation [3] proposes that for concentrated (in our case Fe-Ga) alloys the relaxation magnitude ($\delta J$) can be ascribed by the Eq. (4.7). Consequently, an increase in the Ga concentration in the Fe-Ga alloys up to ~19 at.%, i.e. until the value of the parameter $f(\chi_o, C_{Ga}) \approx 1$ in Eq. (4.7), leads to an increase in the Zener peak height. At higher Ga content, the effect of ordering plays a more powerful role as compared with the effect of the concentration, and the Zener peak height decreases pronouncedly. Similar dependence was observed in for Fe-Al alloys which also has a tendency to atomic ordering above the critical Al concentration (Fig. 4.9).

### 4.3.3 Transient effects due to phase transitions

Careful analysis of the TDIF curves shows that Zener effect coexists for several Fe-(13-28)%Ga alloys with non-thermally activated IF contribution in the same temperature-frequency range (Fig. 4.37). In some cases, e.g. for the Fe-27Ga alloys this non-thermally activated IF contribution is very powerful (Fig. 4.37a), and it is accompanied with an increase in the Young modulus. These effects – both the non-thermally activated IF peak and the increase in the modulus – directly relate to the first order phase transition in Fe-27Ga type alloys from the ordered *b.c.c.* to the ordered fcc structure as studied at instant heating recently by *in situ* XRD [147] and neutron diffraction [149], and at isothermal conditions [102, 109]. For Fe-Ga alloys with 13-19%Ga a non-thermally activated IF contribution gives only a minor contribution to the total damping (Figs. 4.37b,c) similar to that of in the Fe-Al alloys with the second order phase transitions [149].

The most powerful transient IF peak occurs in the Fe-27Ga type alloys at instant heating at around 425-525 °C due to the phase transition from D0$_3$ (a BiF$_3$-type structure with Fe and metal (Me) atoms partially ordered, sp. gr. *Fm3m*) to L1$_2$ (a Cu$_3$Au-type structure with Fe and Ga atoms partially ordered, sp. gr. *Pm3m*) phase (Fig. 4.37a). Using isothermal tests, we demonstrated [109] that the phase transition from an ordered *b.c.c.*-derivative D0$_3$ phase to an fcc-derivative ordered L1$_2$ phase first leads to disordering of the D0$_3$ phase to obtain an A2 structure followed by an A2 to A1 transition with final A1 phase ordering to achieve the L1$_2$ structure. The temperature of this IF peak does not depend on measuring frequency while the peak height is reasonably ascribed by a well-known equation (1.12) for transient peaks for first order transitions.

On the left side from the transient peak in the Fe-27Ga alloy (lines with experimental points), there is a thermally activated peak at 300-450 °C with the activation energy from 1.65 to 1.91 eV [57]. This peak disappears at cooling after heating to 600 °C (lines). Heating to 600 °C with a heating rate of 1 or 2 K/min leads to the formation of the equilibrium fcc-originated ordered $L1_2$ phase, which remains stable at cooling to room temperature as confirmed by *in situ* neutron diffractions [57]. Consequently, the thermally activated peak observed at heating at 300-450 °C disappeared at cooling in the sample with the $L1_2$ ordered structure. Above 540 °C, one can see another thermally activated peak on the TDIF curves measured at low frequencies. Unfortunately, it is impossible to evaluate its activation energy as the temperature position of this peak is above the range of our measurements. The situation with the Zener relaxation in Fe-27Ga remains complex and needs further studying.

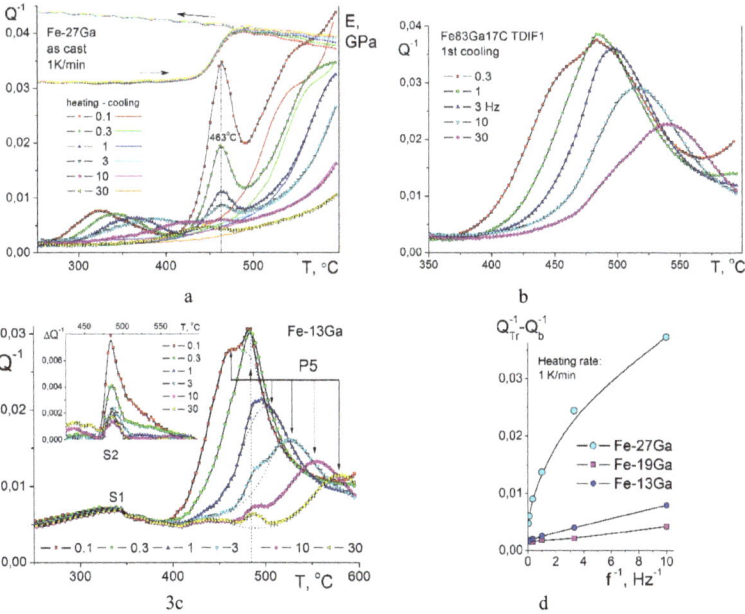

Fig. 4.37 TDIF curves for Fe-Ga alloys: (a) 27%Ga, heating, (b) 17%Ga, cooling, (c) 13%Ga (inset: the transient component of the peak, (d) height of transient peak ($P_{Tr}$) as a function of inverse measuring frequency (1/f) for heating rate of 1 K/min and alloys with 13, 19 and 27Ga. DMA Q800, forced bending vibrations, single cantilever.

Another frequency-independent transient effects, but hardly with the same mechanism, are recorded in Fe-17Ga (Fig. 4.37b) and Fe-13Ga (Fig. 4.37c) alloys at about 480-490 °C. The key difference is that in the alloys with 13 and 17% Ga the transient effects are observed both at heating and at cooling, as well as these effects remain practically unchanged in the subsequent heating-cooling tests, whereas the transient peak in the alloys with 21 and 27% Ga is irreversible and is observed only during the first heating, and then disappears at cooling. These effects have different relaxation strength and dependence on the measuring frequency (Fig. 4.37d), and they manifest themselves at different temperatures. These temperature according to our in situ neutron diffraction tests corresponds well to the second order transition from *b.c.c.*-derivative ordered D03 to A2 disordered structures.

The theory of anelasticity for first- and second order phase transitions invented by L. Landau is given in the book by Nowick and Berry [6], W. Benoit [23] and G. Fantozzi [24]. Maximum of internal friction for both first-order transition and second-order (often called as lambda, $\lambda$, peak) transition according to Landau's theory can be in general described by equation (1.10).

The first order $D0_3$ to $L1_2$ transition takes place in Fe-27Ga at instant heating above 450 °C. At cooling the equilibrium $L1_2$ phase is stable and, consequently, it does not contribute to any transient effects in agreement with our experimental results. This phase transformation explains completely the transient effect in Fe-27Ga at heating and its absence at cooling, see heating and cooling TDIF curves in Fig. 4.37a. The $L1_2$ phase does not exist in Fe-13Ga and Fe-17Ga alloys according to existing phase diagrams, and we were able to detect it only one week annealing at 500 °C in Fe-19Ga samples.. Nevertheless transient IF peaks are recorded in both alloy but with much lower magnitude in the same temperature range (Figs. 4.37:b,c,d). Analysis of parameters of these transient effects at about 450-500 °C in all alloys (Fig. 4.37) gives inverse dependence between peak height and measuring frequency, i.e. $Q_m^{-1} \sim 1/f^m$, where m < 1 for first order transition in Fe-27Ga and m $\approx$ 1 for alloys with lower Ga content, underlying their similarity in studied alloys. Thus, we can suggest that appearance of the D03 phase at furnace cooling and $L1_2$ at much longer isothermal annealing phase in Fe-Ga alloys takes place at lower Ga concentrations as compared with the existing equilibrium phase diagrams.

Test of Fe-17.4Ga single crystal (Fig. 4.38) not only confirms the main regularities discussed above for polycrystalline Fe-Ga samples but also helps to exclude contribution grain boundary relaxation in this temperature - frequency range [256]. This confirms non-GB origin of the thermally-activated peaks under discussion. Activation parameters for the Zener relaxation peak in Fe-17Ga single crystal are at heating $H_Z$ = 2.54 eV and $\tau_0$ =

$2 \times 10^{-18}$ s, and at cooling $H_Z = 2.50$ eV and $\tau_0 = 3 \times 10^{-18}$ s. One can also see a frequency independent contribution to TDIF in the single crystal at heating at about 550 °C and cooling about 520 °C. If the Zener peak temperature is far from the transient effect due to low measuring frequency, TDIF curve exhibits two IF peaks at 450 (Zener peak at 0.1 Hz) and 550 °C (transient peak). If frequency is higher (e.g., 10 or 30 Hz) the Zener peak temperature and temperature of transient effect practically coincide and the total peak height increases around 550 °C at heating or at a little bit lower temperature at cooling. This transient effect as discussed above is due to reversible order-disorder transition.

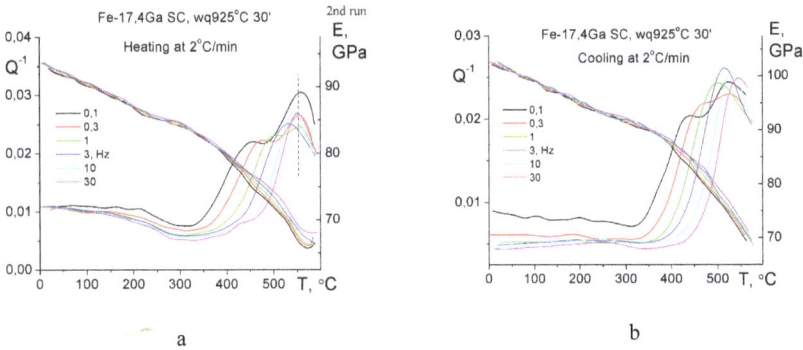

a                                 b

*Fig. 4.38 TDIF curves for single crystal Fe-17,4%Ga (water quenched sample from 925 °C), a) heating and b) cooling. Thermally activated peak (Zener peak) and a frequency independent transient effect at about 550 °C (heating) or 520 °C (cooling) co-exist.*

**Transient effects at higher temperature.** Three transient IF effects or peaks are recorded in the studied Fe-Ga alloys. These peaks correspond to the $D0_3 \rightarrow L1_2 \rightarrow D0_{19} \rightarrow A2$ (or B2) phase transitions. In addition to discussed above transient effect at 450-500 °C in Fe-27Ga alloys, another two transient effects were recorded at higher temperatures using torsion pendulum. This study was performed in cooperation with Prof. Daniele Mari at EPFL Lausanne and presented as a joint talk at ICIFMS conference (Foz do Iguaçu, Brazil: September 12-15, 2017 [283, 284]). Three transient effects - Tr1, Tr2, and Tr3 – as measured at torsion are presented in Fig. 4.39a and they correspond to three phase transitions (Fig. 4.39b). Each time the transient IF peak is associated with a modulus dip. The presence of a drop in the elastic shear modulus witnesses a strong coupling between the stress and the anelastic strain: similarly to martensitic transformation, phase interfaces in Fe-Ga alloys propagate during the transformation. Misfit dislocations migration may also contribute.

**The $D0_3$ to $L1_2$ transition (Tr1 peak).** The relatively low-temperature $D0_3$ to $L1_2$ transition (Tr1) was mainly studied using DMA equipment with varying frequencies from 0.1 to 30Hz. An example of these tests is given in Fig. 4.37a. The dependence of the Tr1-peak height on frequency and the heating rate is plotted in Fig. 4.37d. As discusses, the $D0_3$ to $L1_2$ transition shows that this transition goes through diffusion controlled disordering $D0_3$ to A2 followed by A2 to A1 transition and $L1_2$ ordering of the A1 phase.

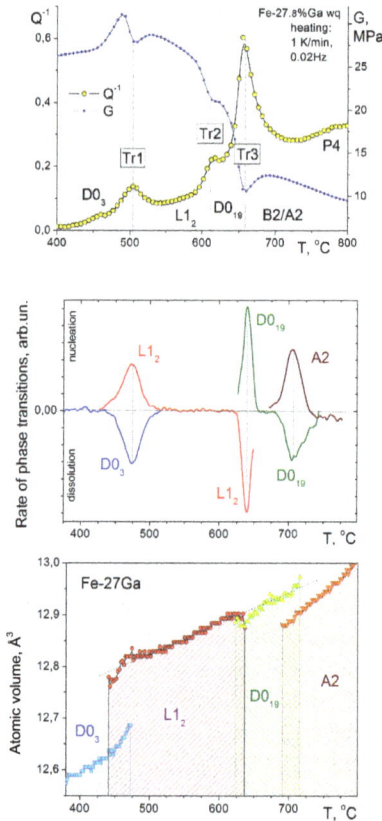

Fig. 4.39. Phase transitions in the Fe-27Ga (a) samples heated with a rate of 1 K/min as measured by internal friction. (b) The rate of the phase transitions—illustrates the rate of growth and the dissolution of different phases. (c) Atomic volumes of the phases in the same temperature range.

The change in the atomic volume (Fig. 4.39c) in the range of co-existing phases ($DO_3$ and $L1_2$ or $DO_{19}$ and A2) leads to an increase in micro stresses in the alloy. The $DO_3 \rightarrow L1_2$ and the $DO_{19} \rightarrow A2$ (or B2 dependently on a heating rate) phase transitions occur with a change in the atomic volume and a rise of the local stresses and strains, which lead to Tr1 and Tr3 anelastic peaks. An increase of the volume fraction of the $L1_2$ phase leads to an increase of the stress mainly in the $DO_3$ phase, whereas the stress in the $L1_2$ phase remains at the same level, which is relatively high from the beginning of the appearance of this phase.

**The $L1_2$ to $DO_{19}$ transition (Tr2 peak).** According to the in situ neutron diffraction, the phase transition between two closed packed phases is not accompanied by a change in the atomic volume and a pronounced increase in internal stresses. Moreover, heating of the sample above the $DO_3$ to $L1_2$ transition leads to a smooth decrease in microstreasses in the sample. Thus, the IF peak (Tr2) in the case of the $L1_2 \rightarrow DO_{19}$ (or fcc $\rightarrow$ hcp) phase transition is due to a long-range motion of the Shockley dislocation, which assists transition between these two closed packed phases. In accordance with neutron diffraction data, the $L1_2 \rightarrow DO_{19}$ transition also proceeds via disordering: $L1_2 \rightarrow A1 \rightarrow A3 \rightarrow DO_{19}$.

Tests with five frequencies (Fig. 4.40a) and four heating rates (Fig. 4.40b) show that the T2 and T3 peaks height changing roughly in agreement with Eq. (1.12). An increase in the heating rate expectably leads to a shift of these diffusion controlled phase transitions to a higher temperature: the red dash lines for 0.5 K/min and the blue dot lines for 3 K/min show the range of existence of $DO_{19}$ phase in Fig. 4.40b.

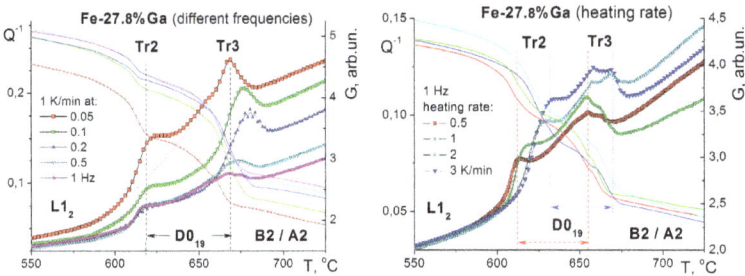

*Fig. 4.40. Second and third transition peaks in the Fe-27.8Ga sample: a) influence of the frequency of vibrations (from 0.05 to 1 Hz, heating rate 1 K/min), b) influence of the heating rate (from 0.5 to 3 K/min, 1 Hz: D. Mari).*

**The $D0_{19}$ to A2 transition (Tr3 peak).** The effect of this first-order phase transition is similar to that of discussed for the $D0_3 \to L1_2$ transition. Here, the $D0_{19} \to B2 \to A2$ transition takes place. According to the *in situ* neutron diffraction study, if the heating rate is 2 K/min, there is not enough time for the $B2 \to A2$ reaction, whereas in the case of heating with a rate of 1 K/min, the neutron diffraction detects weak B2 ordering of the A2 phase. The Tr3 peak splitting with an increase in the heating rate (Fig. 4.40b) can be related to this effect, i.e., to the formation of the A2 phase from $D0_{19}$ followed by a $B2 \to$ A2 reaction.

### 4.3.4 Isothermal frequency dependent damping

Isothermal experiments as a function of the frequency ranging between 0.0001 and 300 Hz were performed in collaboration with Prof. Andre Rivière using a forced pendulum between room temperature and 600 °C, firstly at various increasing temperatures, then during successive cooling for Fe-13Ga alloy [129]. Using this experimental method, the comparison of the internal friction spectra obtained at the same measurement temperature, during the first heating and after *in situ* high temperature annealing was carried out in both cases all transient effects are excluded as the temperature for each tests is constant..

Fig. 4.41a exhibits typical FDIF curves measured at different temperatures (between 425 °C and 500 °C; preliminary annealed at 600 °C) for the Fe-13Ga alloy during the first heating. A thermally activated peak is directly observed above 425 °C. Its height and the width of the peak are stable for various measurement temperatures. The activation parameters deduced from the Arrhenius plot are $H_Z = 2.9$ eV and $\tau_0 = 3 \times 10^{-20}$ s. By comparison of its activation parameters with the TDIF measurement, this peak at the FDIF curves corresponds to the Zener peak for this alloy.

After an *in-situ* annealing at 600 °C, a second peak is exhibited at a lower frequency than the Zener one (Fig. 4.42b). Such a peak is observed at a higher temperature than the Zener peak in the TDIF measurements and it is called "HT (high temperature) peak". This new HT peak is evidenced during the first measurement at 525 °C. Its height is about 0.002. Then, it develops progressively during the high temperature annealing and its height is close to about 0.02 after annealing at 600 °C. This HT peak is thermally activated as shown in Fig. 4.42a. The activation parameters deduced from the Arrhenius plot (Fig. 4.41c) are: $H \approx 3.1$ eV and $\tau_0 \approx 3 \times 10^{-19}$ s.

Activation parameters of HT peak suggest a grain boundary mechanism of this peak. As grain boundary dislocations play an important role in grains sliding, their mobility is different in ordered and disordered state. In FDIF tests at "heating" the material is partly ordered, while at "cooling" tests start from disordered state, and this, similarly to the Fe-Al-Si alloys [126], might be the reason for increase in the HT peak height at cooling.

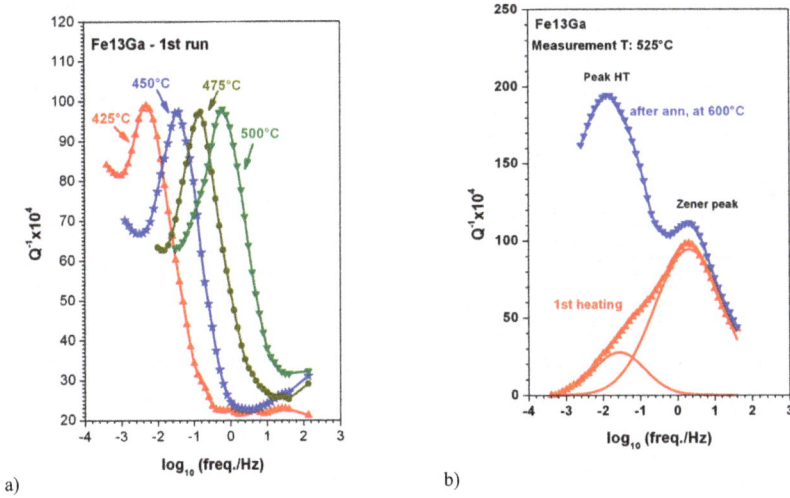

a)
b)

*Fig. 4.41 Fe-13Ga, forced torsion pendulum: (a) Frequency dependent internal friction spectra obtained at several temperatures during the first step-by-step heating. (b). Frequency dependent internal friction spectra obtained at the same temperature (525 °C) during the first heating and after an in-situ annealing at 600 °C with the Zener peak (A. Riviere).*

Figure 4.42b shows the spectra obtained at various temperatures after annealing at 600°C. Only some part of the experimental spectra corresponding to the Zener peaks is plotted for allowing a better comparison with the peaks obtained during the first heating. After annealing, the Zener peak is shifted towards a higher frequency. This shift is more significant for the measurements between 425 and 475 °C. For instance, the frequency $f$ of the maximum of the Zener peak measured at 425 °C during heating is 0.004 Hz ($\log(f)$ = -2.4 in Fig. 4.29.a) and 0.008 Hz ($\log(f)$ = -2.1 in Fig. 4.42.b) after annealing.

This shift is confirmed in Fig. 4.42c where the Napieran logarithms of the frequency of the maximum of the peaks are plotted as a function of the inverse of the absolute temperature of measurement (Table 4.3) for the Zener peak before and after annealing and for the HT peak. The activation parameters for the Zener peak after annealing at 600 °C are H= 2.7 eV and $\tau_0$ = $4\times10^{-19}$ s. Such difference between the first heating and successive cooling was also observed by TDIF. After annealing at a higher temperature (as measured at cooling), the activation energy is slightly lower than that of during heating. This difference can be explained by some atomic rearrangement 'order-disorder' at heating. For obtaining the true value of the relaxation parameters, the measurements at cooling are more reliable.

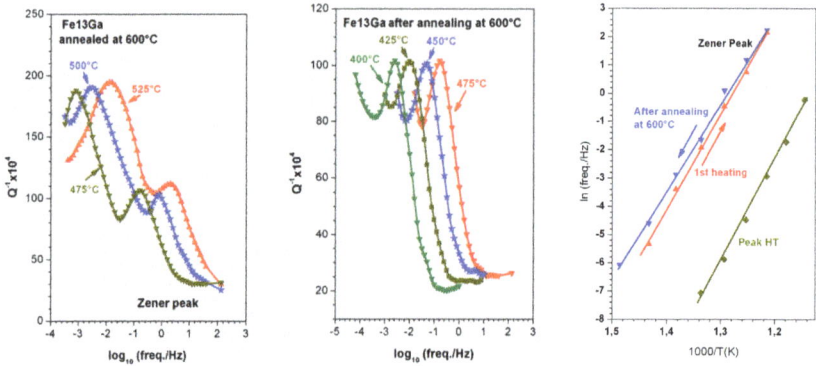

*Fig. 4.42   Fe-13Ga - (a) Frequency dependent internal friction spectra obtained respectively at 525 °C, 500 °C and 475 °C after annealing at 600 °C. - (b) the Zener peak at several measurement temperatures after annealing at 600 °C. - (c) the Arrhenius plots corresponding respectively to the Zener peak during the first heating and after annealing at 600°C, and to the HT peak (cooling). Forced torsion pendulum (A. Rivière).*

The main features of the Zener relaxation in Fe-Ga alloys are rather similar to those in Fe-Al binary alloys. It also can overlap with transient effects due to second order transitions in the alloys.

The FDIF data can be re-plotted as TDIF curves (Fig. 4.43). The difference is that in fact each point at these curves is measured not at instant heating or cooling but as constant temperature. The $P_S$ peaks in Fig. 4.43 consist of two peaks between room temperature and 200 °C: $P_{S1}$ and $P_{S2}$. The activation parameters for these peaks determined from Arrhenius plots are $\tau_{01} \approx 10^{-16}$ s and $H_{S1} = 1.1$ eV for $P_{S1}$, $\tau_{02} \approx 6\times10^{-14}$ s and $H_{S2} = 1.04$ eV for $P_{S2}$. Annealing at 100 °C does not pronouncedly influence these peaks while annealing at 150 °C decreases the $P_{S2}$ peak (see the arrow which starts from the corresponding temperature and shows the peak height measured in step by step cooling). One-day annealing at 350 °C leads to the disappearance of $P_{S1}$ and $P_{S2}$ peaks.

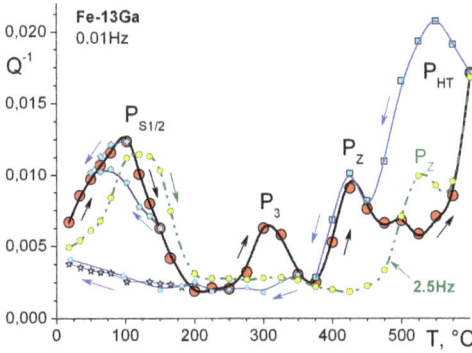

*Fig. 4.43 TDIF spectrum as reconstructed from isothermal FDIF curves for f = 0.01 and 2.5 Hz. Arrows indicate whether FDIF tests were carried out with a step by step increase (→, also red circles for experimental points) or a decrease (←) in the test temperature. Similar reconstruction for f = 2.5 Hz is given only for heating (in olive colour) for the purpose of comparison.*

The $P_3$ peak is situated at about 300 °C if the frequency is 0.01 Hz. The $P_3$ peak, which is very sensitive to heating and disappears during 24 hours of isothermal annealing, preceding the FDIF tests, decreases drastically. The activation characteristics of the $P_3$ peak according to the "temperature – frequency shift" in FDIF tests are: $\tau_{03} \approx 4 \times 10^{-6}$ s and $H_3 = 0.85$ eV.

The thermally activated $P_Z$ peak at about 425 °C is stable with respect to heating. Activation parameters of $P_Z$ peak from FDIF tests are given in the Fig. 4.42c. The P4 peak appears after annealing at about 500 °C and increases strongly with the annealing temperature. The activation parameters are $H_4 \approx 3.1$ eV and $\tau_{04} \approx 3 \times 10^{-19}$ s.

In spite of some quantitative uncertainties in the activation parameters obtained from temperature dependent and frequency dependent tests, several mechanisms of anelasticity acting in the Fe-Ga alloys are proved. The carbon Snoek-type (Fe-C-Fe and Fe-C-Ga) and Zener relaxations are recognized among them. Two subzero ($P_{LT}$) peaks and P3 peak, which are unstable with respect to heating, are addressed to dislocation induced relaxation mechanisms ($P_{LT}$ are probably the Hasiguti peaks similarly to the "D" peak in Fe-Al alloys). The HT peak occurs due to grain boundary relaxation controlled by motion of grain boundary dislocations. All these internal friction peaks are thermally activated relaxation peaks.

## 4.4 Anelastic relaxation in binary Fe-Ge alloys

We have studied Fe-Ge-C alloys with compositions Fe–3Ge, Fe–8Ge, Fe–12Ge, Fe–19Ge, Fe–22Ge, Fe–27Ge and Fe–38Ge (all in at.%) according to spectral and chemical analysis, and checked by X-ray diffraction and DSC calorimetry. All alloys except Fe-8Ge were produced by induction melting of 99.98% Fe and 99.99% Ge under argon atmosphere [63]. A small amount of carbon (~0.01–0.04 at.%, in form of powder embedded into Fe-foil to avoid fogging during melting) was added in order to check the appearance of the Snoek relaxation; for simplicity the indication of carbon will be dropped in the following. For the production of the Fe-8Ge alloy a plasma arc furnace and high purity raw materials (Fe: 99.99%, Ge: 99.999%, C: 99.9999%) were used with argon 5.0 to prevent oxidation. The Fe–38Ge alloy was too brittle for cutting, so that in this alloy only the supplementary methods as mentioned below, e.g. for checking the Curie point in the β phase, could be applied.

The internal friction, $Q^{-1}$, and the relative modulus change were measured as a function of temperature $T$ (from 100 to 875 K), at an amplitude of about $\gamma_0 \approx 5 \cdot 10^{-5}$, in (i) an inverted torsion pendulum at $f$ = 0.5-3 Hz [257] and (ii) two vibrating-reed set-ups at $f$ = 150-3000 Hz [154, 174, 258] and (iii) DMA Q800 [57]. Heating rates from 0.5 to 2 K/min up to a chosen temperature were usually applied, followed by cooling with nearly the same, less controlled rate.

The experimental results [259, 260] are presented and discussed in the following order according to their structures: (1) disordered alloys with *b.c.c.* structure and a low amount of germanium (Fe-3Ge and Fe-8Ge), (2) alloys belonging to the *b.c.c.*-originated ordered B2 or DO$_3$ ranges of the Fe-Ge phase diagram (Fe-12Ge and Fe-19Ge), and (3) alloys with ordered hexagonal phases (Fe-22Ge and Fe-27Ge). Although this classification is not very precise for the Fe-19Ge and Fe-22Ge alloys with mixed microstructures to be discussed below, it is helpful for a better understanding of the experimental results. Heat flow data for several studied compositions are presented in Fig. 4.44.

### 4.4.1 Disordered *b.c.c.* structures (Fe-3Ge and Fe-8Ge)

*Fe–3at.%Ge.* DSC indicates $\alpha \leftrightarrow \gamma$ and $\gamma \leftrightarrow \alpha$ transition temperatures of 1075/1035 °C and 1301/1279 °C (on heating/cooling, Fig. 4.44), respectively, in reasonable agreement with the literature data, as well as a Curie point of $T_C$ = 770 °C. XRD pattern for specimens after water quenching from 1000 K (5 h) shows a *b.c.c.* lattice parameter of 2.8758 ± 0.0004 Å.

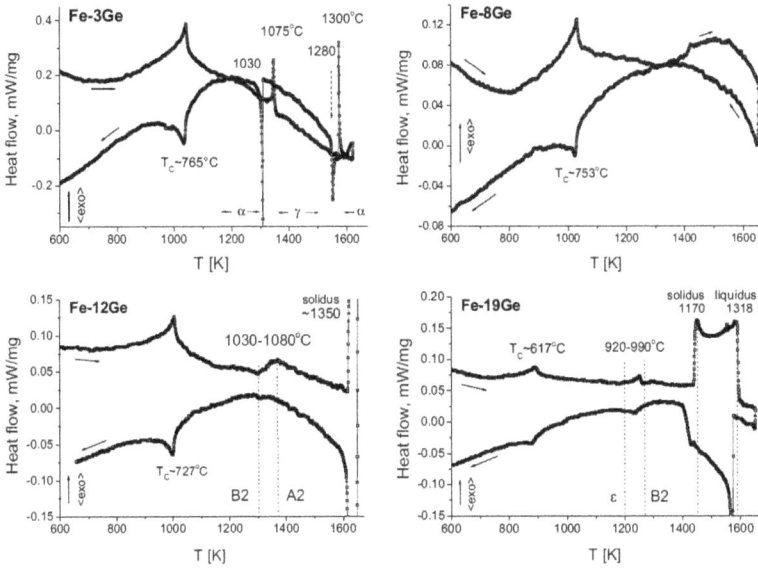

*Fig. 4.44   Heat flow measurements for Fe-3Ge, Fe-8Ge, Fe-12Ge and Fe-19Ge at heating and subsequent cooling (10 K/min) (C. Siemers).*

With a carbon content of about 0.03 at.%, an IF peak (P1 in Fig. 4.45) occurs practically at the same temperature as the Snoek peak in "pure" $\alpha$-Fe [11], with activation parameters $H = 0.86$ eV and $\tau_0 = 2\times10^{-15}$ s from the frequency-temperature shift [166]. Based on these values the peak is classified as a carbon Snoek-type peak which is wider than a Debye peak: the parameter for the relaxation time distribution amounts to $0.4 < \beta < 1.4$, and also the related modulus defect is wider on the temperature scale. The slight broadening of the peak from the high-temperature side suggests the appearance of a weak second peak (P2) at a higher temperature (inset in Fig. 4.45). This effect is typical for C in Fe-Me alloys (Me: metal without strong carbide formation tendency) [11], and has been already discussed in this chapter for Fe-Al and Fe-Ga. Two partial peaks have been distinguished in the range of Snoek relaxation in dilute alloys, respectively, from carbon jumps within Fe-C-Fe and Fe-C-Me surroundings, indicating a long-range elastic interaction between interstitial C and substitutional atoms in iron. When the Fe-3Ge alloy is annealed step by step in the furnace, the Snoek peak decreases with increasing annealing temperature due to trapping of carbon atoms by defects.

151

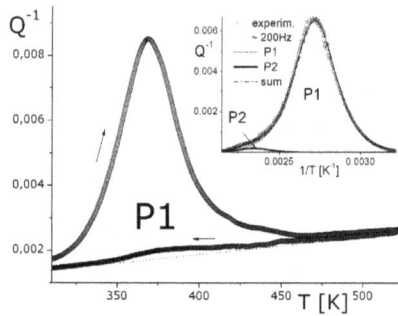

*Fig. 4.45 Temperature dependence of damping $Q^{-1}$ at ~480 Hz in Fe-3Ge-0.03C (water quenched after 3 h at 1000 K) showing the Snoek-type peak (P1). Inset: fit of Eq. (6) to the Snoek-type peak (at ~ 200 Hz) in 1/T coordinates; a second small peak P2 can be distinguished (fit parameters: $Q_1^{-1} = 0.0065$, $H_1 = 0.86$ eV, $\beta_1 = 0.75$; $Q_2^{-1} = 0.0002$, $H_2 = 1.26$ eV, $\beta_2 = 0.98$).*

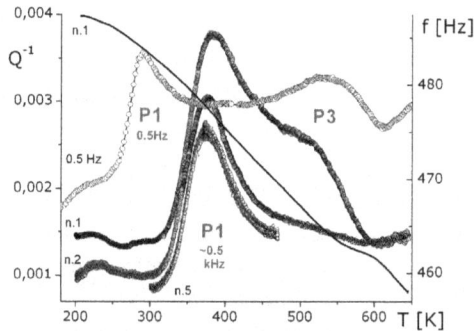

*Fig. 4.46 Temperature-dependent IF of water-quenched Fe-8Ge measured at 0.5 Hz (forced torsion vibrations, upper curve), as well as at ~ 480 Hz (free decay bending) during first (curves n.1 for both $Q^{-1}$ and frequency), second (n.2), and fifth (n.5) heating up to 700 K (see text). The heating rate was 0.5 K/min, except for the fifth run with 1 K/min, where 24 h isothermal annealing was performed at the peak temperature 100 °C (373 K).*

**Fe–8at.%Ge.** By DSC, the Curie point is found at 755 °C; the solidus temperature exceeds 1380 °C, the highest temperature reached by our DSC equipment (Fig. 4.44). Anelasticity at this composition was additionally studied in the subresonant mode at $f =$

0.5 Hz using a forced torsion pendulum at EPFL Lausanne, where the Snoek-type peak (P1) at first heating is found at $T_m$ = 14 °C (Fig. 4.46) (and at 23 °C at 2.8 Hz in the free-decay mode, not shown in Fig. 3). P1 is again broadened at the high-temperature side by a hidden peak P2 (see below), and then followed by another broad peak or plateau with maximum at 500-550 K (0.5 Hz), denoted here as P3 (a similar but much weaker effect is also seen in Fe-3Ge if measured at low frequency).

Bending tests at about 480 Hz show P1 at $T_m$ = 376 K (Fig. 4.46); in subsequent runs with a step-by-step increase of the highest temperature, P3 decreases in each run while P1 becomes more and more "stable". If, alternatively, repeated tests up to 420 °C are performed, P3 vanishes more or less in the first two runs, while P1 does not change any more, neither in height nor in width, after three to five runs. In Fig. 4.46 one can see the first heating (n.1) followed by the cooling curve (n.2), and the fifth run (n.5) where a heating rate of only 0.5 K/min was applied. Within this fifth run an annealing stage for 24 h was carried out at the peak maximum (100 °C), and no change in the peak height was detected. Also, the cooling curve completely coincides with the heating one demonstrating the peak stability. This behavior of P1 indicates a relatively stable concentration of mobile carbon atoms in solid solution, while the disappearance of P3 is attributed to the annealing of thermal vacancies.

The mean activation parameters for this "stable" P1 peak are $H \approx 0.8$ eV and $\tau_0 = 6 \times 10^{-14}$ s. Similar to the situation in Fe-3Ge, a P2 peak can be distinguished at its shoulder by computer fit, using two Debye-type peaks with free parameters. The relative height of P2 is higher in Fe-8Ge than in Fe-3Ge, while the temperature position is not very different. The comparison of the IF spectra measured at low and high frequencies demonstrates that in contrast to P1, the P3 peak plateau location is not connected with a thermally activated relaxation process, although its level changes due to a thermally activated recovery process. Its height increases with Ge content from 3 to 8% (i.e., ordering increases the vacancy concentration), and is very sensitive to the history of the specimen (P3 increases for higher quenching temperatures and cooling rates). At cooling the P3 peak-plateau gets smoother and lower, depends on the cooling rate after measuring TDIF at heating, and can be annealed out during subsequent IF tests. Thus, the IF spectra depend on measuring frequency: at low frequency these effects (P1+P2 Snoek-type and P3 vacancy related peaks) are well separated while they overlap in the kHz range.

### 4.4.2 Disordered and ordered *b.c.c.* structures (Fe-12Ge and Fe-19Ge)

According to the phase diagram [37, 261] these alloys include the B2 ($\alpha_2$) and D0$_3$ ($\alpha_1$) ordered structures, originating from the *b.c.c.* lattice, in a wide range of temperatures.

***Fe-12at.%Ge.*** As this composition is close to the boundary to disordered *b.c.c.* alloys, ordering is weak. A phase transition interval from B2 to A2 was detected at 1300–1350 K [260], while DSC and magnetometry yield a Curie point $T_C \sim 727$ °C and a solidus temperature $T_S = 1347$ °C; these points are in reasonable agreement with the literature data. After homogenizing for 24 h at 1000 °C and air cooling to room temperature, weak ordering in the *b.c.c.* solution (both $DO_3$ and B2) was detected by XRD in one specimen, in contrast to two others where only the disordered A2 phase was recorded [260].

In the "as quenched" state two IF peaks, P1 and P3 in Fig. 4.47, were recorded between 300 and 550 K, together with a new double peak below room temperature ($P_{LT}$ in Fig. 4.47a). The latter, nearly disappearing after heating to 400 °C, could be attributed to dislocations and will be briefly discussed below.

P1 is the carbon Snoek-type peak, but now significantly broader ($\beta \sim 3$) and consequently reduced in height (even in relation to the lower C content as compared to Fe-3Ge). Its asymmetry suggests that it again consists of two partial peaks (P1 and P2) from Fe-C-Fe and Fe-C-Ge surroundings, similar to the cases discussed above. By gradually annealing out the more unstable peak P3 (see below), the combined P1+P2 peak (with P1 dominating) is distinguished very clearly, so that an apparent activation parameters for Snoek relaxation for this composition can be determined: in case of Fig. 4.47b (with $Q_m^{-1}$ $\sim 0.0005$ above background), $H = 0.87$ eV and $\tau_0 \sim 10\text{-}16$ s. These values are very close to those of P1 in Fe-3Ge.

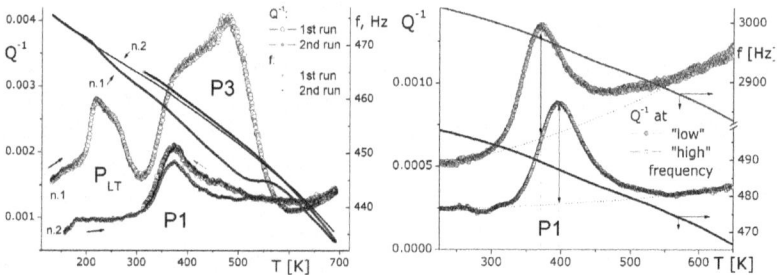

*Fig 4.47 Temperature-dependent IF of Fe-12Ge-0.01C (water-quenched after 3 h at 900 °C, heating rate 2 K/min): (a) heating-cooling cycle from the as-quenched state (n.1), and second heating (n.2); (b) the P1 peaks in the fourth run of a stepwise heating program, after previous heating to 300 °C, using two different frequencies (right scale); the peak heights after background subtraction (dotted lines) are indicated as double arrows.*

The P3 peak, on the other hand, is accompanied by an *increase* in resonance frequency (or elastic modulus, Fig. 4.47a), which is not expected for reversible thermally activated relaxation but more typical for recovery or recrystallisation processes known as Köster effect. The corresponding annealing behaviour of both P1 and P3 was studied in more detail, using two different procedures: a stepwise repeated heating with a rate of 2 K/min up to successively increasing temperatures (Fig. 4.48a), and an isothermal treatment at 167 °C (Fig. 4.48b) which is close to the not very well defined maximum temperature of P3. In the latter case, damping at 440 K was taken as a direct measure of the height of P3, while that of P1 was determined after repeated cooling to 100 °C during short interruptions of the isothermal treatment. During both of these annealing procedures, P3 is significantly less stable than P1, and completely disappears after heating to ~420 °C (Fig. 4.47a). In summary, these findings exclude P3 to result from a thermally activated relaxation process (as argued for Fe-8Ge above). Most probably it represents an effect linked to vacancy annealing, which is additionally supported by the fact that P3 is more pronounced in specimens quenched from higher temperatures [260]; this should be checked by positron annihilation experiments. Ordering in Fe-12Ge probably enhances this effect, as can be seen from the comparison with Fe-8Ge not subjected to ordering.

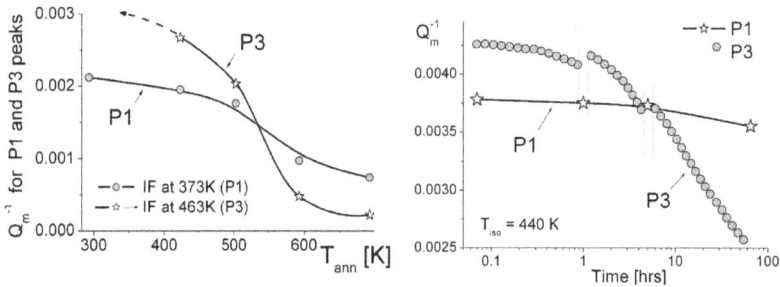

*Fig. 4.48 Reduction of the peak heights $Q_m^{-1}$ of the damping peaks P1 and P3 in Fe-12Ge by annealing: (a) after stepwise heating with 2 K/min, plotted as a function of the maximum temperature reached in the previous heating cycle (background subtracted as in Fig. 4.45b); (b) during isothermal annealing at 440 K (see text; no background subtracted).*

A further IF peak (P4) is found in Fe-12Ge at 775 K in low-frequency tests (Fig. 4.49), corresponding to earlier results [145] where parameters $H = 2.05$ eV and $\tau_0 = 10^{-13.5}$ s have been reported. We determine from our data the same activation energy and frequency factor, with a width parameter $\beta \approx 1$, and relaxation strengths $\Delta = 0.006$ in as-cast and $\Delta = 0.01$ in water-quenched (from both 1000 and 1273 K) states, respectively. These parameters suggest the Zener relaxation [10, 11] to operate, i.e. stress-assisted thermally activated reorientations of solute atom pairs in the present alloy with some low degree of B2 order [260]. A slight indication of another peak, around 670 K for f = 1.5 Hz, can be observed only during first heating after quenching from 1000 K, but nearly completely disappears in the second run after heating to 873 K. It should be noted that the interpretation of P4 as a Zener peak is supported by recent Fe59 tracer diffusion results in Fe-Ge where an activation enthalpy of about 2.5 eV has been found.

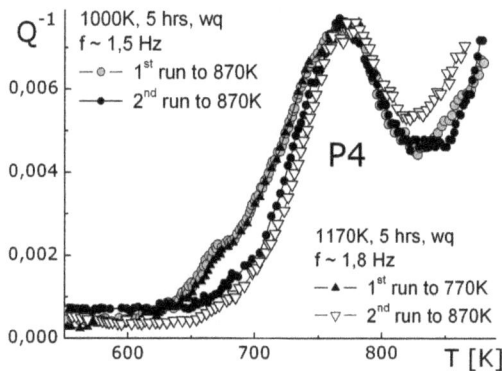

*Fig. 4.49 Low frequency (1.5–1.8 Hz) internal friction at elevated temperatures in Fe-12Ge (water-quenched, first and second heating runs as indicated), showing the Zener peak (P4) around 770 K, and at first heating an unstable peak at 670 K.*

**Fe-19at.%Ge.** This alloy is close to the two-phase range (crossing the solvus line twice [37], and at about 900 °C it consists of two phases due to the minimal solubility of Ge in the b.c.c. iron-based lattice. At a heating rate of 10 K/min, DSC indicates the Curie point at 614 °C, the ε→α transition at 869 °C, the solidus at 1173 °C and the liquidus at 1924 °C, respectively. After most heat treatments the alloy consists of a single ordered α phase ($\alpha_1$ or $\alpha_2$), except after quenching from 950 °C where it is characterized by a two-phase structure with b.c.c. (α) and hexagonal (ε) phases [154, 260] as shown in Fig. 4.50 (a).

After water-quenching from 727 °C, Fe-19Ge exhibits pronounced P1 and P3 internal friction peaks (Fig. 4.51), where P3 again disappears after first heating to 700 K. This instability of P3 supports its relation to vacancies and ordering in the B2 b.c.c. structure as discussed for Fe-8 and -12Ge above.

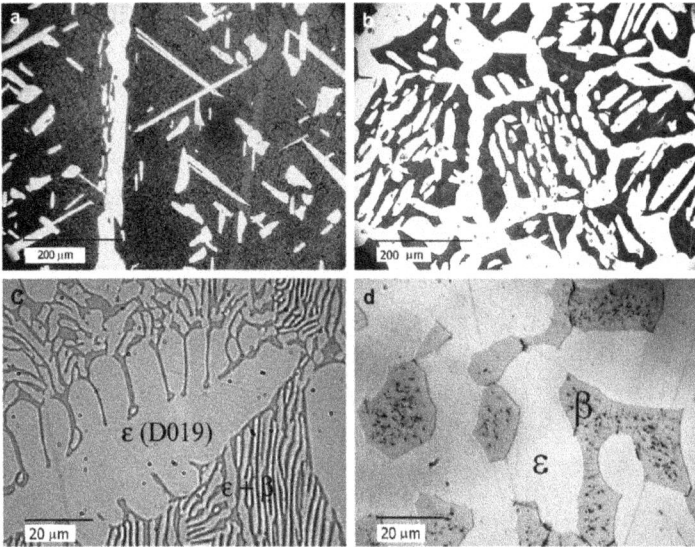

*Fig. 4.50 Optical micrographs of Fe-Ge alloys with ordered b.c.c.-type and hexagonal phases: (a) Fe-19Ge; (b) Fe-22Ge, both after quenching from 950 °C (dark phase: α; white phase: ε); (c) Fe-27Ge as cast; (d) Fe-27Ge after homogenising for 24 h at 1000 °C and additional annealing for 100 h at 375 °C.*

### 4.4.3 Hexagonal structures (Fe-22Ge and Fe-27Ge)

*Fe-22at.%Ge.* Again the solvus line is crossed twice, with two *b.c.c.* temperature ranges separated by a two-phase field [37]. In agreement with these phase diagrams, DSC at 10 K/min show the Curie point at 605 °C, the ε → α transition at 1085 °C, the solidus at 1159 °C, and the liquidus at 1259 °C. In the medium temperature range this alloy is a mixture of the *b.c.c.*-type ordered (α) phases, i.e. B2 ($\alpha_2$) above ~947 °C or D0$_3$ ($\alpha_1$) below ~847 °C [37], and the hexagonal ε (D0$_{19}$) phase (Fig. 4.50b). The alloy was homogenised at 1000 °C for 24 h, then annealed at 950 °C for 24 h to produce the α/ε phase mixture, and water-quenched; XRD then shows the presence of both phases (Fig. 4.52). After additional annealing at 380 °C for 200 h, only the α phase (with the (110) reflection shown in Fig. 4.52) is found with a lattice parameter a = 2.88542 ± 0.00029 Å, and with a very weak B2 ordering as detected by the (100) and (111) reflections at 31° and 55°, respectively.

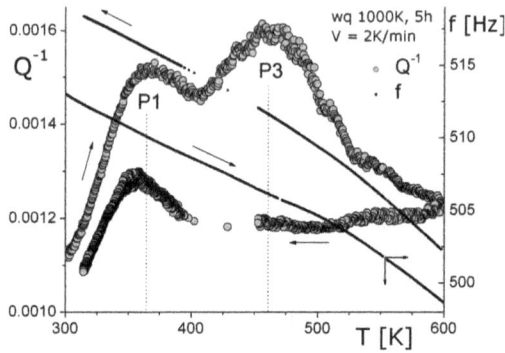

*Fig 4.51 Internal friction in Fe-19Ge during heating and cooling.*

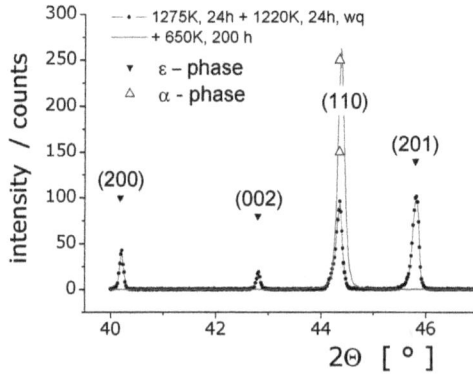

*Fig 4.52 X-ray diffraction patterns of Fe-22Ge homogenized at 1000 °C (24 h) and then quenched from 950 °C, with and without a subsequent annealing for 200 h at 380 °C.*

Internal friction mainly shows a pronounced double peak ($P_{LT1}$ + $P_{LT2}$ in Fig. 4.52, similar to $P_{LT}$ for quenched Fe-12Ge in Fig. 4.47a), with activation enthalpies roughly of $H_{PLT1} \approx 0.76$ eV and $H_{PLT2} \approx 0.96$ eV.

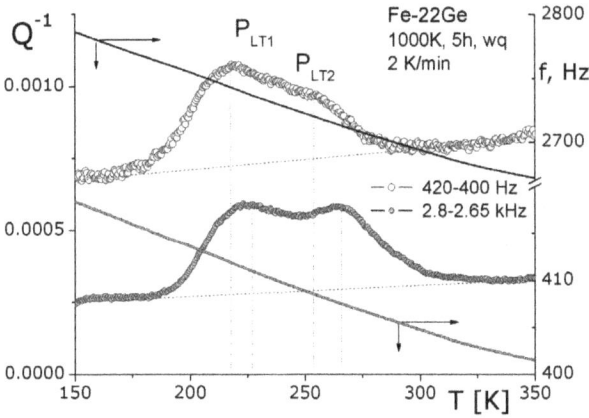

*Fig. 4.53 Temperature-dependent IF in Fe-22Ge quenched from 727 °C, showing the low-temperature double peak ($P_{LT1}$ + $P_{LT2}$) at two different frequencies (right scale).*

We consider two possible interpretations for both peaks: (1) They may belong to the class of $\beta$ and $\gamma$ relaxations in the *b.c.c.* structure, which result from the generation of double kinks in $a_0\{111\}/2$ screw dislocations on $\{110\}$ and $\{112\}$ slip planes, respectively, where the core is modified by solute atoms, as suggested by Seeger [262, 263] for pure and doped *b.c.c.* metals. This assignment is supported by the activation enthalpy values, which may be slightly modified due to possible order and existence of superdislocations. (2) They may represent Hasiguti-type peaks [264] which are often double- or triple-headed, probably due to dislocations relaxing in the vicinity of self lattice defects. Typically Hasiguti peaks are observed in deformed metals and "anneal out" at the recovery stage, while the $\gamma$ peak is known to consist of reversible and irreversible components. Unfortunately, a direct check of this hypothesis by definite plastic deformations is hindered by the extreme brittleness of the alloys.

***Fe-(25-27)at.%Ge with Fe$_3$Ge-type order.*** Contrary to the *b.c.c.*-type alloys discussed above, Fe-25Ge (see also in situ neutron diffractions in Chapter 3) and Fe-27Ge is basically of hexagonal structure. Three melting transformations were identified by DSC: the eutectic reaction $\varepsilon + \beta \to L$ at 1104 °C, the transformation $\varepsilon \to \alpha + L$ at 1125 °C, and the melting of the remaining $\alpha$ phase at 1127 °C [63].

Light microscopy and EDX in the as-quenched state show a two-phase structure with hexagonal $D0_{11}$-ordered $\varepsilon$ dendrites [63] (from the reaction $L + \alpha_2 \rightarrow \varepsilon$) and eutectic $\varepsilon +$ $\beta$, where $\beta$ is hexagonal $B8_1$-ordered (Fig. 4.50c). While short treatments at 1000 °C do not cause a distinct change, long-time (24 h) annealing leads to homogenization and disappearance of the dendrites. The $\varepsilon + \beta$ two-phase structure reappears after low-temperature annealing (100 h at 375 °C) of the homogenized sample, as confirmed by VSM and EDX [63], contrary to the expected $\alpha_1$ ($D0_3$) + $\beta$ phases (Fig. 4.50d). The existence of these phases is essentially corroborated by XRD which after 24 h at 1000 °C shows mainly the $\varepsilon$ phase, possibly some $\beta$ and perhaps some $L1_2$-ordered $\varepsilon'$, while after the following 100 h at 375 °C $\varepsilon$ is still present, together with $\beta$ and with only weak signs of $\varepsilon'$; in both cases no evidence for $\alpha$ is found.

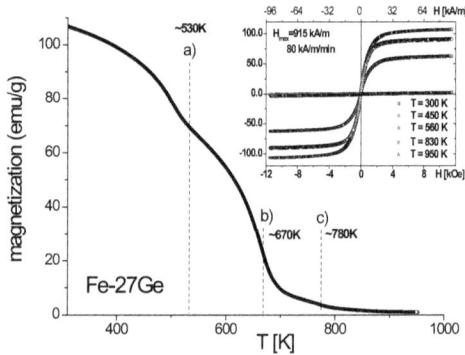

*Fig. 4.54 Temperature dependence of the magnetization for Fe-27at %Ge (1000 °C, 24 h, in air) at external field of 800 kA/m. Inset: field dependence of the magnetization at different temperatures.*

The magnetic properties also support these results from DSC and XRD: the Curie temperatures determined by VSM (Fig. 4.54) are 530 K for $\beta$ (in small amount), 638 K for the dominating $\varepsilon$ phase, and 750 K for the small amount of $\varepsilon'$. Some small collective magnetism for T > 780 K indicates a low fraction of $\alpha$ (not detected by DSC and XRD) with a Curie temperature of ~870 K [63, 260]. The small remaining magnetization above 780 K may arise from a very low volume fraction of $\alpha$ phase ($T_C$ = 870 K) not detectable by XRD.

The average magnetic moment of the iron atoms as calculated from the saturation magnetization of the hysteresis loops measured at different temperatures (Fig. 4.54, inset) decreases with increasing temperature and nearly vanishes for temperatures exceeding 780 K. The saturation value of the magnetic moment of iron in the Fe-27Ge alloy at $T \rightarrow$ 0 K amounts to about 1.7 $\mu_B$ which is in good agreement with the results of Konygin et al. [178] and values obtained by Mössbauer spectroscopy [65, 265].

Because of the dominating hexagonal phases, the Snoek relaxation (P1) is nearly fully suppressed in Fe-27Ge (Fig. 4.55; a small amount of residual $\alpha$ may cause the weak indication of P1 at 425K (152 °C)). A large IF peak is however recorded around 670 K (400 °C) at $f$ = 500 Hz (or 570 K (300 °C) at $f$ = 2 Hz) during both heating and cooling, which does not change with heat treatment. The relaxation strength $\Delta$ = 0.0036, activation enthalpy $H$ = 1.78 eV, and pre-exponential relaxation time $\tau_0 = 2 \times 10^{-17}$ s indicate that the peak is due to the Zener mechanism [63, 260], analogous to the P4 peak described above for Fe-12Ge. This is possible because of local deviations from $D0_{19}$ order, e.g., near defects like grain boundaries, dislocations, or vacancies. The distinct broadening of the peak and its asymmetry can be explained by the Zener effect in both the $\epsilon$ and $\beta$ phases present in this alloy [63]. This interpretation is also supported by recent measurements of the activation energy for Fe59 tracer diffusion [262] which yields a value of 1.57 eV.

The damping background is generally high after quenching and decreases with annealing, which suggests apart from dislocation recovery an annihilation of quenched-in vacancies connected with ordering. The latter is probably also the reason for the observed increase in the elastic modulus around 600-700 K [63]).

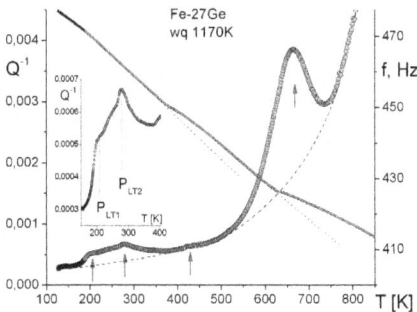

*Fig. 4.55 Temperature-dependent IF in Fe-27Ge after annealing at 900 °C. Inset: low-temperature range with two $P_{LT}$ peaks.*

In the range of low temperatures (200 – 280 K) a pair of peaks ($P_{LT1}$ and $P_{LT2}$) is observed which are supposed to be related to dislocations similar to the $P_{LT1,2}$ peaks described before for the other alloys Fe-(8-22)Ge. The fact that they appear also here in this hexagonal structure may support the Hasiguti relaxation [264] rather than the kink mechanism ($\gamma$, $\beta$ relaxation [262, 263]) which should only occur in *b.c.c.* phases.

**Hysteretic (amplitude dependent) effects.** The increase of Ge content in Fe lowers the Curie point and the ferromagnetism at room temperature. The amplitude dependent IF in case of ferromagnetic alloys consists of two main components, i.e. a magnetic (magnetomechanical hysteresis) and a non-magnetic one (as a rule of thumb – contribution from dislocations and vacancies). These components can be easily distinguished by applying of an external magnetic field. Indeed, an increase of Ge content in Fe decreases magnetomechanical damping (Fig. 4.56). This source of damping plays the main role in *b.c.c.* Fe-3Ge and Fe-12Ge alloys and decreases drastically in Fe-21Ge and is negligible in case of Fe-27Ge. Magnetomechanical damping is also sensitive to the heat treatment regime (Fig. 4.56b) and decreases with increasing temperature of measurements according to the decrease in magnetism. At the same time the absolute IF changes non-monotonously: while it decays in the range of *b.c.c.*-born structures, it increases slightly in the hcp alloy.

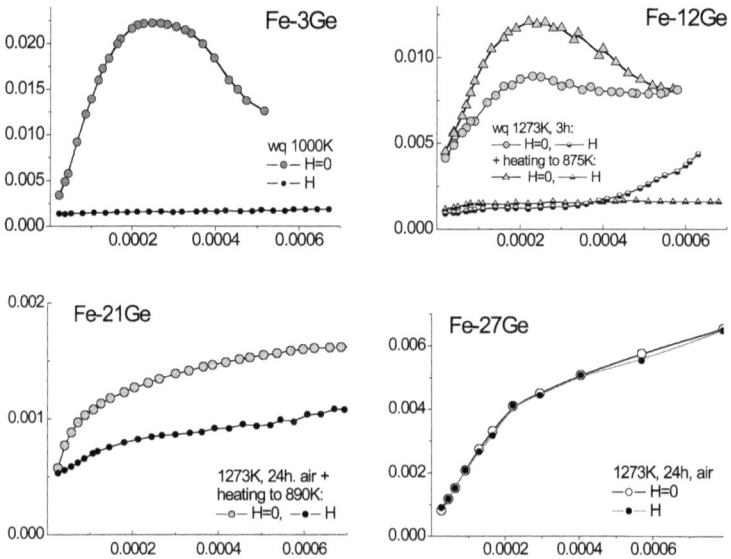

*Fig. 4.56 ADIF curves in several Fe-Ge alloys as measured with external magnetic field (H ($\sim 2 \times 10^4$ A/m)) and without magnetic field (H=0), frequency $\sim 2$ Hz.*

Thus, various kinds of internal friction peaks have been observed in the investigated Fe-(3-27)Ge alloys; the proposed mechanisms are related to the various crystallographic structures. In the *b.c.c.*-originated structures (Fe-3Ge, Fe-8Ge, Fe-12Ge, Fe-19Ge) the added small amount of carbon gives rise to a rather stable relaxation peak around 115 °C ($f \approx 500$ Hz), which is divided into a main Snoek peak (P1) and a smaller Snoek-type peak (P2) corresponding to Fe-C-Fe and Fe-C-Ge surroundings, respectively. Around 210 °C a rather unstable peak (P3), most probably not of thermally activated relaxation nature, is presumably connected with ordering and vacancy annealing.

The hexagonal structure (Fe-27Ge) does not show a Snoek peak as expected from crystallography, but a pronounced Zener relaxation (P4) around 400 °C for $f \approx 500$ Hz. The Zener relaxation was also found in *b.c.c.* Fe-12Ge, but at a higher temperature (770 K for $f \approx 1.5$ Hz). The mixed *b.c.c.* + hexagonal structures of Fe-19...22Ge exhibit a broadening of the Zener peak corresponding to the Zener effects in both structures.

For temperatures below room temperature a pair of peaks ($P_{LT1} + P_{LT2}$) is found in Fe-8....27Ge, supposed to correspond either to the β and γ relaxations in *b.c.c.* structures, i.e. kink pair formation in $a_0\{111\}/2$ screw dislocations on $\{110\}$ and $\{112\}$ slip planes, respectively, or to Hasiguti-type relaxations of dislocations with neighbouring self lattice defects. The latter interpretation is supported by the observation of $P_{LT}$ in the hexagonal structures.

### 4.5. Concluding remarks

The Fe-Al, Fe-Ga and Fe-Ge alloys have some similarities not only from the viewpoint of anelasticity but also in their structures. All these binary alloys have a disordered *b.c.c.* solid solution (A2) structure up to a certain solute concentration, which is roughly 11-12 at.% Al, 9-10 at.% Ge and 17-18 at.% Ga. The A2 range is followed by D0$_3$ or/and B2 order with an increase in the concentration of substitute atoms (with some intermediate, well-known short-range ordered "*K1*" state in case of Fe-Al and less studied "*nano-sized D0$_3$ domains*" in case of Fe-Ga). Al, Ga and Ge are not carbide forming elements in iron. In Fe-Al alloys κ-carbides were detected only for high-Al alloys and practically no data is available on carbide formation by Ga and Ge. From our internal friction results about the carbon Snoek (P1) and Snoek-type (P2) peaks, it is possible to conclude that neither Ga nor Ge trap carbon in carbides in iron but they influence the carbon atoms diffusivity via elastic interaction with carbon atoms. C atoms may prefer octahedral positions in A2, B2 and D0$_3$ structures around Al, Ga, Ge atoms.

The occurrence of the Snoek-type internal friction peak, sometimes also called as i-s (interstitial-substitution) peak or Fe-C-Me peak is the result of C atoms distribution near substitute atoms. The elastic distortions around the solute atoms due to their size misfit seem to play a decisive role. When comparing the respective peaks with a constant, relatively small content of 3 at% Me, the strongest "shoulder" produced by P2 at the high-temperature side of P1 was recorded in Fe-3Al [166, 174], then in Fe-3Si [166], followed by Fe-3Ga, while only a little second peak can be seen for Fe-3Ge, and practically no second P2 peak was detected in Fe-3Co [166].

Taking the atomic (Goldschmidt) radii for Fe, Al, Si, Ga, Ge and Co as 128, 143, 117, 135, 139, and 126 pm, correspondingly [266], the relative distortions ($\Delta r/r_{Fe}$) are Al/Fe = 11.2, Ga/Fe = 5.5, Ge/Fe = 8.6, Co/Fe = -1.6%. While only a rough correlation to the relative strength of the P2 can be seen, it is remarkable that the C-Me attraction does not seem to depend on the sign of elastic distortion: the low-energy sites appear likewise near both the big Al and the small Si atoms. P2 is typically broader than P1, and most pronounced in Fe-Al due to the high lattice distortion by Al: above ~6 at.% Al P2 dominates over P1 [166, 174], and above 12 at.%Al P1 is hardly distinguished. The activation energy is slightly higher in Fe-Al than in Fe-Si [166]. The relative heights of P1 and P2 are influenced by different heat treatments leading or prevention of segregation of C atoms around Me atoms.

The P3 internal friction peak, observed in Fe-Ge only after quenching, is not a completely thermally activated peak. Its height increases with the increase in Ge content towards $b.c.c.$-derivative superstructures (D0$_3$, B2), but it is not found for ordered hexagonal phases. Vacancies are expected to be involved. P3 in Fe-Ge may be analogous to the discussed in 4.2.2 so-called X-peak in Fe-Al alloys [137, 140]. However, the X-peak was found to be a thermally activated relaxation peak with an activation energy about 1.6-1.7 eV. It has been explained by C-vacancy [139] or vacancy-vacancy complexes [135], or with both types of defects [193]. Both peaks – the X-peak in Fe-Al and the P3 in Fe-Ga - decrease due to annealing much faster than the Snoek-type (P1+P2) peaks.

Zener peak is produced by re-orientation of pairs of substitution atoms. The activation energy is close to that of diffusion, which was well demonstrated for Fe-Al alloys (Table 3.3). Notable is the rather big difference in the peak parameters between the $b.c.c.$ Fe-12Ge and the hexagonal Fe-27Ge alloys. The Zener peak for Fe-Al and Fe-Si alloys has long been known [267]. The Zener peak in Fe-Ga may overlap with phase transition from ordered $b.c.c.$ (D0$_3$) to ordered fcc (L1$_2$) lattice. A common feature is the decrease in the Zener peak height with long-range ordering. Thus the Zener peak in disordered Fe-Al and Fe-Ga alloys can be observed up to 22 at.% Al and 19 at.% Ga, correspondingly, and only up to 11 at.% Si in Fe-Si alloys. It is largely suppressed in ternary Fe-Al-Si alloys due to a mutual compensation of the distortions produced by the big Al and small Si atoms, respectively [254].

The two "low temperature" ($P_{LT1}+P_{LT2}$) peaks, measured in both *b.c.c.*-type and hexagonal Fe-Ge and Fe-Ga alloys, which were not studied in detail, are most probably linked with dislocations motion. $P_{LT}$ in Fe-Ge and Fe-Ga is analogous to the so-called D 'dislocation' peak in Fe-Al [268]. The D-peak grows by one order in case of severe deformation [269], where it is suggested to consist of at least three Hasiguti-type [263] peaks.

Transient internal friction effects due to second order transitions ($\lambda$ - peaks): ordering - disordering were recorder and analyzed in Fe-Al and Fe-Ga poly and single crystals.

Three transient IF effects are recorded due to first order transitions in the studied Fe-Ga alloys. These peaks correspond to the $D0_3 \rightarrow L1_2 \rightarrow D0_{19} \rightarrow A2$ (or B2) phase transitions. The transient peaks has similar features with transient peaks at thermoelastic martensitic transition. Nevertheless, *in situ* neutron diffraction does not support the idea about the appearance of tetragonal phases by shear deformation. The first order $D0_3 \rightarrow L1_2$ and the $D0_{19} \rightarrow A2$(or B2) phase transitions occur with a change in the atomic volume and a rise of the local stresses and strains, which lead to these anelastic peaks. The $L1_2 \rightarrow D0_{19}$ phase transition is accompanied neither with a change in the atomic volume nor with a significant rise of the local strains. In this case the IF peak is due to a long-range motion of the Shockley dislocation that assists this transition.

Alloys of all three systems: Fe-Al, Fe-Ga and Fe-Ge exhibit magneto-mechanical damping which can be optimized by alloy composition and heat treatment. Fe-Al alloys are well-known with respect to their application as high damping alloys.

### References

[154] Golovin, I.S. Chapter "Anelasticity of iron-based ordered alloys and intermetallic compounds" to the Book "Intermetallics Research Trends" (ISBN 978-1-60021-982-5) Editor: Y.N. Berdovsky, Nova Science Publishers, Inc, 2008, 65-133.

[155] Woodruff, E. A study of the effect of temperature upon a tuning fork. Phys Rev., 16, 1903, 321–355. https://doi.org/10.1103/PhysRevSeriesI.16.325

[156] Weller, M. Point defect relaxation. In Mechanical Spectroscopy $Q^{-1}$ 2001 with Applications to Materials Science; Schaller, R., Fantozzi, G., Gremaud, G., Eds.; Trans Tech Publications: Switzerland, 2001, 95-140.

[157] Doernberg, E. Ermittlung des gelösten Kohlenstoffgehaltes mittels verschiedener Methoden und seine Auswirkungen auf das Bake-Hardening- und Alterungsverhalten von Stählen; Diplomarbeit; Technische Universität Clausthal: Clausthal-Zellerfeld, März, 2007.

[158] Wert, C. In Thermodynamics in Physical Metallurgy; ASM: Cleveland, 1950; Chapter "Phenomena Accompanying Precipitation from Solid Solution of C and N in α-Iron"

[159] Dijkstra, L. J.; Sladek, R. J. Trans Metall Soc AIME 1953, 197, 69-72.

[160] Koiwa, M. Theory of the snoek effect in ternary b.c.c. alloys. Philosophical Magazine, 1971, 24:187, 81-106. https://doi.org/10.1080/14786437108216426

[161] Wert, C. Trans Metall Soc AIME 1952, 194, 602-603; Internal friction of an alloy of 16 wt% Al in iron (1954) 640-641.

[162] Langerberg, G. Acta Met 1959, 7, 137-. https://doi.org/10.1016/0001-6160(59)90124-5

[163] Krishtal, J. A. Fiz Metallov Metallovedenie, 19 (1965) 111-115 (in Russian).

[164] Kuzka, R.; Moron, J.; Wroblewski, J. Arch Naucki Materialach, 4 (1983) 161-203.

[165] Golovin, S.A. Snoek effect in alloyed ferrite. In: Thermal treatment and physics of metals; UrPI: Sverdlovsk, 4 (1978) 67-75.

[166] Golovin, S. A.; Morozyuk, A. A; Ageev, V. S. Influence of quenching temperature and alloying elements on Snoek relaxation in iron. In: Internal Friction in Metals, Semiconductors, Dielectrics and Ferromagnets; Nauka: Moscow, 1978, 41-45.

[167] Ogi, H.; Ledbetter, H.; Kim, S. Snoek relaxation in a copper-precipitated alloy steel. J All Comp, 310 (2000) 432-435; and Snoek Relaxation and Dislocation Damping in Aged Fe-Cu-Ni Steel. Metall Trans A, 32 (2001) 1671-1677. https://doi.org/10.1007/s11661-001-0145-3

[168] Saitoh, H. S.; Yoshinaga, N.; Ushioda, K. Influence of substitutional atoms on the Snoek peak of carbon in b.c.c. iron. Acta Mater, 52 (2004) 1255-1261. https://doi.org/10.1016/j.actamat.2003.11.009

[169] Numakura, H; Koiwa, M. J Phys (Paris) IV, 6 (1996) C8-97-102.

[170] Ritchie, I. G.; Rawlings, R. Acta Metall, 15 (1967) 491-496. https://doi.org/10.1016/0001-6160(67)90081-8

[171] Kruk, A.; Pietrzyk, J.; Kapusta, C. Scr Metall, 31 (1994) 1679-1683. https://doi.org/10.1016/0956-716X(94)90463-4

[172] Semin, V.A.; Golovin, S.A.; Golovin, I.S. Program for analyses of temperature (i) and frequency (ii) dependent internal friction data (Russian registration number № 2005611581 (i) and № 2006610674 (ii)).

[173] Ruiz, D.; Rivera-Tovar, J. L.; Segers, D.; Vandenberghe, R. E.; Houbaert, Y. Aging phenomena in high-Si steels studied by internal friction. Mat Sci Eng A, 442 (2006) 462-465. https://doi.org/10.1016/j.msea.2006.05.164

[174] Strahl, A.; Golovin, I.S.; Neuhäuser, H.; Golovina, S.B.; Pavlova, T.S.; Sinning, H.-R. Influence of Al concentration on parameters of the short and long range carbon diffusion in Fe-Al alloys. Mater Sci Eng A, 442 (2006) 128–132. https://doi.org/10.1016/j.msea.2006.05.160

[175] Golovin, I.S.; Golovina, S.B.; Sokolova O.A. Effect of thermal aging on the temperature spectrum of internal friction of alloyed Fe–Si–Al–C ferrite. The Physics of Metals and Metallography, 105 (2008)193-201. https://doi.org/10.1134/S0031918X08020129

[176] Koiwa, M.; Numakura, H. The Snoek Effect in Ternary BCC Alloys. A Review. Solid State Phenomena, 115 (2006) 37-40. https://doi.org/10.4028/www.scientific.net/SSP.115.37

[177] Wern-bin, L.; Guo-ping, Y.; Cheng-hsiu, L.; Zheng-qun, L. J Phys (Paris) 42 (1981) C5-463-474

[178] Konygin, G. N.; Yelsukov, E. P.; Porsev, V. E. The structure and magnetic properties of the non-equilibrium $Fe_{100-x}Ge_x$ (x=5-40 at%) system produced by mechanical alloying. J Magn Magn Mat, 288 (2005) 27-36. https://doi.org/10.1016/j.jmmm.2004.07.052

[179] Jäniche, V.W.; Braunen, J.; Heller, W. Archiv Eisenhüttenwesen, 37 (1966) 719-728. https://doi.org/10.1002/srin.196604573

[180] Kulish, N. P.; Mandrika, V.M.; Petrenko, P. V. Fiz Metallov Metallovedenie, 51 (1981) 1229-1237 (in Russian).

[181] Strahl, A. Anelastische Relaxationen durch Punktdefekte und Versetzungen in Fe-Al-Legierungen; Verlag im Internet GmbH, Berlin-Braunschweig, 2006.

[182] Pozdova, T.V.; Golovin, I.S. Mechanical spectroscopy of Fe-Al-C alloys ordering Solid State Phenomena, 89/90 (2003) 279-286. https://doi.org/10.4028/www.scientific.net/SSP.89.279

[183] Damson, B. Innere Reibung in Fe-Al mit B2-Struktur; Dissertation; Universität Stuttgart, 1998.

[184] Nagy A. Mechanische Spektroskopie an Eisen-Aluminium und Polymerschichten; Dissertation; Technische Universität Braunschweig, 2002.

[185] Wu, J.; Han, F. S.; Gao, Z. Y.; Hao, G. L.; Wang, Q. Z. Internal friction peak associated with the ordering process in B2 Fe-Al alloys, Phys Stat Sol A, 203 (2006) 485-492. https://doi.org/10.1002/pssa.200521404

[186] Rokhmanov, N.Ya. Relaxation spectrum of ordering carbon-containing alloys Fe-(25-31)%Al. Functional Materials. 7 (2000) 235-240.

[187] Rokhmanov, N.Ya.; Hamana, D. Visnyk Kharkov Nat University 516: Fizika, 5 (2001) 104 (in Russian).

[188] Zhou, Z.C.; Han, F.S.; Gao, Z.Y. The internal friction peaks correlated to the relaxation of Al atoms in Fe-Al alloys. Acta Mat, 52 (2004) 4049-4054. https://doi.org/10.1016/j.actamat.2004.05.020

[189] Boufenghour, M.; Hamana, D.; Rokhmanov, N.; Andronov, V. Identification of a new tetragonal phase q in an Fe-31.5 at.% Al alloy, using internal friction and X-ray diffraction. La Revue Metallurgie, (2004) 663-669 (in French). https://doi.org/10.1051/metal:2004114

[190] Rokhmanov, N.Ya.; Golovin, I.S. Mechanism of X-relaxation in Fe-Al-C alloys. Reports Kharkov National University: Physics, 558 (2002) 158-167.

[191] Golovin, I.S.; Pozdova, T.V.; Zharkov, R.V.; Golovin, S.A. Relaxation processes in Fe-Al alloys. Metal Science and Heat Treatment, 6 (2002) 16-22.

[192] Golovin, I.S. Proc Int Conference Imperfections Interaction and Anelasticity Phenomena in Solids; 13-15. Nov. 2001; TSU: Tula (2002) 47-55.

[193] Wu, J.; Han, F.S.; Wang, Q.Z.; Hao, G.L.; Gao, Z.Y. The internal friction peaks correlated to the relaxation of atomic defects in $Fe_{47}Al_{53}$ alloy. Intermetallics, 15 (2007) 838-844. https://doi.org/10.1016/j.intermet.2006.10.037

[194] Blanter, M.S.; Golovin, I.S.; Sinning, H.-R. Simulation of the X relaxation in Fe-Al-Me (Me = Co, Cr, Mn, Si) alloys. Mat Sci Eng A (2006) 438, 442/1-2, 133-137. https://doi.org/10.1016/j.msea.2006.05.162

[195] Tanaka, K.; Sahashi, K., The Zener relaxation in Fe-alloys and its application to diffusion problems, Trans. Jpn Inst. Metals, 12 (1971) 130-135. https://doi.org/10.2320/matertrans1960.12.130

[196] Golovin, I.S.; Strahl, A.; Neuhäuser, H. Anelastic relaxations and structure of ternary Fe-Al-Me alloys with Me = Co, Cr, Ge, Mn, Nb, Si, Ta, Ti, Zr. J Mat Res (formerly Z Metallkde), 42 (2006) 1078-1092. https://doi.org/10.3139/146.101341

[197] Golovin, I.S.; Blanter, M.S.; Schaller, R. Snoek relaxation in Fe-Cr alloys and interstitial-substitutional interaction. Phys Stat Sol A, 160 (1997) 49-. https://doi.org/10.1002/1521-396X(199703)160:1%3C49::AID-PSSA49%3E3.0.CO;2-P

[198] Golovin, I.S.; Anelastic relaxation in ternary Fe-Al-Me alloys: Me = Co, Cr, Ge, Mn, Nb, Si, Ta, Ti, Zr. Mat Sci Eng A, 442 (2006) 92-98. https://doi.org/10.1016/j.msea.2006.03.118

[199] Pavlova, T.S.; Golovin, I.S.; Sinning, H.-R.; Golovin, S.A.; Siemers C. Internal friction in Fe-Al-Si alloys at elevated temperatures. Intermetallics, 14 (2006), 1238-1249. https://doi.org/10.1016/j.intermet.2005.12.015

[200] Golovin, I.S.; Rivière, A. Mechanical spectroscopy of Fe-25Al-Cr alloys in medium temperature range. Solid State Phenomena, 137 (2008) 99-108. https://doi.org/10.4028/www.scientific.net/SSP.137.99

[201] Mosher, D.R.; Raj, R. Use of the internal friction technique to measure rates of grain boundary sliding. Acta Metall., 22(12) (1974) 1469–1474.

[202] Gremaud, G.; Kustov, S. Theory of dislocation-solute atom interactions in solid solutions and related nonlinear anelasticity. Phys. Rev. B, 60(13) (1999) 9353–9364. https://doi.org/10.1103/PhysRevB.60.9353

[203] Rivière, A. In Mechanical spectroscopy $Q^{-1}$ 2001 with applications to materials science; Schaller, R; Fantozzi, G.; Gremaud, G. (Eds.); Trans Tech Publications: Switzerland, 2001; pp 635-651; and Mater Sci Forum 2001, 366/368, 635-651.

[204] Wert, C. Internal friction of an alloy of 16 percent aluminum in iron. J Appl Phys., 26 (1955) 640-641. https://doi.org/10.1063/1.1722059

[205] Bonetti, E.; Scipione, G.; Frattini, R.; Enzo, S.; Schiffini, L. J Appl Phys., 79 (1996) 7537-7544. https://doi.org/10.1063/1.362408

[206] San Juan, J.; No, M. L.; Lacaze, J.; Viguier, B.; Fournier, D. Internal friction in advanced Fe–Al intermetallics. Mat Sci Eng A, 442 (2006) 492-495. https://doi.org/10.1016/j.msea.2006.06.064

[207] Lambri, O.A.; Pérez-Landazábal, J.I.; Recarte, V.; Cuello, G.J.; Golovin, I.S. Order controlled dislocations and grain boundary mobility in Fe-Al-Cr alloys. Journal of Alloys and Compounds, 537 (2012) 117–122. https://doi.org/10.1016/j.jallcom.2012.04.119

[208] Gargicevich, D., Lambri, O.A., Pérez-Landazábal, J.I., Recarte, V., Bonifacich, F.G., Cuello, G.J., Sánchez Alarcos, V. Mobility of dislocations and grain

boundaries controlled by the order degree in iron-based alloys. Journal of Physics: Conference Series, 663 (2015) 012013. https://doi.org/10.1088/1742-6596/663/1/012013

[209] Gargicevich, D., Galvan Josa, V.M., Blanco C., Lambri, O.A., Cuello, G.J. Structure determination of Fe-Al-Ge alloys. Journal of Physics: Conference Series 663 (2015) 012004. https://doi.org/10.1088/1742-6596/663/1/012004

[210] Ritchie, I.G.; Fantozzi, G. In Dislocations in Solids; Nabarro, F. R. N. (Ed.), 9 (1992) 57-133.

[211] Mizubayashi, H.; Kronmuller, H.; Seeger, A. J Phys (Paris), 46 (1985) C10-C309.

[212] Brinck, A.; Neuhäuser, H. Intermetallics, 8 (2000) 1019-1024. https://doi.org/10.1016/S0966-9795(00)00042-X

[213] Sestak, B.; Seeger, A. Scripta Metall, 5 (1971) 875-882. https://doi.org/10.1016/0036-9748(71)90062-7

[214] Hivert, V.; Groh, P.; Moser, P.; Frank, W. Phys Stat Sol A, 42 (1977) 511-518. https://doi.org/10.1002/pssa.2210420212

[215] D'Anna, G.; Benoit, W. Mater Sci Forum, 119 (1993) 165-170. https://doi.org/10.4028/www.scientific.net/MSF.119-121.165

[216] Hehenkamp, Th.; Scholy, P.; Köhler, B.; Kerl, R. Defect Diffusion Forum 2001, 194- 199, 389-394.

[217] T. S. Pavlova, I. S. Golovin, D. V. Gunderov, C. Siemers. Effect of Severe Plastic Deformation on the Structure and Low-Temperature Internal Friction of Fe3Al and (Fe,Cr)3Al. The Physics of Metals and Metallography, 2008, Vol. 105, No. 1, pp. 36–44.

[218] Golovin, I.S.; Pavlova, T.S.; Grusewski, C.; Ivanisenko, Y.; Gunderov, D. V. In Nanomaterials by Severe Plastic Deformation; Trans Tech Publications: 503-504 (2005) 745-750.

[219] T. S. Pavlova, I. S. Golovin, D. V. Gunderov, C. Siemers. Effect of Severe Plastic Deformation on the Structure and Low-Temperature Internal Friction of Fe3Al and (Fe,Cr)3Al. The Physics of Metals and Metallography, 2008, Vol. 105, No. 1, pp. 36–44

[220] Bruver, R.E.; Glikman, E.A.; Krasov, A.A.; Trubin, S.V.; Krieger, O.S.; Chelystev, N.A. Interaction of phosphorus with grain boundary internal friction of iron-phosphorus solid solutions. Phys. Metals Metallogr. 42(4) (1976) 187–190.

[221] Jiles, D.C. Introduction to Magnetism and Magnetic Materials; Addison-Wesley: London, U.K., 1998.

[222] Pal-Val, P.P., Pal-Val, L.N., Golovina, S.B., Golovin, I.S. Effect of heat treatment on acoustic properties of chromium polycrystals at low temperatures. Solid State Phenomena, 137 (2008) 43-48. https://doi.org/10.4028/www.scientific.net/SSP.137.43

[223] Udovenko V.A., Tishaev S.I., Chudakov I.B. Structure and properties special features of high damping alloys on the base of Fe. Russian Metallurgy (Metalli), 1994, No 1, pp. 64-69 (engl).

[224] Udovenko, V.A.; Chudakov, I.B.; Polyakova, N.A. In $M^3D$ III, Mechanics and Mechanisms of Material Damping, ASTM STP 1304; Wolfenden, A.; Kinra, V. (Eds.); ASTM: Philadelphia, 1997; 204-213. https://doi.org/10.1520/STP11749S

[225] Udovenko, V.A.; Chudakov, I.B. Specific properties of industrial Fe-Al high damping steels. Solid State Phenomena, 115 (2006) 57-62.

[226] Mielczarek, A.; Riehemann, W.; Sokolova, O.A.; Golovin, I.S. Influence of heat treatment on structure, magnetic and damping properties of the Fe-11at.%Al alloys. Solid State Phenomena, 137 (2008) 129-136. https://doi.org/10.4028/www.scientific.net/SSP.137.129

[227] Chudakov, I.B., Polyakova, N.A., Mackushev, S.Y., Udovenko, V.A., On the formation of high damping state in Fe-Al and Fe-Cr alloys", Solid State Phenomena, 137 (2008) 109-118. https://doi.org/10.4028/www.scientific.net/SSP.137.109

[228] Emdadi A., Nartey, M. A., Xu, Y., Golovin, I.S. Study of damping capacity of Fe-5.4Al-0.05Ti alloy. J. All. and Comp., 653 (2015) 460-46. https://doi.org/10.1016/j.jallcom.2015.09.031

[229] Seo, Y.I., Lee, B.-H., Kim, Y.D., Lee, K.H., Grain size effects on magnetomechanical damping properties of ferromagnetic Fe-5 wt.% Al alloy. Mater. Sci. Eng. A, 431 (2006) 80. https://doi.org/10.1016/j.msea.2006.05.083

[230] Wang, H., Tian, X.F., Yin, C.G., Huang, Z.H. The effect of heat treatment and grain size on magnetomechanical damping properties of Fe-13Cr-2Al-1Si alloy. Mater. Sci. Eng. A, 619 (2014) 199. https://doi.org/10.1016/j.msea.2014.09.089

[231] Wang, H., Huang, H.W., Hong, X.F., Yin, C.G., Huang, Z.H., Chen, L. Synthesis and characterization of $Ho^{3+}$-doped strontium titanate downconversion

nanocrystals and its application in dye-sensitized solar cells. J. Alloy. Comp. 622 (2015) 17. https://doi.org/10.1016/j.jallcom.2014.10.019

[232] Li, N., Xu, Y., Yu, X., et al., Mech. Eng. Mater. 28 (2004) 4.

[233] Wang, W., Zhou, B., Zheng, Z., Damping capacity of Fe-Cr-Al-Si alloys. Acta Metall. Sin. 34 (1998) 1039.

[234] Xu, Y.G., Li, N., Shen, B.L., Hua, H.X., Effect of annealing treatment on damping capacity of Fe–7Al–0.5Ti alloy. Mater. Sci. Eng. A 447 (2007) 163-166. https://doi.org/10.1016/j.msea.2006.10.009

[235] Golovin, I.S., Mechanisms of damping capacity of high-chromium steels and $\alpha$-Fe and its dependence on some external factors. Metall. Mater. Trans. A, 24 (1994) 111-124. https://doi.org/10.1007/BF02646680

[236] O'Handley, R.C., Modern Magnetic Materials, first ed., Wiley, NewYork, NY, (2000) 218-259.

[237] Engdahl, G. Handbook of Giant Magnetostrictive Materials, First ed., Academic Press, London, U.K, (2000) 69-81.

[238] Hubert, A.; Schafer, R.; Magnetic Domains, third ed., Springer, Berlin, Germany (2009) 125-138.

[239] Udovenko, V.A.; Chudakov, I.B.; Alexandrova, N.M.; Kakabadze, R.V.; Perevalov, N.N. On the formation of high damping state and optimization of structure of industrial damping steels. Solid State Phenom., 137 (2008) 119. https://doi.org/10.4028/www.scientific.net/SSP.137.119

[240] Smith, G.W.; Birchak, J.R. Internal stress distribution theory of magnetomechanical hysteresis - An extension to include effects of magnetic field and applied stress. J. Appl. Phys., 40 (1969) 5174-5178. https://doi.org/10.1063/1.1657370

[241] Udovenko, V.A.; Tishaev, S.I.; Chudakov, I.B. Magnetic domain structure and damping in Fe-Al system alloys. Dokl. Akad. Nauk, 329 (1993) 585 ([in Russian]).

[242] Chudakov, I.B.; Alexandrova, N.M.; Makushev, C.U. Influence of external stresses on magnetic and damping properties of Fe-Al alloys. Probl. Chernay Metall. Mater. 4 (2012) 75 ([in Russian]).

[243] Xiaoxiao, M.; Ning, L.; Jiazhen, Y. Investigation of constant prestress in effecting damping capacity of Fe-15Cr-2.5Mo-1.0Ni casting ferromagnetic alloy. J. Alloy. Comp., 650 (2015) 92. https://doi.org/10.1016/j.jallcom.2015.07.273

[244] Pulino-Sagradi, D.; Sagradi, M.; Karimi, A.; Martin, J.L. Damping capacity of Fe-Cr-X high-damping alloys and its dependence on magnetic domain structure. Scr. Mater., 39 (1998) 131-138. https://doi.org/10.1016/S1359-6462(98)00157-2

[245] Schneider, W.; Schrey, P.; Hausch, G.; E. J. Phys. Colloq. 42 (1981) C5-C635.

[246] H. Xiao-Feng, L. Shuwei, L. Xiuyan, R. Lijian, Chin. J. Mater. Res. 27 (2013) 225.

[247] Hu, X.F.; Liu, S.W.; Li, X.Y.; Zhang, B.; Rong, L.J.; Yin, F.X.; Li, Y.Y. Influence of static stress on damping behavior in Fe–15Cr and Fe–8Al ferromagnetic alloys. Mater. Sci. Eng. 528 (2011) 5491. https://doi.org/10.1016/j.msea.2011.03.062

[248] Yin, F.; Takamori, S.; Ohsawa, Y.; Sato, A.; Kawahara, K. The effects of static strain on the damping capacity of high damping alloys. Mater. Trans., 43 (2002) 466. https://doi.org/10.2320/matertrans.43.466

[249] Golovin, I.S.; Emdadi, A.; Balagurov, A.M.; Bobrikov, I.A.; Cifre, J.; Zadorozhnyy, M.Yu.; Rivière, A. Anelasticity of $Fe_3Al$ type single and polycrystals. J. Alloy. Comp. (JALCOM-D-17-02390) (2017) accepted, in press.

[250] Becker, R.; Doring, W. Ferromagnetismus, Springer, Berlin, (1939) 336. https://doi.org/10.1007/978-3-642-47366-1_5

[251] Leamy, H.; Gibson, E. D.; Kayser, F. X. Acta Metall., 15 (1967) 1827-1838. https://doi.org/10.1016/0001-6160(67)90047-8

[252] Ruffoni, M.P.; Pascarelli, S.; Grossinger, R. et al. Direct measurement of intrinsic atomic scale magnetostriction. Phys. Rev. Lett., 101 (2008) 147202. https://doi.org/10.1103/PhysRevLett.101.147202

[253] H. Cao, F. Bai, J. Li, D.D. Viehland, T.A. Lograsso, P.M. Gehring. Structural studies of decomposition in Fe–x at.%Ga alloys. J. All. Comp., 465 (2008) 244-249. https://doi.org/10.1016/j.jallcom.2007.10.080

[254] Sinning, H.-R., Golovin, I.S., Strahl, A., Sokolova, O.A., Sazonova, T. Interactions between solute atoms in Fe-Si-Al-C alloys as studied by mechanical spectroscopy. Mat. Sc.Eng. A, 521–522 (2009) 63–66. https://doi.org/10.1016/j.msea.2008.09.110

[255] Golovin, S.A., Golovin, I.S. Mechanical spectroscopy of Snoek type relaxation. Metal Science and Heat Treatment, 54 (2012) 208-216. https://doi.org/10.1007/s11041-012-9483-6

[256] Boyer, S.A.E.; Gerland, M. ; Rivière, A.; Cifre, J.; Palacheva, V.V. ; Mikhaylovskaya, A.V.; Golovin, I. S. Anelasticity of the Fe-Ga alloys in the range of Zener relaxation. J. All. Comp., 730 (2018) 424-433. https://doi.org/10.1016/j.jallcom.2017.09.275

[257] Golovin, S.A.; Arkhangelskij, S.I. Relaxator for amplitude, temperature and frequency internal friction tests. Strength of Materials, 5 (1971) 120-124 (in Russian).

[258] Nagy, A.; Harms, U.; Neuhäuser, H. Def. Diff. Forum 203-205 (2002) 257-260. https://doi.org/10.4028/www.scientific.net/DDF.203-205.257

[259] Golovin, I.S.; Ivleva, T.S.; Jäger, S.; Neuhäuser, H.; Redfern, S.A.T.; Siemers, C. Structure and anelasticity of Fe-Ge alloys. Solid State Phenomena, 137 (2008) 59-68. https://doi.org/10.4028/www.scientific.net/SSP.137.59

[260] Golovin, I.S.; Neuhäuser, H.; Redfern, S.A.T.; Sinning, H.-R. Mechanisms of anelasticity in Fe-Ge-based alloys. Materials Science and Engineering A 521–522 (2009) 55–58. https://doi.org/10.1016/j.msea.2008.09.099

[261] Massalski, T.B. Binary Phase Diagrams, 2nd ed., ASM, OH 2006, p.1740.

[262] Seeger, A. The mechanism of the β-relaxation in high-purity bcc metals. Phil Mag Lett 83 (2003) 107-115. https://doi.org/10.1080/0950083021000053329

[263] Seeger, A. Dislocation relaxation processes in bcc metals: β versus γ relaxation. Phil Mag Lett 84 (2004) 79-86. https://doi.org/10.1080/09500830310001628239

[264] Hasiguti, R.R.; Igata, N.; Kamoshita, G. Internal friction peaks in cold worked metals. Acta Metall 10 (1962) 442-447. https://doi.org/10.1016/0001-6160(62)90023-8

[265] Adelson, E.; Austin, A. E. J. Phys. Chem. Solids. 1965, 26, 1795-1804. https://doi.org/10.1016/0022-3697(65)90212-X

[266] Brandes, R.A.; G.B. Brook (eds.), Smithells Metals Reference Book, 7th edition, Butterworth-Heinemann, Oxford 1992.

[267] Tanaka, K. The Zener relaxation effect in ferrous alloy systems. Trans Japan Inst Met 16 (1975) 199-206. https://doi.org/10.2320/matertrans1960.16.199

[268] Strahl, A.; Golovina, S.B.; Golovin, I.S.; Neuhäuser, H. Anelasticity of $Fe_3Al$ intermetallic compounds. J. Alloys Comp 370 (1-2) (2004) 268-273. https://doi.org/10.1016/j.jallcom.2003.10.066

[269] Golovin, I.S.; Pavlova, T.S.; Golovina, S.B.; Sinning, H.-R.; Golovin, S.A. Effect of severe plastic deformation on internal friction of an Fe-26 at.% Al alloy and titanium. Mat Sci Eng A 442/1-2 (2006) 165-169. https://doi.org/10.1016/j.msea.2005.12.081

Chapter 5

# Ternary Fe-based Alloys: Structure Induced Anelasticity

## Contents

### 5.1 Ternary Fe-Ga based alloys

After pioneering studies of Fe-Ga alloys mainly by US research groups in the early 2000s (A.E. Clark, T.A. Lograsso, A. Flatau, M. Wittig, A. Khachaturyan et al), in the last few years an enormous research activity in this field has taken place in different research groups in China (C. Jiang, X. Ren, T. Ma, J. Zhu, X. Gao, S.U. Jen et al). They introduced several fruitful ideas to improve functionality of Galfenols including a 'ferromagnetic strain glass' approach, alloying Fe-Ga by rare earth elements, annealing in a magnetic field, formation of special texture by rolling etc. First internal friction studies of Fe-Ga alloys were reported in papers by M. Wuttig and H. Numakura, then by our research group. We also applied for the first time systematical studies of *in situ* neutron diffractions to characterize phase transitions in Fe-Ga based bulk alloys and to interpret anelastic phenomena measured by different mechanical spectroscopy techniques in the same heating and cooling or annealing regimes. This chapter is mainly based on the results of our own studies with some references to relevant findings by other research groups.

### 5.1.1 Fe-Ga-Tb alloys

Fe-Ga alloys exhibit high values of magnetostriction under very low magnetic fields $\sim 100$ Oe (8000 A×m$^{-1}$); and apart from having very low magnetic hysteresis, they demonstrate rather high tensile strength ($\sim 500$ MPa) and limited variation in magnetomechanical properties for temperatures between -20 and 80 °C [270,271, 272]. The magnetostriction of Fe100$^{-x}$Ga$^x$ was found to vary as $x^2$. This suggested that the magnetostriction is associated with Ga-Ga pairs similarly to the Zener effect, which takes place in the alloys with Ga content from 13 at.% [127]. Recently, a remarkable improvement in magnetostriction of binary Fe-Ga alloys by doping them with trace amounts of rare earth elements like Tb, Dy, and Ce was reported by Chengbao Jiang et al [273,274,275,276]. A significant improvement of magnetostriction by trace dopants of the rare earth elements is mainly linked with two phenomena: (i) the increase in local strains produced by Ga-Ga pairs and the strong localized magnetocrystalline anisotropy induced by the dopants, and (ii) the contribution of a single-ion crystal field connected to the quadrupole moment Q of the magnetic rare earths [274]. First principal calculation also proved that random substitution of the Tb for Fe or Ga sites favors the formation of the tetragonal symmetry [277]. C. Meng et al. [278,279] reported that doping with trace amount of Tb and directional solidification could yield optimized magnetostriction and enhanced tensile properties in Fe-17Ga-xTb (x≤0.2) alloys as compared with the binary alloy.

In our studies we do not aim on achieving extra ordinary magnetostriction values but rather on structural changes produced by adding rare earth elements. Here we provide tangible results and reasonable interpretation on the effect of Tb, on the type and kinetics of the phase transitions in the Fe-27Ga alloys, as well as on the functional properties of the Fe-19Ga and Fe-27Ga alloys. Fe-19Ga and Fe-27Ga type alloys with 0.1-0.5 at% Tb were produced by rapid solidification in a copper mould using pure Fe and Ga by arc melting under protection of high-purity inert argon gas using an vacuum furnace.

Using energy dispersive spectroscopy, the chemical compositions of the cast buttons were measured with ±0.2% accuracy. The analysis reveals compositions of the Fe-19Ga based alloys as Fe-18.8Ga, Fe-19.1Ga-0.15Tb, Fe-19.6Ga-0.37Tb (denoted as Fe-19Ga, Fe-19Ga-0.15, and Fe-19Ga-0.37Tb throughout the text, respectively), and the Fe-27Ga based alloys as Fe-27.4Ga-xTb (x = 0.0, 0.15, 0.3, 0.4, and 0.5).

**Fe-19Ga based alloys**

**Study of the microstructure.** Doping with small amounts of Tb affects the microstructure of Fe-Ga alloys. Fig. 5.1a displays typical back scattered electron images of the as-cast Fe-19Ga-xTb samples doped with different amounts of Tb. In the un-doped

Fe-19Ga alloy, single-phase morphology with coarse equiaxed grains is observed. In the Tb-doped ternary alloys, the bright phase is observed both at the grain boundaries and inside the grains, and the volume fraction of the bright phases clearly increases with Tb content.

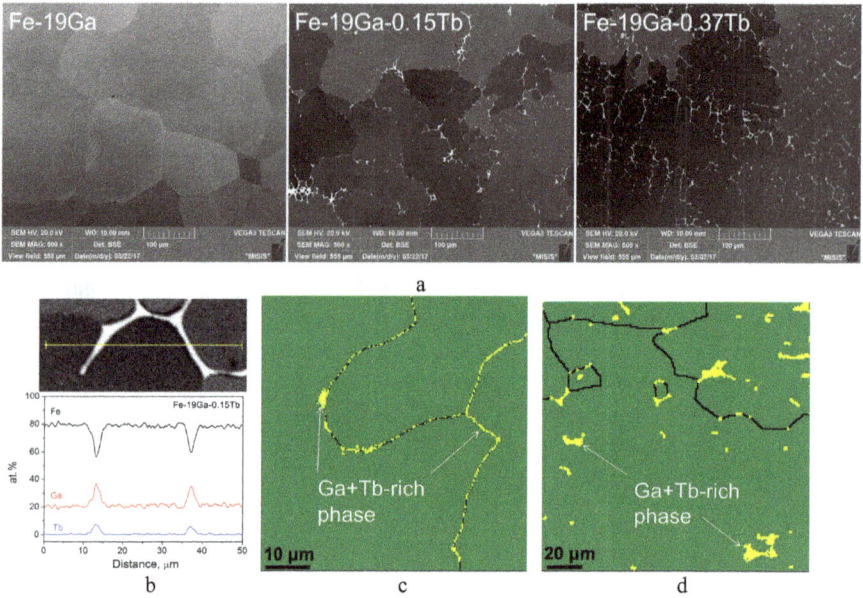

*Fig. 5.1 Typical back scattered electron images of the as-cast Fe-19Ga-xTb samples doped with different amounts of Tb (a), and EDX line mapping for Fe-19Ga-0.15Tb sample with corresponding SEM image (b), EBSD analysis for Fe-19Ga-0.15Tb (c) and Fe-19Ga-0.37Tb (d) (green, black, and yellow regions represent several grains, grain boundaries, and Ga+Tb-rich phase, respectively) (V. Cheverikin et al).*

The EDX line mapping for the Fe-19Ga-0.15Tb sample is displayed in Fig. 5.1b. In the grain body, Fe and Ga elements and a trace amount of Tb (not more that 0.1 at%) were recorded; whereas the grain boundary consists of all three elements: Fe, Ga, and Tb. The chemical composition of the matrix and the bright color precipitates obtained by EDX demonstrates that the grain boundaries are enriched in Tb (on average, 7 at% Tb) compared with the grains. These results confirm that the Tb atoms are accumulated at the grain boundaries and form precipitates enriched in Ga and Tb whose chemical formula is not recognized in this paper. Fig 5.1c and d exhibit the EBSD analysis for Fe-19Ga-0.15Tb and Fe-19Ga-0.37Tb samples, respectively. Regions of different colors correspond to several grains with different orientations, black regions represent the grain boundaries, and white regions indicate Ga+Tb-rich precipitates. In Fe-19Ga-0.15Tb sample, the Ga+Tb-rich phase is observed mainly at the grain boundaries of the matrix phase, whereas in Fe-19Ga-0.37Tb sample the precipitates form not only at the grain boundaries but also inside the grains decorating dendrides. From Fig. 5.1a, we can also conclude that the solubility limit of Tb in the Fe-19Ga alloy is below 0.15at.%. Beyond the solubility limit, a Tb- and Ga-rich ternary phase forms at the grain boindaries of the parent phase, which is detectable by the XRD results. Fig. 5.2 exhibits X-ray diffraction patterns of the Tb-free and the Tb-doped Fe-19Ga alloys at room temeprature. All the main diffraction peaks are indexed on a body centered cubic (b.c.c.) structure demonstrating that the matrix possesses the same crystal structure as the un-doped binary alloy. There is a slight deviation from symmetry of the (310) peak at the high angle side in all the alloys, which may be attributed to the weak contribution from the $D0_3$ reflections [59]. For the alloy doped with 0.37at.%Tb, a weak undefined diffraction peak at $2\theta \approx 41.4°$, marked by arrow, is observed, which may result from the Tb-rich precipitates [278, 280]. Although the SEM images discovered the presence of a Tb-rich phase for the alloy containing 0.15at.% Tb as well, it was, however, not detected by the XRD presumably due to a very low volume fraction of the precipitates in this sample.

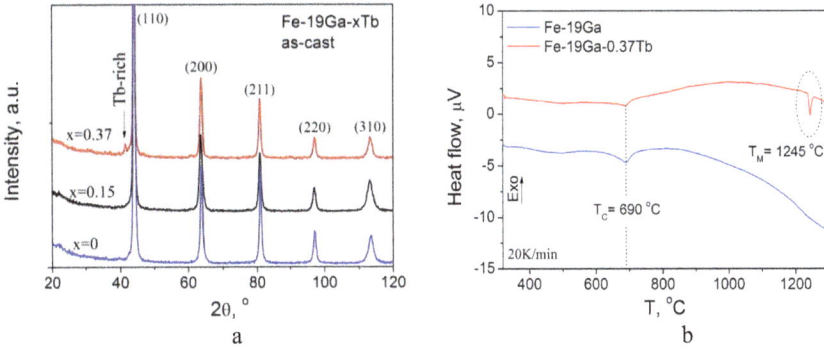

*Fig. 5.2  Room temperature X-ray diffraction spectra (a) and DSC curves (b) of the parent and the doped Fe-19Ga-xTb alloys with different Tb content in as-cast state. $T_C$ and $T_M$ are referred to the Curie point and melting temperature of the studied samples.*

The DSC measurements were carried out for the studied specimens to find out the temperature at which Tb and Ga-rich phase may appear in the studied alloys (Fig. 5.2b). The endothermic peak observed only for Tb-doped alloys at ~1245 °C is due to the melting of Tb-rich precipitates, denoted as $T_M$ in Fig. 5.2b. We can, therefore, conclude that the solidification of binary alloy proceeds with the formation of dendrite structures in the primary stage of the solidification. With a further decrease in the temperature, segregation of Tb-rich phases into interdendiritic regions occurs. Therefore, many dispersed regions form inside the matrix after solidification as proven by EBSD results in Fig. 5.1d. Doping with Tb keeps the Curie point of the binary alloy unchanged, which is denoted as $T_C$ in Fig. 5.2b.

**Magnetization and magnetostriction.** Our results demonstrate that the addition of Tb to the Fe-19Ga alloys remarkably alters both the temperature dependence of the magnetization and the magnetic field dependence of the magnetostriction [149]. The saturation magnetostriction proved to be higher after water quenching as compared with the furnace cooled samples [149]. Fig. 5.3a illustrates the magnetization vs. temperature, M(T), and parallel magnetostriction vs. the applied magnetic field, λ(H), curves at room temperature for the as-cast Fe-19Ga-xTb samples doped with different amounts of Tb. The inset in Fig. 5.3b illustrates the dependence of saturation magnetostriction, $λ_S$, on the Tb content. For all three specimens, magnetization curves exhibit typical behavior against temperature, i.e., the magnetization smoothly decreases with temperature until it approaches zero value at the Curie temperature. Doping with Tb does not appreciably influence the overall temperature dependence of the magnetization and the Curie temperature of the un-doped binary alloy. However, a significant enhancement of the saturation magnetostriction, $λ_S$, of the ternary alloys occurs by doping with trace amounts of Tb. For a sample of the binary Fe-19Ga alloy, the $λ_S$ is only 73 ppm, which is in good agreement with [149]. The $λ_S$ increases up to 210 ppm for the sample doped with 0.15at% Tb, and further increases to 235 ppm for the sample doped with 0.37 at.% Tb, which are about three times higher than those of the un-doped Fe-19Ga alloy. A similar effect of Tb on the structure and properties we also observed for Fe-12.5Al-xTb alloys with x=0.1 and 0.5Tb. The effect of rare earth elements dissolved in the solid solution on magnetostriction in Fe-Ga alloys is explained in several papers by group of Prof. C. Jiang by significant enhance of the nano-sized tetragonal distortion of the lattice produces by groups of Ga-Ga pairs. VSM and DSC (Fig. 5.2b) results do not demonstrate a pronounced change in the Curie temperature of the binary alloy with Tb doping, which does not support the results of Ref. [281] showing the Curie temperature rise of ~150 °C by addition of 0.3 at.% Tb to Fe-19Ga alloy.

A certain deviation from monotonous decrease in magnetization with temperature in the range from 300 to 500 °C is most probably the result of change from week $D0_3$ order to disorder as discussed in Chapter 4. This effect can also be seen at temperature dependent internal friction. The exact temperature position of the transient peak caused by disordering depends strongly on the alloy composition and can vary from 300 to 600 °C according to steep line connecting single phase A2 range and two phase range $A2+L1_2$ at equilibrium phase diagram (Fig. 3.1b). As the formation of the $L1_2$ equilibrium phase in Fe-19Ga is practically impossible, the two phase range is a mixture of A2 and $D0_3$ phases. The boundary between single A2 phase and non-equilibrium $A2+D0_3$ phases is not well defined and it depends not only on the alloy composition but also on heating rate and another technological factors. We have reported this effect at ~350°C for Fe-18Ga [282] and at ~500 °C for Fe-19Ga [114] alloys.

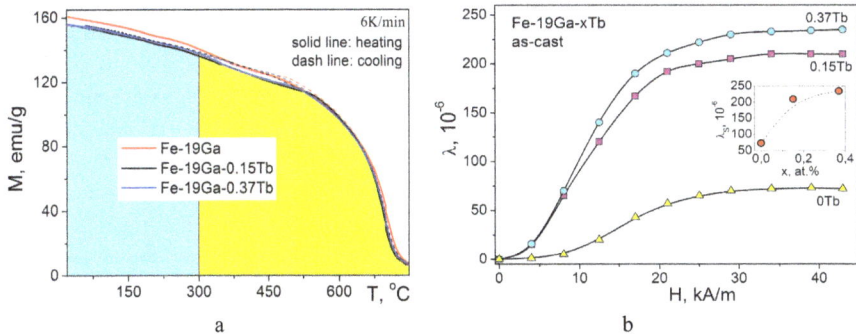

*Fig. 5.3  Magnetization vs. temperature, M(T), (a), parallel magnetostriction vs. applied magnetic field, λ(H), (b), for the Fe-19Ga-xTb (x = 0, 0.15, and 0.37 at. %) alloys in as-cast state. Inset illustrates the dependence of saturation magnetostriction, λ$_S$, on the Tb content, x. (E. Zanaeva, A. Emdadi).*

### Fe-27Ga-Tb alloys

An important difference between alloys with 19 and 27Ga is that alloys with about 27%Ga have metastable $D0_3$ structure after casting and they have several phase transitions both of the first and the second order at heating. In contrast alloys with 19%Ga only have second order transition – A2 to $D0_3$ structure. According to the neutron diffraction results, the initial state after direct solidification in a copper mould for the parent Fe-27Ga and the doped Fe-27.4Ga-xTb (x = 0.15, 0.30, 0.40 and 0.50 at.% Tb) samples is represented by the $D0_3$ structure. As already discussed, this conclusion is different from the XRD results showing dominating disordered A2 phase in the as-cast Fe-27.5Ga alloy [101]. This difference is related to the fact that the XRD diffractions are recorded from most rapidly cooled surfaces of the sample; whereas the neutron diffraction is recorded from the bulk samples in which the cooling rate is lower. Thus, only a thin surface area of the samples has a disordered A2 structure, whereas, the $D0_3$ ordering takes place in the bulk sample. It is noteworthy to mention that structural similarity of A2 and $D0_3$ structures and identical X-ray scattering factors between Fe and Ga also plays an important role in complication of the structural characterization in Fe–Ga alloys by conventional X-ray diffraction. On the other hand, neutron diffraction results still cannot give a clear answer about size of D03 ordered domains due to relatively low intercity of superlattice peaks. HRTEM studies suggest nano-size of ordered domains.

**Microstructure and phase transitions.** Doping Fe-27.4Ga alloy with trace amounts of Tb has a significant effect not only on its saturation magnetostriction but also on the phase fraction of different phases in the studied samples, as shown in Fig. 5.4a-c. The addition of Tb increases stability of phases with A2, $D0_3$, and B2 structures and reduces the amounts of ordered phases with $L1_2$ (*f.c.c.*, $Cu_3Au$-type structure, Pm3m) and $D0_{19}$ (hcp, $MgCd_3$-type structure, $P6_3/mmc$) structures. Fig. 5.4d illustrates the dependence of the amount of $L1_2$ on the Tb content appeared at instant heating with a heating rate of 2 K/min at 580 °C. It is observed that the volume fraction of the phase with structure $L1_2$ decreases from 100% for the un-doped Fe-27.4Ga alloy to 75% for the x=0.15Tb sample, and further to 33% for the x=0.3Tb sample. Beyond 0.3Tb, the volume fraction of $L1_2$ remains mainly unchanged around ~30%. The results confirm that doping with Tb remarkably suppresses the formation of the ordered close-packed structures $L1_2$ and $D0_{19}$ and stabilizes the phases with structures A2, $D0_3$, and B2 during heating as compared to the un-doped Fe-27.4Ga alloy. The temperature range of the consecutive phase transformations of A2/$D0_3$/B2 → $L1_2$ → $D0_{19}$ marginally changes in the samples of different amounts of Tb, and they take place at the interval of 455-620 °C and 600-655 °C, respectively.

Temperature dependence of magnetization of the as cast ternary Fe-26Ga-0.15Tb alloy (Fig. 5.5.a, red curve) is qualitatively similar to the binary alloy, but the qualitatively magnetization due to the $L1_2$ phase is about four times lower than that of in the binary alloy: <40 emu/g instead of ~130 emu/g in the binary alloy. The magnetization at 500-600 °C, i.e., in the range of the $L1_2$ phase existance, increases in several subsequent heating-cooling cycles, proving an increase in the $L1_2$ phase volume fraction in each cycle of heating-and-cooling. A certain deviation from a smooth behaviour of the magnetization at about 200 °C both at heating and cooling is not yet completely clear: This effect at first heating of the as cast or water quenched sample might be explained by A2 → $D0_3$ ordering and annihilation of as quenched vacancies recorded in [59, 84], but hardly in the next subsequent heating and cooling cycles.

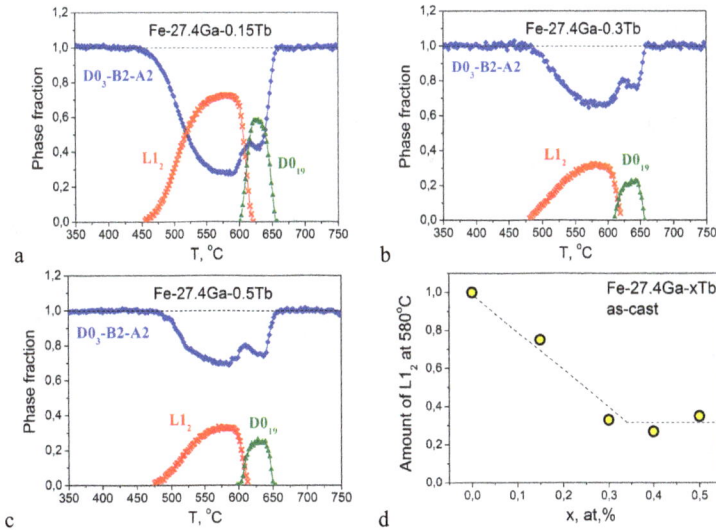

Fig. 5.4    The changes in the fractions of the phases during heating (a-c) and the dependence of the amount of the $L1_2$ appeared at instant heating with a heating rate of 2 K/min on the Tb content at $580^oC$ (d) in the Fe-27.4Ga-xTb (x = 0, 0.15, 0.30, 0.40, and 0.50 at.% Tb) alloys.

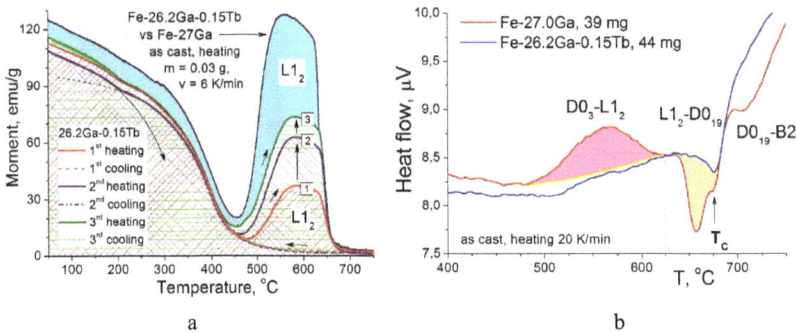

Fig. 5.5    Temperature dependent magnetization with a heating/cooling rate of 6 K/min (a) and a heat flow with a heating rate of 20 K/min (b) for the Fe-27Ga and Fe-26.2Ga-0.15Tb as cast alloys. The magnetization curves for the Fe-26.2Ga-0.15Tb alloy are shown for three subsequent heating-and-cooling cycles (A. Churyumov).

Heat flow tests also demonstrate effects for the $D0_3 \rightarrow L1_2 \rightarrow D0_{19} \rightarrow B2$ transitions in the binary alloys whereas in the ternary Fe-26Ga-0.15Tb alloy, they are practically not detectable (Fig. 5.5b).

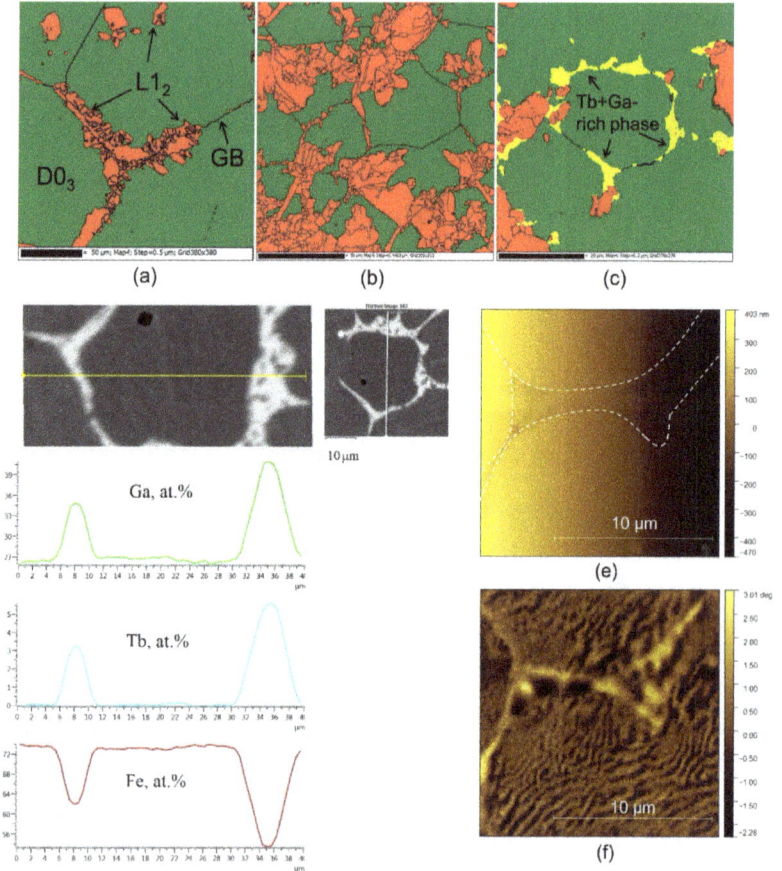

Fig. 5.6 EBSD analysis displaying nucleation (a) and growth (b) of the $L1_2$ phase on the grain boundaries of $D0_3$ phase in binary Fe-27.4Ga alloy; Tb-rich phase (yellow) on the $D0_3$ grain boundaries of Fe-27.4Ga-0.5Tb sample after annealing at 500 °C, 30 min ($D0_3$ – green, $L1_2$ – red, Tb-rich phase – yellow) (c); elements concentration in the same ternary alloy (d) obtained from EDX line scan; topography (e) and MFM image (f) of Fe-27.4Ga-0.5Tb quenched in water after annealing at 1000 °C (Cheverikin, Emdadi [283]).

The mechanism of the terbium influence on the kinetics of the $A1/L1_2$ phase growth is as follows [283]: in the Tb doped Fe-27Ga alloy, a Tb- and Ga-rich phase (about 3-5%, sometimes up to 7-9 at.% Tb) is already formed at casting from a mould state on the grain boundaries of the metastable $D0_3$ phase. At heating of the as cast Fe-27Ga alloy, the equilibrium $L1_2$ phase nucleates mainly at the grain boundaries of the metastable $D0_3/A2$ phase. In ternary Fe-27Ga-Tb alloy, the grain boundaries are already occupied by the Tb-rich phase, and the $L1_2$ phase must find other less favourable places to nucleate and grow (Fig. 5.6 a-c). The EBSD parameter MAD is 0.25 which indicates a reliable identification of the phases (Fig. 5.6d). When a significant volume fraction of a sample is transformed from D03 to L12 phase, a Tb-rich phase can be detected inside the L12 phase by concentration profiles. Presence of Tb- and Ga-rich precipitations finds also reflection in magnetic structure (Fig. 5.6e and f): the $D0_3$ matrix shows fine stripe domain with some degree of alignment, while Tb- and Ga-rich precipitates at the grain boundaries (region in topography image marked by white dotted line) does not show any distinctive magnetic structure.

In the Chapter 4 we have shown that the phase transition in Fe-Ga alloys from the ordered phase with *b.c.c.*-derivative $D0_3$ structure to the ordered phase with *f.c.c.*-derivative $L1_2$ structure takes place through intermediate steps: first, the $D0_3$ is subjected to a disordering transition to obtain an A2 structure followed by an A2 to A1 transition with final A1 ordering to achieve the $L1_2$ structure. The order – disorder - order type of structural transitions (the sequence $D0_3 \rightarrow A2 \rightarrow A1 \rightarrow L1_2$) was first suggested for the Fe-27Ga alloy [101] and then proven the isothermal transitions in this alloy [109]. Fig. 5.7 shows that the same transition sequence during instant heating takes place not only in binary Fe-27Ga alloys but also in the ternary Tb-doped alloys.

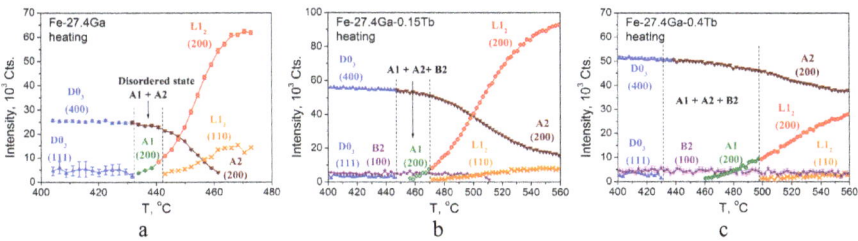

*Fig. 5.7 Behavior of the intensities of the characteristic diffraction peaks in the Fe-27.4Ga (a), Fe-27.4Ga-0.15Tb (b), and Fe-27.4Ga-0.4Tb (c) in the $D0_3 \rightarrow L1_2$ transition region during heating. There are no superstructural peaks in the temperature range indicated by the vertical lines.*

There is a temperature gap (431-443 °C) indicated by the vertical lines between the disappearance of the superstructural peaks of the $D0_3$ structure and the appearance of the superstructural peaks of the $L1_2$ structure (Fig. 5.7a). The absence of the superstructural peaks within this range suggests that the $D0_3$ has already transformed to a disordered state through the transition of $D0_3 \rightarrow A2$. Ordering in the $L1_2$ has not yet occurred, the (2 0 0) $L1_2$ reflection actually corresponds to the (2 0 0) reflection of the disordered A1 structure. Accordingly, the sequence of the transformations that take place in the Fe-27.4Ga alloy is: $D0_3 \rightarrow A2 \rightarrow A1 \rightarrow L1_2$ - the structural transformation of one ordered structure into another goes through a disordered state.

For the Tb-doped Fe-27.4Ga-xTb (x = 0.15, 0.30, 0.40 and 0.50 at.% Tb) alloys, a similar temperature gap (445-472 °C for x = 0.15 and 430-499 °C for x = 0.4Tb alloys) but in a wider range occurs, which is indicated by the vertical lines in Fig. 5.7 b and c. The situation is different from that of in the parent Fe-27.4Ga alloy: in contrast to the Fe-27.4Ga sample, the (1 0 0) reflection of the B2 structure is observed in the transition region for the Tb-doped Fe-27.4Ga-xTb alloys. This means that in a significant part of the volume of the sample, partial B2 ordering of the atoms was preserved, i.e., not all the volume occupied by the $D0_3$ has transformed into the disordered A2. We assume that the transition of $D0_3 \rightarrow B2$ in the transition region prevents the formation of the $L1_2$ in addition to the occupation of the grain boundaries by the Tb-enriched phase. Subsequently, the formation of the A3 structure in the entire volume of the sample is also prohibited. Similar conditions, i. e., the presence of the B2 in the transition region, were detected for x = 0.30 and 0.5Tb alloys. Therefore, in the samples doped with Tb, the $D0_3 \rightarrow A2 \rightarrow A1 \rightarrow L1_2$ transition occurs, but not in the entire volume of the sample. In a significant part of the volume, the transition of $D0_3 \rightarrow B2$ takes place, and the $L1_2$ and A3 structures do not appear in that part of the volume. The aforementioned situation of the phase transformations in the transition region accounts for the small volume fraction of the $L1_2$ and A3 for x > 0.15at.%Tb samples as compared with the un-doped Fe-27.4Ga sample.

**Magnetization and magnetostriction.** Our results prove that the doping of Tb to the Fe-27Ga alloys remarkably alters both the temperature dependence of the magnetization and the magnetic field dependence of the magnetostriction. The saturation magnetostriction of the water quenched samples is higher than that of the furnace cooled samples, which contains a small fraction of the $L1_2$ phase [149].

Microalloying with Tb significantly increases the saturation magnetostriction, $\lambda_S$, of the as-cast Fe-27.4Ga-xTb (x = 0, 0.15, 0.3, and 0.5at.%) alloys, as shown in Fig. 5.7. The $\lambda_S$ increases from 130 ppm for the un-doped Fe-27.4Ga alloy to 186 ppm for the x = 0.15at.% specimen, and further increases to 242 ppm for the x = 0.3 at.% Tb specimen.

Beyond x = 0.3 at.% Tb, the $\lambda_S$ decreases to 147 ppm, which is still higher than that of the un-doped sample. The main reason for magnetostriction decrease with increase in Tb content above 0.3% is formation of non-magnetic Tb-and-Ga rich phase. The inset in Fig. 5.8 presents the dependence of the $\lambda_S$ on the Tb content. There is a monotonous increase in the $\lambda_S$ with Tb addition until the highest value at Tb = 0.3 at.% is achieved. Therefore, it can be concluded that as long as the Tb is dissolved in the Fe-Ga solid solution, the electronic effects of Tb including the localized strain and the induced magnetocrystalline anisotropy enhance the $\lambda_S$. However, beyond the solubility limit of Tb in the Fe-27.4Ga (~0.3at.%), the formation of the non-magnetic Tb-rich phase mainly at the grain boundaries and dendrides, and its opposing effect to magnetic domain wall motions cause a reduction in $\lambda_S$ from the maximum value and prevent the $D0_3 \rightarrow L1_2$ structural transition.

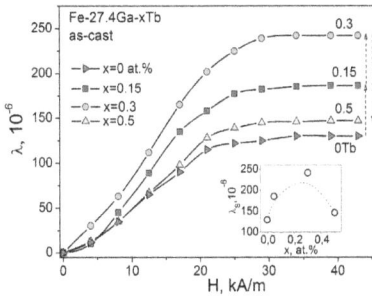

*Fig. 5.8 Parallel magnetostriction vs. applied magnetic field, λ(H), curves for the as-cast Fe-27.4Ga-xTb (x=0.15, 0.3, and 0.5at.%) samples. Inset shows the dependence of the saturation magnetostriction, $\lambda_S$, on the Tb content at room temperature.*

**Internal friction.** Dopping of the Fe-27.4Ga alloy with Tb decreases the rate of the *b.c.c.* to *f.c.c.* phase transition as measured by internal friction. This result is in agreement with the aforementioned approach. Fig. 5.9. represents a temperature dependent internal friction (TDIF) and elastic modulus in arbitrary units (TDEM) DMA test of the un-doped Fe-27Ga (a) and Tb-doped Fe-27.4Ga-xTb (x = 0.15 and 0.5 at.%) alloys (b and c) at heating up to 600 °C. By using six frequencies from 0.1 to 30 Hz, we can easily distinguish the effects related to different phase transitions ($P_{Tr}$) between thermally activated effects at about 460 °C (Fig. 5.9a). This traisient effect is not recorded in Tb-doped alloys (Fig. 5.9b,c). For thermally activated effects, the peak position increases with an increase in the frequency of the forced vibrations (P1, P2 and P3). In the case of phase transition, it is the peak height (or relaxation strength) rather than the peak temperature that depends on the frequency.

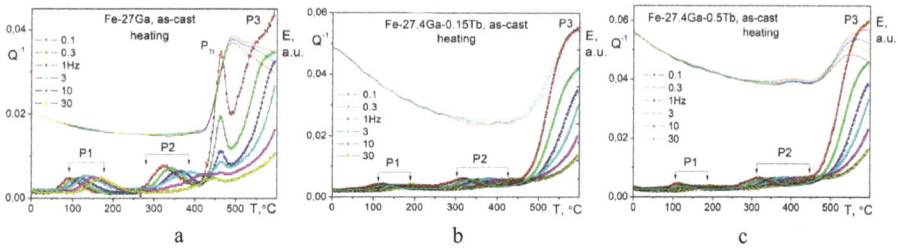

*Fig. 5.9  TDIF and TDEM (in arbitrary units) curves at different frequencies from 0.1 to 30 Hz for continuous heating (1 K/min) for the parent Fe-27Ga (a), and doped Fe-27.4Ga-0.15Tb (b) and Fe-27.4Ga-0.5Tb (c) in as-cast state.*

Thermally activated anealstic effects at 80-180 °C (the P1 peak), at 300-420 °C (P2), and around 600 °C (P3) are similar for the parent and doped alloys. The temperatures of the thermally activated P1, P2, and P3 peaks depend on the measuring frequency of the forced vibrations; the peak positions move to higher temperatures with an increase in the frequency. This allows calculating the activation parameters for the corresponding thermally activated relaxation processes. The activation energy of the P1 and P2 are 1.08 ± 0.01 eV and 1.91 ± 0.07 eV for Fe-27Ga alloy, 1.13±0.02 eV and 1.77±0.05 eV for Fe-27.4Ga-0.15Tb alloy, and 1.16±0.03 eV and 1.72±0.08 eV for Fe-27.4Ga-0.5Tb alloy, correspondingly. These activation paramentrs support the peaks interpretation as the P1 is the Snoek-type relaxation and the P2 is most probably the Zener relaxation [41, 57, 59, 147]. The main difference between the TDIF curves for the parent and doped alloys is the frequency independent peak at about 460 °C, which is recorded only in the binary Fe-27Ga alloy.

Transient internal friction $P_{Tr}$ peak temperature at about 460 °C that is accompanied with an increase of the Young's modulus and its temperature position does not depend on the frequency, whereas its height is strongly affected by the measurement frequency and the heating rate. The peak temperature increases with an increase in the heating rate: it is 458, 479, and 512 °C for the heating rates of 0.5, 1, and 2 K/min, respectively [109]. According to *in situ* neutron diffraction results, a significant atomic volume jump occurs at the $D0_3$ (*b.c.c.*-derivative ordered structure) → $L1_2$ (*f.c.c.*-derivative ordered structure) first order transition. The difference in volumetric thermal expansion coefficients for co-existing *b.c.c.* and *f.c.c.* phases lead to further growth of microstresses at heating.

Microdeformations manifest themselves in neutron diffraction patterns as broadening of the peaks. A lattice shear deformation occurs and leads to a well pronounced internal friction transient peak, $P_{Tr}$. The absence of $P_{Tr}$ in the Tb-doped Fe-27.4Ga-xTb alloys demonstrates that the transition from $D0_3$ to $L1_2$ structure is highly suppressed. This results is in full agreement with our magnetization, dilatometry, and neutron diffraction results which demostrates partial suppressing of this transition in presence of Tb.

Doping Fe-27Ga alloys by Tb significantly slows down the $D0_3$ to $L1_2$ transition both upon isothermal annealing and instant heating. The effect of Tb on the kinetics of the phase transition in the Fe-Ga alloys is also confirmed by neutron scattering. As a result, the Tr peak at 460 °C is not recorded in our tests of the Tb-doped Fe-27Ga alloys including high temperature tests carried out at EPFL torsion pendulum (Fig. 5.10) [284]. In contrast, the transient peaks at higher temperatures, namely at 610 °C (Tr2) and 650 °C (Tr3), are well recorded (Fig. 5.10). Their temperature positions are in exellent agreement with the $L1_2$ to $D0_{19}$ and $D0_{19}$ to B2 transitions, also shown in this figure, similarly to this effect in binary alloys. In contrast with binary alloys, in the Tb-doped samples, the $L1_2$ and $D0_{19}$ phases coexist with the B2 phase (volume fraction of B2 phase is at least twice bigger as compared with these closed packed structures) as the $D0_3$ to $L1_2$ transition is not completed in the alloys doped by Tb.

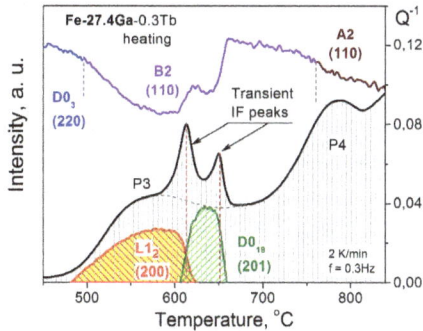

*Fig. 5.10 (a) Amount of $L1_2$ phase determined by X-ray analysis after isothermal annealing in Fe-27.4Ga (yellow circles) and in Fe-27.4Ga-0.5Tb (red circles) samples; (b) Phase transitions in Fe-27.4Ga-0.3Tb sample according to in situ neutron diffraction (left scale) and TDIF (black curve corresponds to right Y-axe) [284].*

The positive effect of Fe-Ga (mainly with 17-19%Ga) alloys doping by Tb was reported in several papers by C. Jiang [273-276] and ourselves [119, 149]. A remarkable enhancement of the saturation magnetostriction of the Fe-19Ga-xTb alloys occurs by doping with trace amounts of Tb. The saturation magnetostriction increases from 73 ppm for Fe-19Ga sample to 210 ppm for the sample doped with 0.15at% Tb, and further increases to 235 ppm for the sample doped with 0.37at.% Tb, which are about three times higher than those of the un-doped Fe-19Ga alloy.

We could establish a microstructural understanding to figure out why doping with the trace amounts of Tb leads to an enhanced magnetostriction in the Fe-27Ga type alloys. The results demonstrated that the microalloying of Fe-27Ga type alloys with Tb significantly suppresses the formation of the phases with $L1_2$ and $D0_{19}$ structures and stabilizes the dominant phase with $D0_3$ structure. The $L1_2$ has negative magnetostriction compared to positive magnetostriction of the $D0_3$ matrix; therefore, the lower the volume fraction of the $L1_2$ is the higher the saturation magnetostriction at room temperature is. It was shown that the transformation of the phase with ordered $D0_3$ structure to the phase with ordered $L1_2$ structure upon heating proceeds through the diffusion-controlled formation of disordered structures according to the scheme of $D0_3 \rightarrow A2 \rightarrow A1 \rightarrow L1_2$.

### 5.1.2 Fe-Ga-Al alloys

Fe-19Ga and Fe-27Ga alloys exhibit the highest magnetostriction not only among Galfenols but probably also among all Fe-based alloys. Nevertheless, they have some disadvantages: high brittleness and low corrosion resistance are among them. That is the reason why researchers are looking for new multicomponents Fe-Ga based alloys. Al is one of these elements: Al increases improves corrosion resistance of Fe-Ga alloys. Al and Ga have similar electronic structures: The ground-state electron configurations of nonmagnetic elements Al and Ga are $1s^22s^22p^63s^23p^1$ and $1s^22s^22p^63s^23p^63d^{10}4s^24p^1$. Their outer-shell electron configurations are similar, both Al and Ga enhance the magnetostriction of *b.c.c.* iron giving a magnetoelastic contribution to damping capacity of these alloys. Atomic ordering in both systems may take place and it will decrease damping due to additional pinning of magnetic domain walls at antiphase boundaries.

In this section we consider influence of additional alloying of Fe-Ga alloys by Al on the alloy structure using in situ neutron diffraction and damping capacity and magnetostriction of several Fe-Ga-Al alloys. Most of the results were obtained on Fe-18(Ga+Al) and Fe-18Ga-8Al alloys and compared with binary Fe-Ga alloys with 19 and 27%Ga, correspondingly.

In the as cast state all peaks at neutron diffraction high resolution spectra correspond to the $D0_3$ phase for Fe-18Ga-8Al alloy and to the A2 phase for Fe-12Ga-5Al alloy (Fig. 5.11)

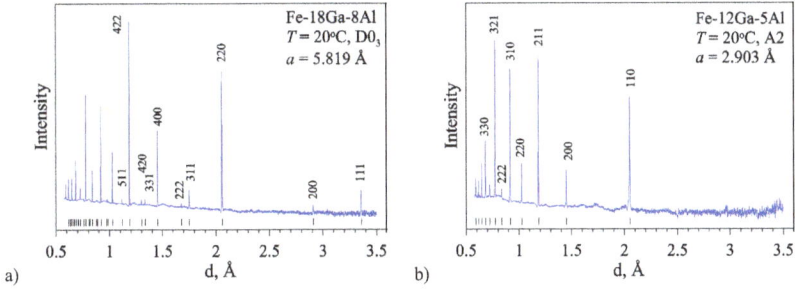

*Fig. 5.11 High-resolution diffraction patterns of Fe-18Ga-8Al (a) and Fe-12Ga-5Al (b), measured at room temperature. The vertical bars indicate the calculated peak positions for the $DO_3$ (the first compound) and A2 (the second compound) unit cells.*

In order to evaluate influence of Al in Fe-Ga, one may consider the Fe-18Ga-8Al alloy as the Fe-26Ga alloy in which 8 at.% of Ga are substituted by 8 at.% Al, and the Fe-12Ga-5Al alloy as the Fe-17Ga alloy in which 5 at.% of Ga are substituted by 5 at.% Al. Structural transitions in Fe-27Ga type alloys at heating and cooling as well as at isothermal annealing were discussed in Chapters 3 and 4. In contrast with them, the Fe-18Ga-8Al alloy structure is more similar to Fe-Al alloys which do not have neither $L1_2$, not $DO_{19}$ phases. Fig. 5.12 proves only second order phase transitions in as cast Fe-18Ga-8Al alloy which take place at instant heating and subsequent cooling. Thus, as generally expected from Fe-Al and Fe-Ga phase diagrams, Al suppresses formation of the closed packed phases in Fe-Ga-Al ternary alloys.

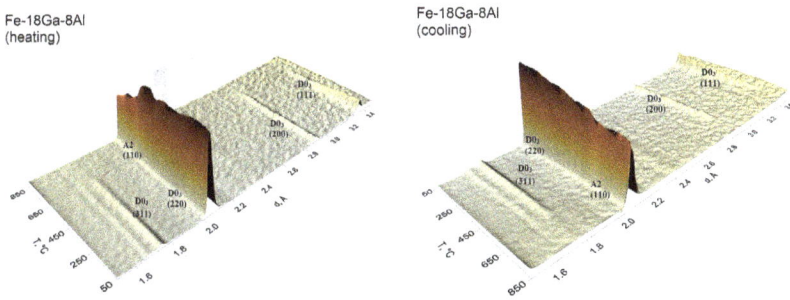

*Fig. 5.12 3D plot for Fe-18Ga-8Al alloy at heating (a) and cooling (b).*

191

Often the formation of the equilibrium $L1_2$ phase in Fe-Ga needs longer time and instant heating or cooling with the rate of 2 K/min may be not enough for even beginning of this transition. To check if this transition might take place during isothermal annealing we checked structure of the Fe-18Ga-8Al after 4 h annealing at 400°C, and then after 6 h annealing at 500 °C (Fig. 5.13): no sign of the $D0_3$ to $L1_2$ transition was found.

On the other hand, TEM analysis [59] allows to detect little amount of the $L1_2$ phase after isothermal annealing. Fig. 5.14 shows a bright field TEM micrograph of the annealed at 1000°C and water quenched Fe-18Ga-5Al alloy and the corresponding selected-area electron diffraction pattern (SAEDP). The microstructure contains dispersed islands of dislocations in a matrix with a slightly blurred contrast (Fig. 5.14a). The diffraction pattern taken along [101] zone axis contains the 111-type superlattice reflections, typical for $D0_3$ order. Some additional diffuse spots are also present in the SAEDP, marked by arrows in Fig. 5.14b, which will be discussed below.

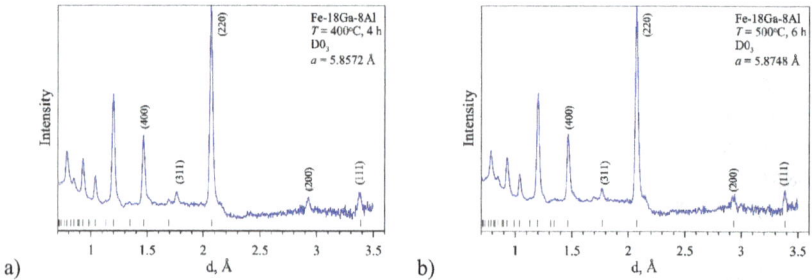

*Fig. 5.13 Diffraction patterns of Fe-18Ga-8Al after 4 h (a) and 6 h (b) annealing at 400°C and 500°C, correspondingly.*

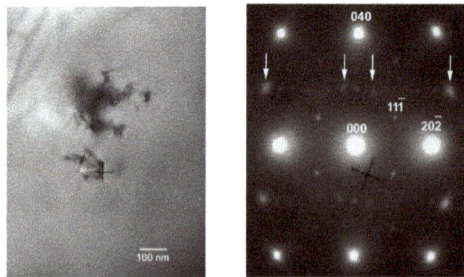

*Fig. 5.14 (a) Bright field micrograph of the water quenched Fe-18Ga-5Al alloy. (b) Corresponding diffraction pattern, indexed according to the $[101]_{D03}$ zone axis. The arrows indicate additional reflections (J. Pons).*

The microstructure of a water quenched sample aged during 4 hrs at 275 °C is shown in Fig. 5.15a, together with the corresponding diffraction pattern (Fig. 5.15b). Fig. 5.15c shows a bright field image of a sample aged for 4 hrs at 400 °C. These images evidence the formation of nanometer-sized precipitates during the ageing treatments, these precipitates being larger in the sample aged at 400 °C.

It is worth to mention that the precipitates interfered with the electropolishing process of the sample preparation. As a result, the TEM foils had a poor quality and only small areas were transparent to the beam, which limited their studies in the TEM. This was especially critical in the alloy aged at 400 °C, which contains a larger amount of the precipitated phase. The SAEDP shown in Fig. 5.15b for the alloy annealed at 275°C was recorded with a concentrated electron beam, in order to increase the visibility of the weak reflections. The pattern contains the $D0_3$ reflections (labeled with black indexes) and the same additional spots present in the as-quenched alloy (marked by arrows in Fig. 5.15b), but with higher intensity. After a careful analysis, the diffraction pattern could be indexed as superposition of the [101] zone axis of a $D0_3$ structure and the [111] zone axis of a $L1_2$ structure. The spots marked with arrows in Fig. 5.15b are due to double diffraction. Then, the precipitates are identified as the $L1_2$ phase in the $D0_3$ matrix. The diffraction pattern indicates the $[101]_{D03} \parallel [111]_{L12}$ orientation relationship, which is consistent with the Bain-type orientation relationship between *b.c.c.* and *f.c.c.* crystals. Moreover, the $040_{D03}$ and $202_{L12}$ spots are nearly coincident, which means that the lattice parameters fulfill the following relationship: $a_{L12} \approx a_{D03}/\sqrt{2}$. The $220_{L12}$ spots appear a bit diffuse in Fig. 5.15b due to the nanometric particle size. As mentioned before, weak $220_{L12}$ spots are also present in the patterns from the quenched alloy (marked by arrows in Fig. 5.14b), indicating that the $L1_2$ phase nucleates during the quenching process. The subsequent ageing treatments at 275°C and 400°C produce a slow growth of this phase up to nanometric particles, due to the slow diffusion at these temperatures. The $L1_2$ nuclei in the quenched sample are too small to be distinguished by TEM, but they are probably responsible of the blurred matrix contrast visible in these samples (Fig. 5.14a). Fig. 5.15d shows an area containing dislocation arrays and $L1_2$ precipitates in the sample annealed at 275 °C. Interestingly, the regions surrounding the dislocation arrays are free of precipitates. This is indicative of a rapid migration of the vacancies towards the dislocations during ageing in agreement with positron annihilation results. Then, the absence of vacancies around the dislocation arrays restricts the atomic diffusion and suppresses the $L1_2$ phase growth in these zones.

Thus, the neutron diffraction of Fe-18Ga-8Al alloys rule out a massive formation of the $L1_2$ phase at 400 °C and confirm the expected role of the Al addition to Fe-Ga alloys in terms of stabilization of the $D0_3$ phase. However, the TEM observations reveal the nucleation and a limited growth of the $L1_2$ phase up to nanometer-size particles upon ageing at 275 °C or 400 °C, which cannot be detected by XRD. This indicates a metastable formation of the $L1_2$ phase at low temperatures (400 °C or below), although it cannot attain a significant volume fraction due to the slow diffusion at these temperatures.

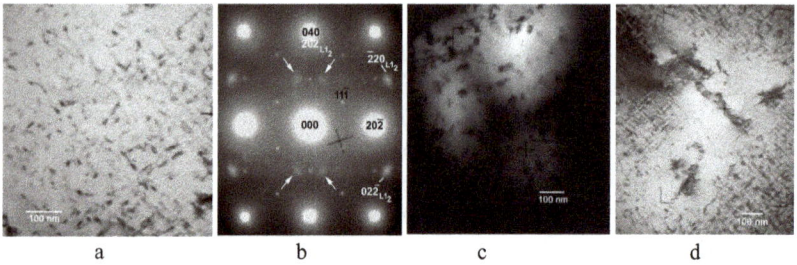

a                b                c                d

*Fig. 5.15 (a) Bright field micrograph of the Fe-18Ga-5Al alloy aged for 4 hrs at 275°C; (b) Corresponding diffraction pattern; (c) Micrograph of the alloy aged for 4 hrs at 400°C; (d) Micrograph showing dislocations and precipitates in the alloy aged for 4 hrs at 275°C (J. Pons).*

Substitution of Ga atoms in Fe-27Ga by Al atoms leads to a decrease in saturation magnetostriction at about 30% (Fig. 5.16a) and suppressing of the $L1_2$ phase formation as judged from VSM curves (Fig. 5.16b).

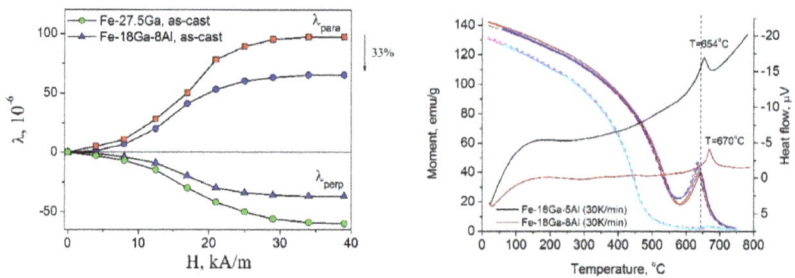

*Fig. 5.16 The $\lambda_{parallel}(H)$ and $\lambda_{perp}(H)$ dependencies for Fe-27Ga and Fe-18Ga-8Al.*

In contrast with Fe-27Ga, no transient internal friction due to the $D0_3$ to $L1_2$ transition was recorded at 450-550°C at TDIF curves for Fe-18Ga-5Al and Fe-18Ga-8Al alloys.

Internal friction tests have been also performed in the low temperature range up to 300 °C, and an IF peak is detected at 100-200 °C in ternary Fe-18Ga-5Al alloys, as shown in Fig. 5.17.

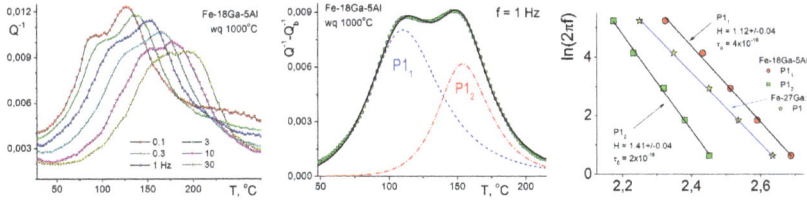

*Fig. 5.17 Temperature dependent IF curves for Fe-18Ga-5Al alloy (a) after water quenching from 1000 °C in the range from 300 to 575 °C: heating rate 1 K/min, frequencies from 0.1 to 30 Hz. Experimental results for 1 Hz fitted by two Debye peaks (b), and activation energies for experimental peaks P11 and P12 compared with single headed P1 peak in Fe-27Ga (c).*

It is a Snoek-type internal friction peak related to the substitutional-interstitial atoms interaction (Ga/Al with C in our alloys). In order to enhance the carbon diffusion effects, the ternary alloy with an intentionally added small amount of carbon (0.05 at.%C) was used. Addition of a certain amount of not carbide forming elements in $\alpha$-Fe may lead to the appearance of a well-pronounced shoulder at the Snoek-type peak [11].

Fig. 5.16a shows the temperature dependent internal friction curves for Fe-18Ga-5Al, where a thermally activated bi-modal internal friction peak (denoted as $P1_1$ and $P1_2$) is clearly seen at 100-200°C. To make sure that the internal friction peak in the Fe-18Ga-5Al alloy is indeed bi-modal we have measured five water quenched samples: even though the relative height of $P1_1$ and $P1_2$ varies slightly, the total '$P1_1$ plus $P1_2$' peak retains always a bi-modal shape. In carbon-free Fe-18Ga-7Al alloy these peaks are at least one order of magnitude lower as C was not added to this alloy.

The bi-modal peak in Fe-18Ga-5Al alloy can be reasonably fitted by two Debye-type peaks in all cases. An example of a bi-modal with practically the same heights of the $P1_1$ and $P1_2$ peaks measured at 1 Hz is given in Fig. 5.17b. The activation energies and pre-exponential factors $\tau_0$ determined from the Arrhenius plot (Fig. 5.17c) are $H_{11}$ =1.12±0.04 eV, $\tau_0 = 4 \cdot 10^{-16}$ s and $H_{12}$ =1.41±0.04 eV, $\tau_0 = 2 \cdot 10^{-18}$ s, for the $P1_1$ and $P1_2$ peaks, respectively. The pre-exponential factor values correspond to point defect relaxation. In turn, the parameters of the P1 peak in Fe-27Ga are H = 1.04±0.02 eV, and $\tau_0 = 10^{-14}$ s.

According to the activation parameters and height of the peaks, they are most probably caused by carbon atoms jumps in solid solution. The substitutional (s) – interstitial (i) atoms interaction often causes stronger modifications of the Snoek relaxation in substitutional alloys, where this type of relaxation is usually called "Snoek-type" relaxation. The bi-modal internal friction peak in ternary alloy is the result of presence of two types of substitutional atoms, Ga and Al, in solid solution. Comparison of activation parameters of the peaks makes it possible to suggest that the $P1_1$ peak in ternary alloy, the parameters of which are similar to the P1 peak in binary alloy, is due to C-atom jumps nearby Ga atoms (this peak can be denoted as the Fe-C-Ga peak). Then the $P1_2$ peak is due to C-atom jumps nearby Al atoms (this peak can be denoted as the Fe-C-Al peak). This hypothesis is reasonable if one considers the size of atoms: the Al atoms (effective Goldschmidt atomic radius equal to 0.143 nm) are larger than Fe (0.128 nm) and Ga (0.135 nm) atoms. Therefore, Al and Ga atoms generate compressive stresses and increase the lattice parameter of iron but Al atoms do this more effectively.

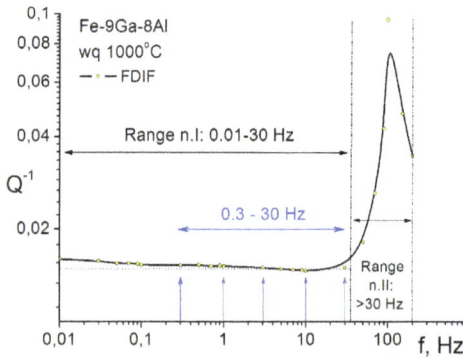

*Fig. 5.18 Frequency dependent internal friction curve for forced vibrations (DMA Q800): choice for frequencies for further tests.*

The height of the $P1_1$ and $P1_2$ peaks decreases with the same rate if the sample is annealed [59]. This is additional evidence that their origin is the same, i.e. their height is controlled by carbon atom jumps. If, for example, one of the peak is due to reorientation of pairs of vacancies – the sensitivity to annealing would be different. Moreover, the positron annihilation study has shown that as quenched vacancies annihilate after two hours annealing at 200-250 °C [283].

Damping capacity was studies using alloys of nominal composition $Fe_{80}Ga_{(18-x)}Al_x$ ($x = 0$, 5, 8, 12). Actual compositions of the alloys were obtained by both energy dispersive spectrometry: Fe-11.9%Ga-5.1%Al, Fe-9.0%Ga-8.0%Al, Fe-5.2%Ga-11.9%Al and chemical analyses: Fe-12.2%Ga-5.1%Al-0.05%C, Fe-9.2%Ga-8.0%Al-0.04%C and Fe-5.0%Ga-12.2%Al-0.04%C. These ternary alloys are denoted below as Fe-12Ga-5Al, Fe-9Ga-8Al and Fe-5Ga-12Al alloys, respectively. Binary Fe-18%Ga alloy, was used in several cases for comparison with ternary alloys.

The internal friction samples were sectioned from the directional solidification rod into shapes of $60 \times 3 \times 0.9$ mm$^3$, then sealed under vacuum into quartz tubes, annealed at 1000°C for one hour followed by a water quench. Internal friction has been measured on a dynamical mechanical analyser DMA Q800 TA Instruments.

The test conditions are illustrated in Fig. 5.18, where typical frequency dependent curve (amplitude of deformation $\varepsilon_0 = 7 \times 10^{-5}$) at room temperature is presented. Dynamical mechanical analyser DMA Q800 TA Instruments operates in the frequency range between 0.01 and 200 Hz but its stiffness with respect to sample stiffness is not sufficient at $f > 30$ Hz (Range II in Fig. 5.18) for chosen geometry of iron based alloys.

Typical amplitude dependent internal friction and elastic modulus (ADIF and ADEM, correspondingly) curves for two Fe-Ga-Al alloys after water quenching from 1000°C have a well pronounced extreme: maximum for IF at $\varepsilon_0 = (1\text{-}2) \times 10^{-4}$ and minimum for elastic modulus values (Fig. 5.19). The nature of this peak-like effect is magnetomechanical damping: This is easy to prove by applying external magnetic field. In the external saturated magnetic field (MF) all magnetic domains inside a sample are fixed by MF and do not contribute to energy dissipation during vibrations in the elastic range of cyclic loading. Construction of DMA Q800 specimen holder does not allow placing a specimen inside electrical coil in order to create saturated magnetic field. Consequently, we used four strong NdFeB magnets (Br = 1.2-1.3 kGs, Hci = 12-15 kOe, $(B_H)_{max}$ = 30-35 MGOe) attached to both wings of the sample mounted in double cantilever configuration in the dynamical mechanical analyser. This method does not allow to fix well all magnetic domains in the magnetic field of attached magnets but even this simple experiment clearly demonstrates that total damping of studied samples supplied with NdFeB magnets is two-three times lower compared to measurements without magnets. The same considerations apply to the relation between modulus defect with and without magnetic field (Fig. 5.19 a and b).

197

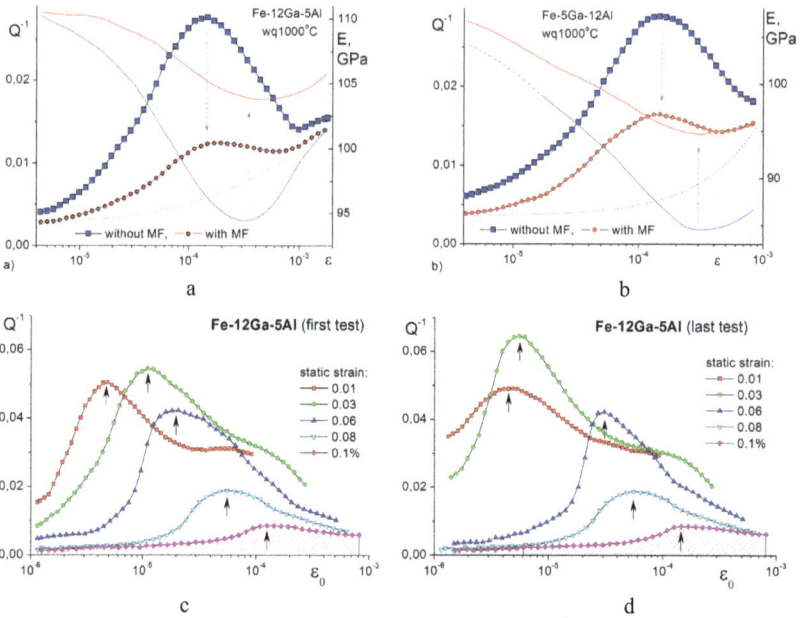

Fig. 5.19. ADIF and ADEM curves measured at single cantilever mode for water quenched Fe-12Ga-5Al (a) and Fe-5Ga-12Al (b) alloys, f = 3 Hz: measurements were carried out first without magnetic field (MF), then in magnetic field. Dotted line – approximation of nonmagnetic IF background. Three point bending mode with different preloading strain for Fe-12Ga-5Al sample in the first test (c) and after several tests (d).

When the maximal value of applied periodic deformation ($\varepsilon_0 = E^{-1}\sigma_0$) is small, one would expect that domain walls execute small motions in the vicinity of their equilibrium position, defined largely by the internal stress distribution. This is the amplitude range where the Smith and Birchak model [31] works. When $\varepsilon_0$ is large, the domain walls are moving considerably further and so the effect of internal stress is averaged over a considerably larger region [285]. For such long-range domain wall motion it is necessary to add pinning effects of domain walls by different nonmagnetic obstacles (Kërsten theory [286]) as well as magnetic inhomogeneities and the inhomogeneities of the structure. That is why the testing of the Smith and Birchak theory by Astie and Deqauque [287, 288] gave unsatisfactory agreement in the vicinity of the damping peak in contrast to the satisfactory situation in the Rayleigh region. In contrast with the Smith and Birchak model and in agreement with the Astie and Deqauque experiments there is no direct proportionality between our experimental data for Fe-Ga-Al alloys: maximal damping level at ADIF curves and magnetostriction of the alloys if we consider different compositions (Table 5.1).

As it was already demonstrated in the Chapter 4 for binary Fe-Al alloys, damping capacity of the alloys with magneto-mechanical damping significantly decreases if they work under external preloading force, which happens very often in many practical applications. Figs. 5.19 c and d show that the sample preloading by step by step increasing of static stress in three point bending configuration leads to a decrease in damping capacity of Fe-Ga-Al samples, too. This decrease is not monotonous (see Fig. 4.25) and it leads to a well pronounce shift of the damping peak to higher amplitudes. The results for three point bending are shown for the first ADIF test (c) and after several loading cycles, i.e. after training of the sample (d). It is also notable, that for not heavy preloaded samples, damping capacity of the samples in three point bending mode is higher than that is for cantilever mode. These experimental facts underlines difficulties to compare damping capacity of the materials measured by different techniques and sensitivity of alloys with enhanced damping to combination of cyclic and static stresses.

*Table 5.1 Saturation magnetostriction and the maximum value of damping at peak position (without/with background subtraction, roughly at $\varepsilon_0 = 0.0002$ for single cantilever mode) for differently treated alloys [40].*

| | Fe-12Ga-5Al | Fe-9Ga-8Al | Fe-5Ga-12Al |
|---|---|---|---|
| | Magnetostriction : | | |
| As-quenched from 1100°C | 80 (A2) | 70 (A2) | 62 (A2+D0$_3$) |
| As-quenched from 730°C | 64 (A2) | 74 (A2) | 38 (A2) |
| Annealed | 20 | 24 | 30 |
| | $Q_{h.max}^{-1}$: total damping / MM-damping | | |
| As-quenched 1100 | 0.028 / 0.019 | 0.022 / 0.011 | 0.029 / 0.018 |
| As-quenched 730 | 0.029 / 0.020 | 0.026 / 0.019 | 0.025 / 0.014 |
| Annealed | 0.014 / 0.0054 | 0.013 / 0.0054 | 0.014 / 0.0046 |

Heat treatment of the Fe-Ga-Al samples plays an important role in the total damping. In Fig. 5.20 one can see ADIF curves for four different states of our alloys: cold-rolled, water quenched from 1000 °C, annealed at 1000 °C cooled down in furnace to 730 °C (3 hrs) and water quenched, and water quenched from 1000 °C plus heated to 400 °C during TDIF tests.

In the cold-rolled state total damping has mainly nonmagnetic sources: there is no magnetomechanical peak at ADIF curves. After water quenching from both 1000 °C and 730 °C, the ADIF curves have clear damping peaks due to magnetomechanical damping. There is no well pronounced difference between the absolute values of ADIF curves following quenching from either temperatures. In contrast, annealing (heating to 400 °C with heating rate 2 K/min in furnace) decreases damping significantly.

The decrease in damping and simultaneous increase in hardness [40] may be a result of short-range ordering of the sample structures. The X-ray structural studies identify texture in as cast samples: after solidification $Fe_{80}Ga_{12}Al_6$ and $Fe_{80}Ga_9Al_8$ samples have better <110> orientation, $Fe_{80}Ga_5Al_{12}$ have <211> orientation. The X-ray peak intensity changes after annealed at 1000°C. The $Fe_{80}Ga_9Al_8$ sample got strong <110> orientation, $Fe_{80}Ga_{12}Al_5$ and change to <100> and $Fe_{80}Ga_5Al_{12}$ change to <110> orientation but we did not find clear evidences of ordering.

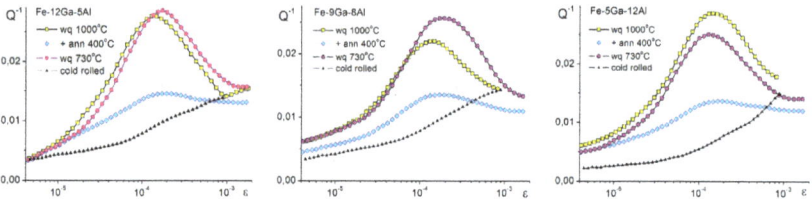

*Fig. 5.20 ADIF curves for Fe-12Ga-5Al (a), Fe-9Ga-8Al (b) and Fe-5Ga-12Al (c) alloys in three different states: cold-rolled, as-quenched (two different annealing temperatures before quenching) and as annealed. All tests without magnetic field.*

These results demonstrate the complicated behavior of the damping capacity of Fe-Ga-Al ferromagnetic materials possessing enhanced level of magnetomechanical damping. The understanding of this complex behavior requires to account for all the changes in both the fine crystalline structure and in the magnetic domain structure. It is also clear that heat treatment influences damping greatly.

Main contribution to total damping in as-quenched Fe-Ga-based alloys at room temperature comes from magnetomechanical damping. According to the damping index $\Psi = 2\pi Q_m^{-1} \approx 15\%$ (for forced vibrations in cantilever mode) all studied ternary alloys belong to high damping materials (i.e., $\Psi > 15\%$). Nevertheless, in contrast with Smith and Birchak theory, we did not find direct proportionality between saturation magnetostriction and maximal damping of Fe-Ga-Al alloys by varying their chemical composition. This fact may be a result of short range ordering in the alloys.

### 5.1.3  Fe-Ga-(Ni, Co) alloys [*]

In this section the structural, magnetic, magneto-mechanical and damping properties of three ternary functional alloys, namely: $(FeCo)_{81}Ga_{19}$, $Fe_{55}Co_{18}Ga_{27}$, and $Fe_{100-x}Ni_xGa_{19}$ (with x = 0 to 26 at.%Ni) are presented and discussed with respect to $Fe_{81}Ga_{19}$ binary alloy. These studies were mainly carried out at Institute of Physics, Academia Sinica, Taipei, Taiwan.

$Fe_{81}Ga_{19}$ **(Bulk):** Structural, magnetic, and magneto-mechanical properties: The alloy was slowly-cooled in furnace to room temperature after being melted in an induction furnace. The samples used were cut from the center part of the master ingot. From X-ray diffraction (XRD) and transmission electron microscopy (TEM) studies, we found that the alloy contains the A2 and $D0_3$ phases [289]. From the magnetic thermal gravimetric (MTG) analysis, we recorder a structural transition from the ordered $D0_3$ to the disordered A2 phase at about 480 °C, and the Curie temperature of the A2 phase at s $T_C$ = 713 °C.

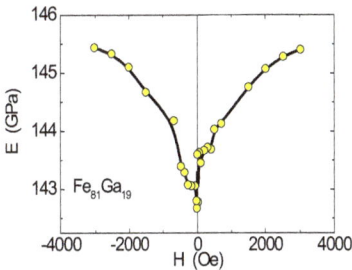

*Fig. 5.21  E(H) dependence: E is the Young's modulus and H is the external magnetic field.*

As to the magnetic property of the slowly-cooled polycrystalline $Fe_{81}Ga_{19}$ alloy at room temperature, saturation magnetization $M_S$ = 160 emu/g; coercivity $H_C$ = 1.1 Oe; and initial permeability $\mu_R$ = 74. The magneto-mechanical property includes the saturation magnetostriction ($\lambda_S$), measured by the strain gauge method, and the $\Delta E$ effect, measured by the Resonant Frequency and Damping Analyzer (RFDA) device. They are $\lambda_S$ = 61 ppm, and $(d\lambda_l/dH)_C$ = 0.64 ppm/Oe, where $(d\lambda_l/dH)_C$ is the longitudinal magnetostriction sensitivity of the alloy after the demagnetization correction [289]. Fig. 5.21 shows the magnetic-induced $\Delta E$ effect of the alloy from the RFDA measurement: the Young's modulus (E) is plotted as a function of the magnetic field (H).

---

[*] Material for section 5.1.3 was kindly prepared by Prof. Dr. S.U. Jen (Institute of Physics, Academia Sinica, Taipei, Taiwan)

By definition, $\Delta E/E_0 = (E_S - E_0)/E_0$, where $E_0 = 142.6$ GPa is the E value at the zero field, and $E_S$ is that at the saturation field. Thus, $\Delta E/E_0 = 1.4\%$ for this SC-alloy. In addition, the magneto-mechanical coupling factor, $K_E$, defined as $(K_E)^2 = \Delta E/E_0$, is 11.7%.

Temperature dependence of damping properties: Total damping capacity, $(\Delta W/W)_T$, of a metallic ferromagnet should consist of four terms (or contributions):

$$\left(\frac{\Delta W}{W}\right)_T = \left(\frac{\Delta W}{W}\right)_h + \left(\frac{\Delta W}{W}\right)_e + \left(\frac{\Delta W}{W}\right)_E + \left(\frac{\Delta W}{W}\right)_{NM} \tag{5.1}$$

where $(\Delta W/W)_h$ is the magneto-elastic hysteresis term, $(\Delta W/W)_e$ is the micro-eddy-current term, $(\Delta W/W)_E$ is the macro-eddy-current term, and $(\Delta W/W)_{NM}$ is the non-magnetic term. Only the $(\Delta W/W)_e$ and the $(\Delta W/W)_{NM}$ terms are important in the case of RFDA measurement [290]. Also, based on the arguments in Ref. [290], one can conclude that Debye peaks of the two curves (with H = 0 and H = 200 Oe) occur at the same inverse peak temperature, $(1/T_{mF})$. For tests at H = 0, it was found $\Delta W/W_{H=0} = 7.1$ % at T $= T_{mF}$, and the total activation energy for the magnetic and non-magnetic relaxations, $E_A$, is 1.03 eV/atom. For tests in H = 200 Oe, the $\Delta W/W_{H=200}$ is equal to 5.4 % at T $= T_{mF}$, and the activation energy for non-magnetic relaxations, $\bar{E}_A$, is 0.81 eV/atom. The magnetic field (H) suppresses the $(\Delta W/W)_e$ contribution, which is mostly related to the magnetic domain wall (MDW) motion. The experimental $(\Delta W/W)_{H=200} = 5.4$ % is equal to $(\Delta W/W)_{NM}$, and (iii) the total damping $(\Delta W/W)_{H=0} = (\Delta W/W)_{eExp} + (\Delta W/W)_{NM}$. Then, the experimental $(\Delta W/W)_{eExp} = 7.1\% - 5.4\% = 1.7\%$ at T $= T_{mF}$, which is only slightly larger than the theoretical $(\Delta W/W)_{eTh}$, 1.2% [290].

**Fe$_{81}$Ga$_{19}$/Si(100) Film**

Magnetic and magnetic domain properties: Fe$_{81}$Ga$_{19}$ film was deposited on a Si(100) substrate by the dc magnetron sputtering technique at room temperature. The film thickness is equal to t = 110 nm. From the curvature measurements, we found the in-plane principal stresses are biaxial and compressive: i.e., $\sigma_{xx} = -420$ MPa (nearly parallel to length, L, of the film), and $\sigma_{yy} = -656$ MPa (nearly parallel to width, w) [290]. Since $\lambda_{100}$ or $\lambda_S$ of the Fe$_{81}$Ga$_{19}$ film is positive (about 40 ppm), the easy-axis defined by the magneto-elastic mechanism should be along L (or the x-axis): e.g., the magneto-elastic anisotropy energy $K_\sigma = 0.27 \times 10^5$ J/m$^3$ > 0. Notice that due to the nano-crystalline nature of the film, the magneto-crystalline anisotropy energy does not play an important role. On the other hand, the (perpendicular) demagnetizing energy of the film is $E_M = 2\pi(M_S)^2$, where $M_S = 1.63$ T is the saturation magnetization. According to [291], the magnetic quality factor ($Q_m$) as $Q_m = K_\sigma/E_M$. From the data shown previously, it is easy to find $Q_m = 8.13 \times 10^{-3}$ for the Fe$_{81}$Ga$_{19}$/Si(100) film. Since $Q_m \ll 1$, it is concluded that the magnetization of the $Fe_{81}Ga_{19}/Si(100)$ film must be lying in-plane, which is also confirmed by the results of in-plane angular dependence of the magnetic hysteresis loop measurements [290].

Due to the fact that the as-deposited $Fe_{81}Ga_{19}$ film is stressed in a highly compressive state, the film must be magnetically dispersive. As a result, the magnetic property of the $Fe_{81}Ga_{19}/Si(100)$ film has the following three features [290]: First, the film is inverted: i.e., coercivity ($H_C$) is larger than the anisotropy field ($H_K$) of the film. Second, after completely ac demagnetizing, the magnetic domain structure exhibits the leaf-like (or ripple-like) pattern. Third, after dc demagnetizing from the maximum field, the magnetic domain structure exhibits the labyrinth pattern at the remanence.

### FeCoGa/Si(100) and FeCoGa/glass (Films)

Magnetostriction property: the Curie temperature ($T_C$) could be increased, and the saturation magnetostriction ($\lambda_S$) slightly increased by replacing 10 at.%Fe with Co atoms in the $Fe_{83}Ga_{17}$ alloy [292]. It should be noticed that the polycrystalline (FeCo)-Ga alloys in Ref. [292] were all in bulk form. Due to the structural differences between the same alloy in a bulk and in a thin-film forms, the magnetostriction behavior or trend as a function of the Co concentration, z, in the $Fe_{81-z}Co_zGa_{19}$ alloy can be totally different.

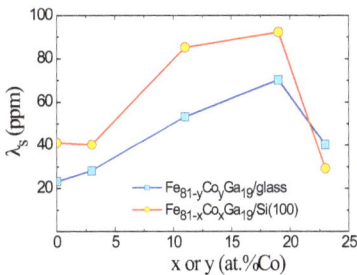

*Fig. 5.22 Saturation magnetostriction of the Fe-Co-Ga films deposited on the glass and Si(100) substrates, respectively.*

Two series of the $(FeCo)_{81}Ga_{19}$ films; one was the $Fe_{81-x}Co_xGa_{19}$ film deposited on the Si(100) substrate ($Fe_{81-x}Co_xGa_{19}/Si(100)$), and the other was $Fe_{81-y}Co_yGa_{19}$ film deposited on the glass substrate ($Fe_{81-y}Co_yGa_{19}/glass$) were prepared. Fig. 5.22 shows that both the 'x' and 'y' dependences of the saturation magnetostriction ($\lambda_S$) for the $(FeCo)_{81}Ga_{19}$ films are similar to each other. Notice, that as x or y increases from zero, $\lambda_S$ increases and reaches the maximum value, when x = y = 19 at.% Co.

In Table 5.2, the experimental data for the $Fe_{81}Ga_{19}$ and $Fe_{62}Co_{19}Ga_{19}$ alloys in a bulk (B) and a thin-film (F) forms are summarized. According to this Table, $\lambda_S$(Exp) of $Fe_{81}Ga_{19}$-B is larger than that of $Fe_{62}Co_{19}Ga_{19}$-B, whereas $\lambda_S$(Exp) of $Fe_{81}Ga_{19}$-F is smaller than that of $Fe_{62}Co_{19}Ga_{19}$-F. This means that the effects of replacing Fe by Co in the $Fe_{81}Ga_{19}$ alloy on $\lambda_S$ are just the opposite for the alloy in a bulk and a thin-film forms. The explanation is given below.

Based on the magnetostriction theory, for a polycrystalline (i.e., with randomly oriented grains) ferromagnetic sample, $\lambda_{S(TH)}$ is expressed as:

$$\lambda_{S(TH)} = (2/5)\lambda_{100} + (3/5)\lambda_{111}, \tag{5.2}$$

where $\lambda_{100}$ and $\lambda_{111}$ are the magnetostriction constants of single crystals. On the other hand, for a highly (110) textured ferromagnetic sample, $\lambda_S^*(TH)$ is expressed as,

$$\lambda_S^*{}_{(TH)} = (1/5)\lambda_{100} + (4/5)\lambda_{111}. \tag{5.3}$$

Table 5.2 also shows the structural properties of the $Fe_{81}Ga_{19}$ and $Fe_{62}Co_{19}Ga_{19}$ alloys (both for bulk and thin-film samples) from the XRD studies. Thus, we can conclude that for the bulk ($Fe_{81}Ga_{19}$-B and $Fe_{62}Co_{19}Ga_{19}$-B) samples, their crystal grains are nearly randomly oriented. Notice that for a polycrystalline $b.c.c.$ sample with truly random orientations, $I_{211}/I_{110} = 80\%$ and $I_{200}/I_{110} = 50\%$, where are $I_{211}$, $I_{200}$, and $I_{110}$ are the diffraction intensities from the (211), (200), and (110) crystal planes, respectively.

*Table 5.2 $I_{211}$, $I_{200}$, and $I_{110}$ are the XRD diffraction intensities from the (211), (200), and (110) crystal planes. $\lambda_S(EXP)$ is the experimental value for saturation magnetostriction. $\lambda_S(TH) = \left(\frac{2}{5}\right)\lambda_{100} + \left(\frac{3}{5}\right)\lambda_{111}$ and $\lambda_S^*(TH) = \left(\frac{1}{5}\right)\lambda_{100} + \left(\frac{4}{5}\right)\lambda_{111}$ are the theoretical values for saturation magnetostriction in the cases of randomly oriented and highly (110) textured grains, respectively.*

| Sample | $I_{211}/I_{110}$ (%) | $I_{200}/I_{110}$ (%) | $\lambda_S(EXP)$ (ppm) | $\lambda_S(TH)$ (ppm) | $\lambda_S^*(TH)$ (ppm) |
|---|---|---|---|---|---|
| $Fe_{81}Ga_{19}$-B | 15.0 | 11.0 | 50 | 64.0 | - |
| $Fe_{62}Co_{19}Ga_{19}$-B | 50.0 | 44.0 | 13 | 28.0 | - |
| $Fe_{81}Ga_{19}$/glass-F | 58.9 | 13.0 | 43 | - | 24.8 |
| $Fe_{62}Co_{19}Ga_{19}$/glass-F | 16.1 | 2.7 | 70 | - | 34.0 |
| $Fe_{81}Ga_{19}$/Si(100)-F | 6.0 | ~0.0 | 57 | - | - |
| $Fe_{62}Co_{19}Ga_{19}$/Si(100)-F | 14.0 | ~0.0 | 90 | - | - |

Further, from Table 5.2, we can also conclude that for the thin-film ($Fe_{81}Ga_{19}$-F and $Fe_{62}Co_{19}Ga_{19}$-F) samples, they are highly (110) textured. Especially, for the $Fe_{81}Ga_{19}$/Si(100)-F and $Fe_{62}Co_{19}Ga_{19}$/Si(100)-F samples, the intensity ratio, $I_{200}/I_{110}$, is practically zero. In turn, it is reasonable that one should use Eq. (5.2) to calculate $\lambda_{S(TH)}$

for the bulk sample, and use Eq. (5.3) to calculate $\lambda_S^*{}_{(TH)}$ for the thin-film sample. The closest existing data for theoretical calculations are from Ref. [293]: we could find $\lambda_{100} = 180$ ppm, and $\lambda_{111} = -14$ ppm for the $Fe_{84}Ga_{16}$ single crystals, and $\lambda_{100} = 10$ ppm, and $\lambda_{111} = 40$ ppm for the $Fe_{65}Co_{19}Ga_{16}$ single crystals. Then, approximately in agreement with the experimental findings that $\lambda_{S(TH)}$ of $Fe_{81}Ga_{19}$-B is larger than that of $Fe_{62}Co_{19}Ga_{19}$-B, whereas $\lambda_S^*{}_{(TH)}$ of $Fe_{81}Ga_{19}$-F is smaller than that of $Fe_{62}Co_{19}Ga_{19}$-F. Finally, from the high-temperature VSM runs, we could find that the Curie temperature ($T_C$) of $Fe_{62}Co_{19}Ga_{19}/Si(100)$-F is 600°C, higher than that (465°C) of $Fe_{81}Ga_{19}$-F.

The direct and inverse magnetoelectric effects in a two-layer planar structure containing mechanically coupled Galfenol and PZT plates was experimentally studied in [318]. It is shown that the direct magnetoelectric effect is lower and the inverse ME effect is equal to the corresponding effects in Terfenol based structures. Along with the low cost of Galfenol, these features make Galfenol–PZT structures promising for designing low frequency ac magnetic field sensors and transducers based on the forward ME effect and electrically controlled sources of ac magnetic fields based on the inverse magnetoelectric effect.

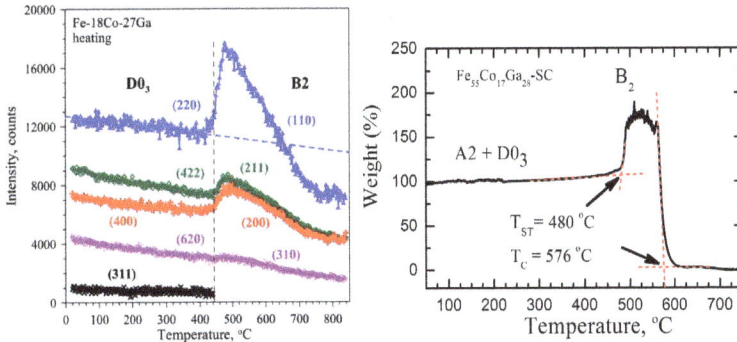

Fig. 5.23 a) Temperature dependencies of neutron diffraction peak intensities in Fe-18Co-27Ga at heating. The (311) peak indicates the presence of the $D0_3$ phase. The vertical dashed line shows a boundary between $D0_3$ and B2 ordering, b) The MTG scan of the slowly-cooled $Fe_{55}Co_{19}Ga_{26}$ alloy: $T_{ST}$ is the structure transformation temperature, and $T_C$ is the Curie temperature.

### $Fe_{55}Co_{18}Ga_{27}$

This alloy was also slowly-cooled. At room temperature, the alloy contains the disordered A2 (major) and ordered $D0_3$ phases. At T ≈ 480°C, there is a structure transition from the $D0_3$ phase to the B2 phase, based on the in situ neutron diffraction data – Fig. 5.23a.

The Curie temperature of the B2 phase of $Fe_{55}Co_{19}Ga_{26}$ is $T_C = 576$ °C (Fig. 5.23b). The magnetic property of the $Fe_{55}Co_{19}Ga_{26}$ (SC) alloy at room temperature is summarized: $M_S$ = 114 emu/g; $H_C = 1.7$ Oe; and $\mu_R = 223$. The mechanical and magneto-mechanical properties of the alloy are: $E_0 = 114.7$ GPa, $\Delta E/E_0 = 4.84\%$, and $K_E = 22\%$ [294].

Substitution of 18 at.% Fe in Fe-27Ga by 18 at.% Co influences the phase transitions at heating and annealing: the main difference is suppressing of the $D0_3$ to $L1_2$ transition and, consequently, the subsequence $L1_2$ to $D0_{19}$ transitions. Thus, the closed packed phases do not appear if 18%Fe in Fe-27Ga are substituted by 18%Co, i.e., in the Fe-18Co-27Ga alloy. The degree of the $D0_3$ atomic order the as-cast state in the Fe-18Co-27Ga alloy is lower compared with the Fe-27Ga alloy if we consider the ordering degree by intensity of super-lattice reflections. These structural peculiarities influence magnetic properties of the alloys. Thus, the high-temperature B2 phase of Fe-18Co-27Ga is ferromagnetic, whereas that of Fe-27Ga is non-ferromagnetic.

**$Fe_{100-X}Ni_XCo_{19}$, with x = 0 to 26 at.%Ni**

These alloys were slowly cooled in furnace. $\Delta E/E_0$ vs. x and $\Delta G/G_0$ vs. x plots for the furnace cooled $Fe_{100-X}Ni_XCo_{19}$ alloys are shown in Fig. 5.24. A maximum $\Delta E/E_0$ is found at x = 11 at.%Ni in Fig. 5.23. This maximum phenomenon maybe related to the similar x dependence of $E_0$ for the alloys; $E_0$(max) also occurs around 11 - 17 at.%Ni [295].

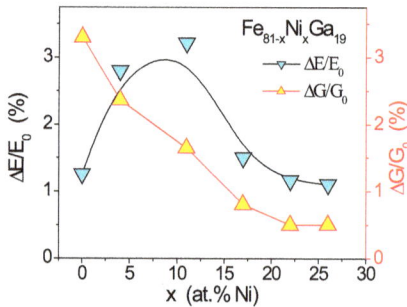

*Fig. 5.24 The $\Delta E$ and $\Delta G$ effects of the (bulk) Fe-Ni-Ga alloy.*

## 5.2 Fe-Al based alloys

Fe-Al based alloys have gained considerable interest owing to their attractive mechanical properties which can be improved by addition of appropriate third elements. Fe-Al alloys with strengthening intermetallic phases are promising candidates for structural applications. Here we provide further information in favour of the mechanisms proposed in Chapters 3 and 4, by checking the changes of the anelastic effects with additions of third elements, e.g. those which trap carbon interstitials into strongly bound carbides. They are grouped in the Sections 5.2.1. Fe-Al-(Nb, Ta, Ti, Zr); 5.2.2. Fe-Al-Cr; 5.2.3. Fe-Al-Si; 5.2.4. Fe-Al-(Co, Ge, Mn), and the results are discussed with respect to their anelastic mechanisms identified in binary Fe-Al alloys.

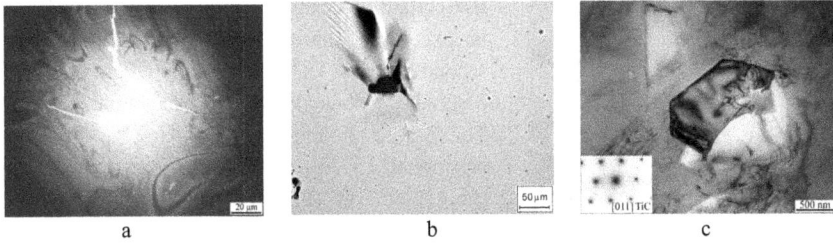

a        b        c

*Fig. 5.25 a) Light-optical micrograph of a TEM-film of Fe-26Al-4Ti showing a large number of TiC precipitates about 1 μm in diameter in the matrix, b) TEM bright field micrograph of the same region showing the TiC precipitates in the thinned region around the TEM hole, and c) enlarged section of the same region showing a typical TiC precipitate and the corresponding electron diffraction pattern which can be clearly indexed with the f.c.c. structure of TiC.*

### 5.2.1 Fe–Al–Me Alloys, Me = Nb, Ta, Ti, Zr

These four systems have some similarities and can be discussed together: the Ti, Nb, Zr and Ta having different solubility in Fe–Al are strong carbide forming elements in iron-based alloys: they produce MeC carbides, e.g. TiC, TaC, and NbC in Fe-(15–26)Al alloys [296]. The Nb, Ti and Ta increase the temperatures of the $D0_3$ to B2 and the B2 to A2 transformations in Fe-Al [297], which agrees with our own DSC data (0.3 at.% Nb: $T_O$ = 550 °C, 2 at.% Ti: $T_O$ = 672 °C, 4 at.% Ti: $T_O$ = 775 °C), and Ti and Ta in alloys with higher concentration lead to the $L2_1$ order, in which not only Fe and Al atoms are $D0_3$-ordered but also Ti and Ta atoms occupy positions of the (4b) sublattice. Nb and Zr have low solubility in Fe-Al and in particular in the $D0_3$ phase.

In all these alloys neither the Snoek-type nor so-called X peaks [196] were recorded: Ti, Nb, Zr, Ta trap free C interstitials into MeC carbides (Fig. 5.25), thus suppressing the Snoek and the X peaks. This effect takes place even in Fe-Al alloys containing only 0.1 at.% Zr or 0.3 at.% Nb: already such small amounts of Zr or Nb erase the S and X peaks. Very little influence of carbide forming elements on the vacancy concentration in these alloys after different regimes of quenching was proved in [139]. The fact that both the

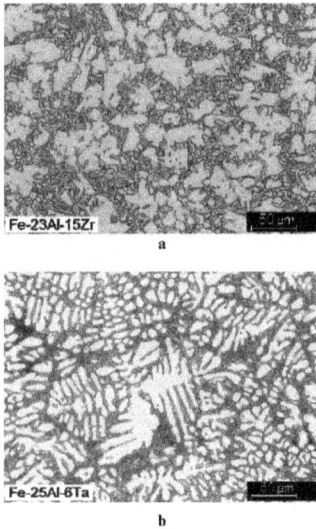

Fe-23Al-15Zr

a

Fe-25Al-6Ta

b

*Fig. 5.26 Structure of Fe-23Al-15Zr (a) and Fe-25Al-6Ta (b) alloys.*

Snoek and the X peaks are suppressed in all alloys containing a strong carbide former supports the conclusion about carbon in solid solution as the decisive ingredient for these relaxations, similar to those discussed for the binary alloys.

If a certain concentration of Ti, Ta or Zr in Fe–Al is exceeded, these elements produce different second phases (Fig. 5.26), e.g. Laves phases: $Zr_2(Fe,Al)$ hexagonal C14 ($\lambda$1) and cubic C15 ($\lambda$2), $ThZr_{12}$-type $\tau_1$ phase in Fe–Al–Zr alloys, and hexagonal C14 Laves phases $(Fe,Al)_2Ti$, $(Fe,Al)_2Nb$ or $(Fe,Al)_2(Ti,Nb)$ in Fe-Al-Ti,Nb alloys, or $(Fe,Al)_2Ta$ precipitates [298,299]. In the examples of Fig. 5.26 the phases could be distinguished by Vickers hardness measurements (*HV*) as follows: For Fe-23%Al-5%Zr (Fig. 5.26a) the dark phase (*HV* = 712) represents a eutectic, the bright one (*HV* = 1085) the Laves phase; for Fe-25%Al-6%Ta (Fig. 5.26b) the dark phase (*HV* = 1145) is the eutectic, and the bright phase (*HV* = 430) mainly Fe–Al solid solution.

Two temperature ranges can be distinguished in such "peak-free" $Q^{-1}(T)$ curves for Fe-20Al-0.1Zr in Fig. 5.27 between 100 and 800 K, (I) below 400 K, and (II) above 450 K. Taking into account the related increase in $f(T)$ and the start of decrease in the heat flow (DSC signal), this effect can be explained by the higher dislocation mobility in quasi-quenched disordered specimens in range "I", while the decrease of mobility and consequently of the damping background between 127 °C and 177 °C is supposed to be the result of ordering. Indeed the effect is smaller in the second heating run of the same specimen, which has been cooled down in the vibrating-reed furnace instead of quenching: in the second run the specimen is better ordered as compared to the quenched one. The increase in Zr content to 4% (in Fe-40%Al) and to 12.5% (in Fe-20%Al) leads to a well pronounced decrease of this effect. This may be due to the Laves phase in the 12.5Zr containing alloy which restricts the dislocation motion in the Fe–Al. Similar effect in Zr-free Fe-25%Al alloy was observed at lower temperature.

Ti (2 and 4 at.%) decreases the Zener peak in Fe-26%Al while no clear effect of the low content of Nb (0.3 at.%) was recorded in Fe-26Al [137], neither 0.1 nor 12.5 % Zr in Fe-20Al change substantially the range of Zener relaxation (Fig. 5.27) due to the low solubility in Fe-Al. The Zener peak decreases if the transformation temperature $T_O$ increases by alloying and if the added metals have reasonable solubility (e.g., Ti); no effect on the Zener peak occurs in case of low solubility (e.g., Zr, Ta, Nb).

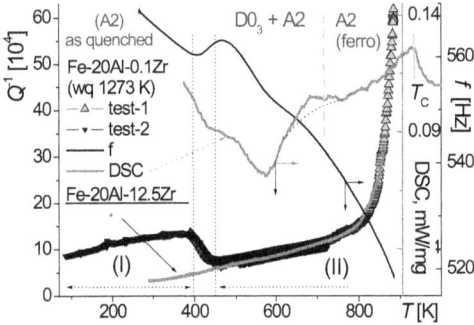

Fig. 5.27 Fe-20Al-0.1Zr (quenched from 1273K): damping $Q^{-1}$ (T), frequency f(T), and DSC curves. $Q^{-1}$(T) for Fe-20Al-12.5 Zr is added for comparison.

Alloying by Zr – contrary to the effect of Ti, Ta or Nb – leads to a prominent effect between 370 and 470 K: the level of IF decreases with increasing temperature in this range, although the increase in temperature does significantly influence neither the volume fraction nor the composition of phases in Fe-Al-Zr alloys [298]: The Zr content of the Laves phase in the Fe-23Al-15Zr alloy is 24.0 (after quenching from $T_q$ = 800 °C), 23.6 ($T_q$ = 1000 °C), and 23.3 at.% ($T_q$ = 1150 °C). At the same time after these treatments the IF differs in the low-temperature range.

## 5.2.2 Fe–Al–Cr Alloys

An important feature of this system is the high solubility of both Al and Cr in b.c.c. iron. Fe-25Al alloys are often additionally alloyed by Cr, in order to increase not only their yield stress but also their ductility and workability. The Cr atoms occupy the 4b or 8c positions with some preference of next nearest neighbour Al positions [300]. The lower dissociation energy $W_{(Cr-Al)}$ = 0.6960 eV for Cr-Al pairs than for Fe–Al pairs ($W_{(Fe-Al)}$ = 0.7457 eV) is responsible for the mentioned decrease of the APB energy. The ordering transition temperature is not changed significantly [301] by Cr addition of <15% (Table 2). The density of vacancies in Fe-25Al-(15-25)Cr alloys quenched from 1000 °C was detected to be about $c_V \approx 3 \times 10^{-5}$ at$^{-1}$ which is three times lower than that in Fe-25Al ($c_V \approx 10^{-4}$ at$^{-1}$ [174]).

Alloying of Fe-Al with Cr complicates the atomic ordering and affects the temperature dependences of internal friction. As in the case of Fe-27Al, it can be organized in the form of antiphase domains (APDs) separated by coherent antiphase boundaries (APBs) or in the form of a matrix of a disordered phase with dispersed clusters (mesoscopic in size) of an ordered phase. We studied by neutron real-time thermodiffractometry a set of samples with the composition close to $(Fe,Cr)_3Al$ with different chromium content, namely, Fe-27Al-3Cr, Fe-25Al-9Cr and Fe-25Al-15Cr.

The measured diffraction patterns show that in the initial state (quenched) all compositions with Cr are in B2 phase (Fig. 5.28). The Miller indices for the $D0_3$ unit cell are indicated in the figure. To be compared with the indices of A2 and B2 phases they must be divided by factor of two. With this notation, peaks with odd $(h, k, l)$ are allowed only in the $D0_3$ phase. Peaks with even $(h, k, l)$ and odd values $(h + k + l)/2$, are allowed in the $D0_3$ and B2 phases. In the disordered phase A2, only intense peaks with even Miller indices $(h, k, l)$ and $(h + k + l) = 4n$ remain. If the sample is in $D0_3$ phase, then the peaks with any combination of odd or even indices, are fundamental or superstructure ones, respectively. If the sample is in phase B2, then the peaks with $(h + k + l) = 4n$ are fundamental, the others are reflections from superstructure.

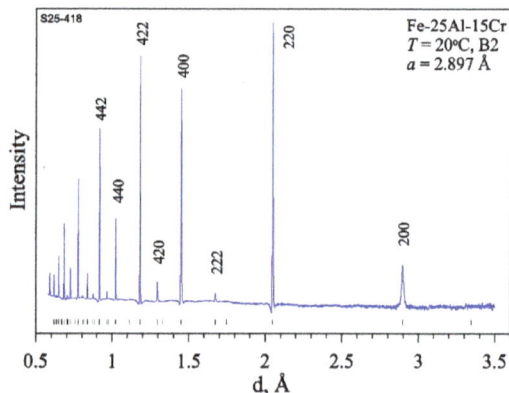

Fig. 5.28.    High-resolution diffraction pattern of Fe-25Al-15Cr, which is in the B2 phase. The vertical bars indicate the calculated peak positions for the $D0_3$ unit cell ($a_{D03} = 2a_{B2}$).

Heating of the samples was carried out directly in a neutron beam up to 850°C with a constant rate of about 2 K/min. Cooling was proceeded linearly with a rate of -2 K/min down to $T \approx 150$ °C. High-resolution diffraction patterns were measured before heating and after cooling to room temperature. In all quenched compositions with Cr, the sequences B2 → $D0_3$ → B2 → A2 at heating and A2 → B2 → $D0_3$ at cooling were observed. The intensities of some particular peaks during heating and cooling and the atomic volume (a unit cell volume per one atom) as a functions of temperature are shown in Fig. 5.29.

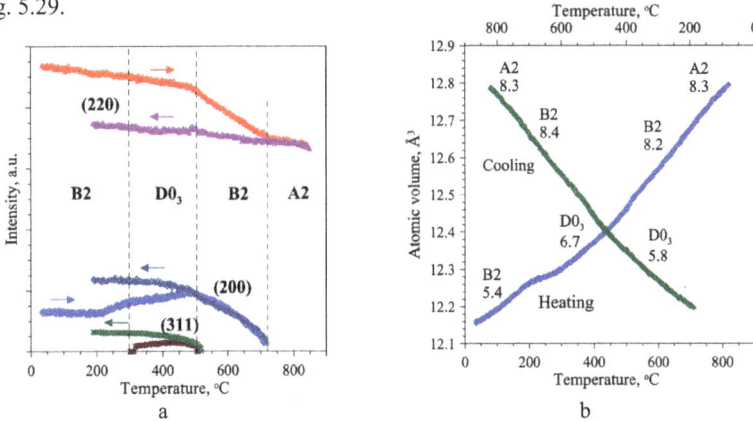

*Fig. 5.29 (a) Intensities of typical diffraction peaks from Fe-25Al-9Cr sample measured upon heating (arrow to the right) and cooling (arrow to the left). Experimental points are shown with statistical uncertainties. The vertical lines show phase boundaries for heating process. (b) Atomic volumes for Fe-25Al-9Cr as functions of temperature at heating (lower scale) and cooling (upper scale). Near curves the volumetric thermal expansion coefficients (in the $10^{-5}$ $K^{-1}$ units) determined over the intervals with linear dependence are indicated.*

One can see the appearance of the (311) peak from the $D0_3$ phase in the range of 300-500 °C and an increase in the intensity of the (200) peak in approximately the same temperature range, while the intensity of the (220) peak decreases monotonically. Similar dependences were obtained for all studied Fe-Al-Cr alloys.The atomic volume varies continuously with temperature including the transition range which distinguishes Fe-Al compositions from Fe-Ga ones. In the latter the transition from $D0_3$ to A2 goes through appearance of the closed packed $L1_2$ and $D0_{19}$ phases with a different lattice symmetry.

The observed changes in the intensities of these peaks characteristic can be quite clearly interpreted. For their analysis, it should be taken into account that the intensities of the fundamental, $I_f$, and superstructure, $I_s$, peaks depend on temperature as follows:

$$I_f \sim |F_f|^2 \cdot \exp[-W(T)], \qquad I_s \sim \xi^2(T) \cdot |F_s|^2 \cdot \exp[-W(T)], \qquad (5.4)$$

where $F_f$ and $F_s$ are structure factors of fundamental and superstructure peaks, $\xi(T)$ is a factor connected with the degree of atomic order, $0 \le \xi(T) \le 1$, $\exp[-W(T)]$ is the Debye-

Waller factor.

The smooth decrease in the intensity of the (220) peak upon heating is associated with the Debye-Waller factor, an increase in the intensity of the (200) peak and the appearance of (311) one are the result of Fe and Al order increasing. A further heating leads to transitions to a partially ordered (B2) and then to a disordered (A2) states. When cooling, the intensities of the (200) and (311) peaks increase due to the increase of $\xi(T)$.

Fig. 5.30 shows the $(\Delta d)^2$ on $d^2$ dependences, measured in the initial state of the Fe-25Al-9Cr alloy and after its cooling. In the initial state, the allowed in A2 dependence for the peaks with $(h + k + l) = 4n$ is linear, in contrast the the $(\Delta d)^2$ on $d^2$ dependence for superstructure peaks is parabolic. After cooling the linear dependence is measured for the peaks with even indices, i.e. allowed in B2, whereas points with odd indices, allowed only in $D0_3$, fit to the parabola. Other clearly visible effects are the absence of any anisotropy in the behaviour of the widths and a noticeable increase in the size of the coherent scattering domains after heating and cooling.

The data presented in Fig. 5.30 are in good agreement with the model of the matrix and dispersed clusters. In the initial as quenched state, the matrix is the disordered A2 phase, in which the clusters of the partially ordered B2 phase, with the mean size of $L \approx 300$ Å, are disperse. After heating to 850°C and subsequent slow cooling, the matrix is B2, in which clusters of the ordered phase $D0_3$, with the average size of $L \approx 930$ Å, are dispersed.

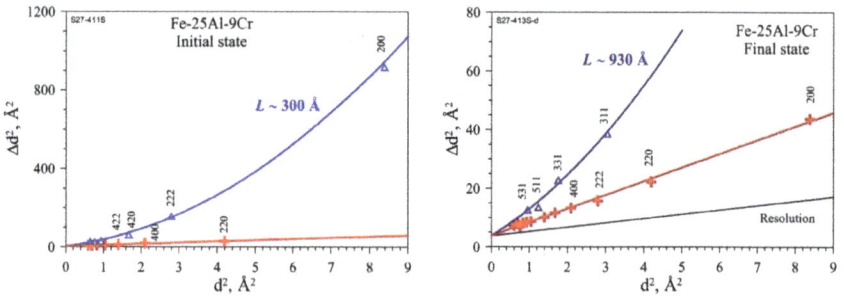

Fig. 5.30 The $(\Delta d)^2$ over $d^2$ dependences for Fe-25Al-9Cr alloy in the initial (before heating) B2 state (a) and after heating-cooling session ($D0_3$ phase) (b). In both cases, the widths of the fundamental peaks (red crosses) fit into a linear dependence, the widths of superstructure peaks (blue triangles) are described by a quadratic dependence. The $(\Delta d)^2$ values are multiplied by $10^6$. Statistical errors of experimental points are about symbol size.

Analysis of the intensities of reflection orders, for instance, (200) and (400), which ratio does not depend on texture effects, makes it possible to obtain temperature dependence of occupancies for a certain type of atoms. i.e. the degree of atomic ordering in the clusters. For Fe-25Al-9Cr alloy, the result of the calculation is shown in Fig. 5.31. It can be seen that during slow heating from $T \approx 220$ °C and up to $T \approx 500$ °C, the occupancy of iron atoms in one of the unit cell positions is growing up to $n_{Fe} \approx 0.93$. If cooled, this parameter increases to $n_{Fe} \approx 0.97$, i.e. almost the maximum possible ordering is achieved.

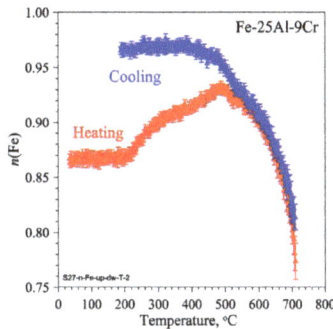

Fig. 5.31   The temperature dependences of the degree of Fe ordering in Fe-9Cr-25Al determined from the ratio between the (200) and (400) intensities. Experimental points and their statistical uncertainties are shown. In the disorder state n(Fe) is equal to 0.66.

Chromium is a carbide forming element both in iron and in Fe-26Al, although not as strong as Ti, Nb, Ta or Zr. Chromium carbides can be dissolved at quenching from high temperatures and some interstitial carbon remains in solid solution. Calculations [197] have demonstrated that the C–Cr 'chemical' plus 'elastic' interaction in iron takes place up to the fourth or fifth coordination shell. This explains the observation that the Snoek-type and so-called X peaks (see Chapter 3 and 4) for alloys Fe-30%(Al+Cr) if the Al content exceeds the Cr content, and the shift of peak positions to higher temperatures with respect to binary Fe-Al alloys due to the additional C–Cr interaction [195]. Substituting Al atoms by Cr in Fe-30(Al+Cr) alloys leads to a decrease in hardness and decrease in Curie and ordering tempeatures (Table 5.3). The Zener peak smears to the direction of higher temperatures since the activation energy of Cr diffusion in Fe₃Al is higher than that of Al. The activation energy of Cr diffusion in iron and also in Fe-27Al ($232 \pm 5$ in A2 and $239 \pm 3$ kJ/mol in B2 phases) is higher than the activation energies of Al diffusion in iron [302].

Table 5.3  Hardness of (Fe,Cr)₃Al alloys after annealing for 48 h at 475°C, D0₃-to-B2 transition temperature and Curie temperature as a function of Cr% (alloys with ≤ 15Cr are D0₃ ordered, the alloy with 25Cr is B2).

|  | 25Al | 27Al-2.5Cr | 26Al-5Cr | 25Al-8Cr | 25Al-15Cr | 25Al-25Cr |
|---|---|---|---|---|---|---|
| HV | 308 | 321 | 297 | 280 | 284 | 361 |
| $T_O$, °C | 546 | 550 | 552 | 552 | 520 | ~465 |
| $T_C$, °C | ~500 |  | ~480 | ~290 | ~150 | ~135 |

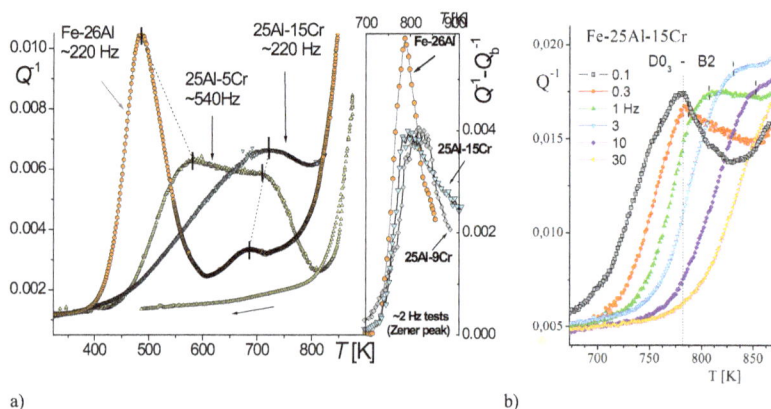

a)                                                          b)

Fig. 5.32 (a) $Q^{-1}(T)$ in Fe-25Al-(5÷15)Cr for 390-540 Hz (vibrating reed), right-hand part of the figure with scale at the top for ~2 Hz (torsion pendulum); (b) $Q^{-1}(T)$ in Fe-25Al-15Cr at 0.1-30 Hz bending (DMA).

If Al content (~25 at. %) is fixed and Cr substitutes Fe, the above mentioned tendencies can also be observed but the relative changes are different (Fig. 5.32). Here we used three different techniques: vibrating reed (left: 220-540 Hz), torsion pendulum (in the middle: about 2 Hz), and DMA (right: 0.1-30 Hz).

The D0₃ order is proved by TEM in Fe-26Al-5Cr, Fe-25Al-8Cr, Fe-25Al-15Cr and Fe-25Al-25Cr composition after annealing at 475 °C. Some selected TEM structures of Fe-25Al-xCr alloys with different Cr content are presented in Fig. 5.33 showing fine and coarse D0₃ domains in all studied alloys after short and long annealing, respectively. B2 order is found in Fe-25Al-15Cr and in Fe-25Al-25Cr after quenching. The fine D0₃ domains form after 48 h annealing at 750 K in Fe-25Al-25Cr which is rather close to the

D0$_3$-to-B2 transition. The temperatures of the A2-to-D0$_3$ and D0$_3$-to-B2 as well as of the ferro- to paramagnetic transformation for Fe-25Al-Cr alloys are shown in Table 5.3. D0$_3$ order is also proved in alloys with 8 and 15%Gr below 500 °C by in situ neutron diffractions.

The decrease in the Snoek-type and X peak heights is due to trapping of carbon atoms in chromium carbides, while some shift of the Snoek peak to higher temperature is the result of the C–Cr interaction in solid solution [197]. The Snoek and X peaks can hardly be distinguished in some cases: only by adding Cr stepwise into Fe–Al, is it possible to see the retaining contribution of these peaks in Fe-Al-Cr alloys with high Cr content, e.g. in Fe-25Al-15 and -25%Cr.

The Zener relaxation is slightly broadened and shifted to higher temperatures: this effect is better seen in the low-frequency tests (Fig. 5.32a: right scale). In case of high frequency tests the Zener peak position is beyond the range of our TDIF measurements. DMA tests (Fig. 5.32b) shows that the transient peak due to D0$_3$-B2 transition (~780 K, dotted vertical line) in the range of the Zener peak (the peak positions are tentatively shoen by vertical lines). The Zener peak parameters in Fe-25Cr-5Al as reported in [303, 304] are: $H$ = 245 kJ/mol ($\tau_0$ = 2.6×10$^{-17}$ s), which is surprisingly more close to Fe-25Al than to the Fe-25Cr alloy if compared with [195, 269]. The peak shape was also recorded to be close to the Debye peak with a single relaxation time.

**The isothermal mechanical spectroscopy** technique was applied in cooperation with Prof. Andre Rivière [200] to study the high temperature relaxation in several Fe-Al-Cr ternary alloys. This technique excludes transient effects as it operates at constant temperature. According to TEM, X-ray and ND examination all studied alloys were ordered. Heat flow tests of as quenched specimens were used to determine the D0$_3$-to-B2 ($T_O$) temperature, vibrating sample magnetometry – to determine the Curie point ($T_C$) (Table 5.3). All the specimens were homogenized at 1000 °C in a quartz ampoule and annealed at 480 °C during 100 hrs to eliminate relaxation effects caused by interstitial atoms and thermal vacancies in solid solution. Two series of experiments were made using each sample. During the first run, the samples have been progressively heated step by step. At each temperature the specimens were annealed for 24 hours, which leads to nearly equilibrium condition of the specimens for each temperature used. In a second stage the specimens were measured at the same temperatures as for heating but after annealing at higher temperatures.

Two IF peaks are recorded in Fe-Al-Cr alloys: the Zener peak denoted as P1 peak in this section and second peak at lower frequency, which is equivalent to a higher temperature at TDIF tests as the P2 peak. The results are presented in Fig. 5.34a after background subtraction as measured at 550 °C during the first run. The procedure of background subtraction is shown in Fig. 5.34b for only one alloy (Fe-25Al-15Cr). There is only peak P1 in Fe-25Al and Fe-28Al-3Cr while in alloys with higher Cr content there are two peaks. Fig. 5.34b shows result for Fe-25Al-15Cr obtained also at 550 °C but after annealing of the specimen at 650 °C. The difference between the two curves corresponding to the Fe-25Al-15Cr is due to the evolution of the peaks during the high temperature annealing.

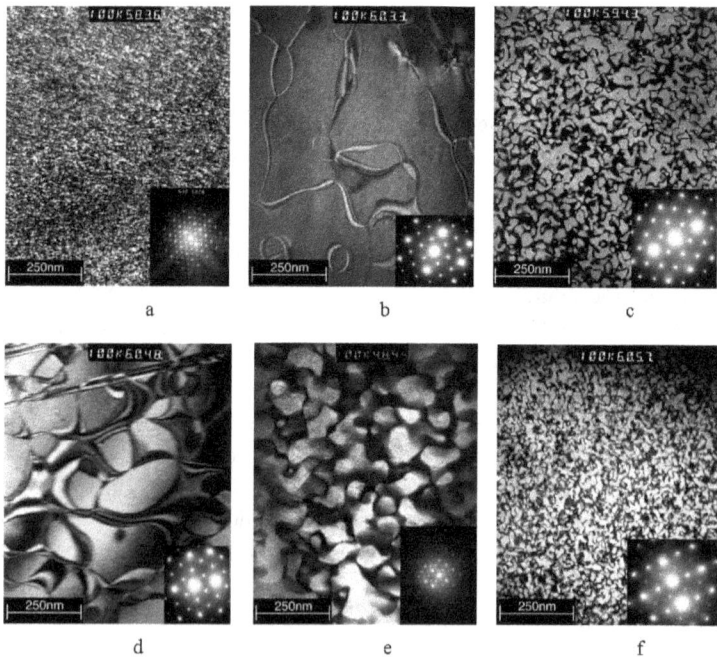

Fig. 5.33 TEM micrographs showing the D0₃ structure in Fe-26Al-5Cr (dark field, [110](111)): quenched from 850°C and annealed 3h at 300°C (fine domains) (a) and quenched from 900 °C and annealed 48 h at 480 °C (coarse domains) (b) Fe-25Al-8Cr quenched from 850 °C and annealed 3 h at 300 °C (c) and 48h at 480 °C (d), Fe-25Al-15Cr, quenched from 900°C and annealed 100 h at 480°C (e), and Fe-25Al-25Cr quenched from 900°C and annealed 48 hrs at 480°C (fine domains) (f).

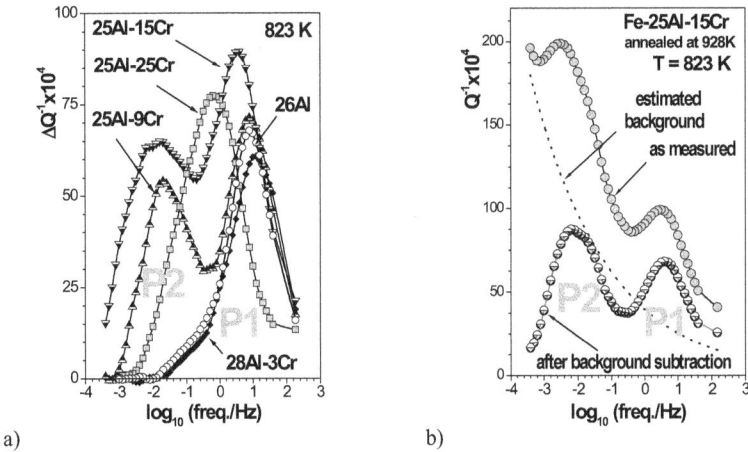

*Fig. 5.34 Overview of two IF peaks in Fe-25Al-(0-25)Cr alloys after low frequency background subtraction (a), procedure of low frequency background subtraction (Fe-25Al-15Cr) (b).*

The $D0_3$ structure was detected in all alloys after annealing at 480 °C. At the same time the B2 order is detected by X-ray studies in Fe-25Al-25Cr alloy: the $D0_3$-to-B2 transition in this composition is close to the annealing temperature [200]. Frequency dependent damping in the Fe-26Al alloy and its structure is discussed in Chapter 4.

***The Fe-28Al-3Cr alloy.*** The $T_0$ transition in this alloy takes place according to DSC tests at lower temperature as compared with Fe-26Al $T_0 = 533$ °C mainly because of a higher Al content. Thus, the Zener peak is mainly recorded in the $D0_3$ range of the phase diagram. The P2 peak was not observed in this alloy. Activation parameters for the P1 Zener peak are found to be $H = 276$ kJ/mol, and $\tau_0 = 3 \times 10^{-19}$ s were found. Only one test belongs to the B2 range (824K), which makes to think that the $D0_3$ structure dominates in the structure of this alloy in the range of our tests.

***The Fe-26Al-8Cr alloy.*** The $T_0$ transition (552 °C) in this alloy takes place according to DSC tests at practically the same temperature as in Fe-26Al, while the Curie point is about 200 °C lower. Both the P1 (Zener) and P2 peaks are recorded (Fig. 5.36a). The P1 peak - in both the $D0_3$ (paramagnetic) and B2 ranges, the P2 peak - above $D0_3$-to-B2 transition. Similarly to the Fe-26Al alloy the P1 peak height increases in Fe-26Al-8Cr with increase in temperature if the temperature of the isothermal test is more than 450 °C (Fig. 5.36b). The same is also true for the P2 peak (Fig. 5.36c). The P1 peak height below

450 °C does not change pronouncedly. Activation parameters for the P1 and P2 peaks were found as follows: the P1 peak: H = 290 kJ/m, $\tau_0 = 5\times10^{-20}$ s, the P2 peak: H = 282 kJ/m, $\tau_0 = 10^{-17}$ s.

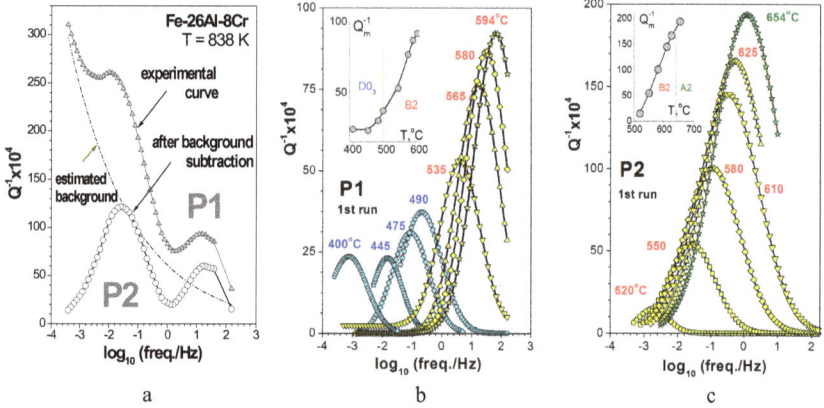

Fig. 5.36 The Fe-27Al-8Cr alloy: experimental data, isothermal test at 565°C in the range of the P1 and P2 peaks (a), P1 (Zener) (b) and P2 (c) peaks measured at different temperatures. Blue color of curves corresponds to $D0_3$, yellow – to B2 and green – to A2 phases.

**The Fe-25Al-15Cr alloy.** The $T_O$ transition (520 °C) in this alloy is 25-30 °C lower as compared with the Fe-26Al alloy, Curie point is at about 150 °C. After 100 h annealing at 475 °C the alloy has $D0_3$ structure and after 100 h annealing at 625 °C – the B2 structure. Both the P1 (Zener) and P2 peaks are recorded (Fig. 5.37). The P1 peak - in both the $D0_3$ (paramagnetic) and B2 ranges, the P2 peak - above $D0_3$-to-B2 transition. The P2 peak height is slightly lower than that is in Fe-26Al-8Cr alloy. Activation parameters for the P1 and P2 peaks in this composition were found as following: for P1 the H = 285 kJ/m (2.9 eV), $\tau_0 = 4\times10^{-20}$ s in cooling, and for P2 peak: H = 290 kJ/m (3 eV), $\tau_0 = 1.3\times10^{-17}$ s (after annealing at 665 °C).

**The Fe-25Al-25Cr alloy.** In contrast with other Fe-Al-Cr alloys only one peak is recorded in this alloy in the studied temperature range. Furthermore, the peak is broader (in all tests $\beta > 1.5$), and its height does not have such a clear dependence on temperature of measurements. The activation parameters of the peak $\tau_0 = 10^{-19}$ s and $H = 288$ kJ/mol.

Experiments were repeated for Fe-26Al-8Cr and Fe-25Al-15Cr alloys at the same temperatures as during heating after annealing at 655 °C. The height, width and frequency of the P1 peak in Fe-25Al-15Cr alloy are the same after and before the annealing. It is similar for the Fe-26Al-8Cr alloy: the frequency of P1 peak is always the same but the height is increased for about $18 \times 10^{-4}$ and the width decreases until $\beta = 1.17$ on average. For P2 peak, the behaviour is the same for both the alloys. The height increases ($\sim 35$ to $40 \times 10^{-4}$) and the peak shifts towards lower frequency after annealing for measurements at the same temperature.

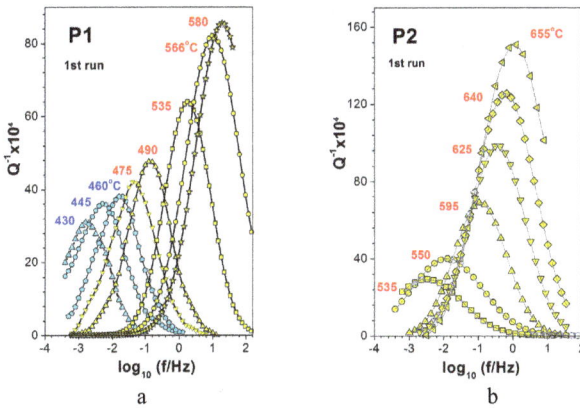

Fig. 5.37  The Fe-27Al-15Cr alloy the P1 (Zener) (a) and P2 (b) peaks measured at different temperatures.

In all alloys except Fe-25Al-25Cr composition an increase in the P1 peak height with temperature of measurements is observed. Increase in the peak height with temperature is decreased in order in the alloy. As it concerns the P2 peak, it is possible to state that in spite its higher temperature (or lower frequency) location of its activation energy is lower than that is for the P1 peak in the same alloys. Possible interpretation for the P2 peak is dislocation creep in the alloys [200]. O. Lambri et al. [207] comes to similar conclusion: in ordered Fe-Al-Ga alloys independently of the type of order, $D0_3$ or $B2$, the mobility of dislocations and grain boundaries is markedly reduced. Addition of Cr in Fe-Al alloys decreases ordering in these alloys at elevated temperatures. If order decreases or disappears the dislocation mobility increases which results in an increase in damping as compared with the ordered state. A relaxation peak (P2) related to the grain boundary or dislocation relaxation during the cooling after annealing at temperatures higher than 700 °C.

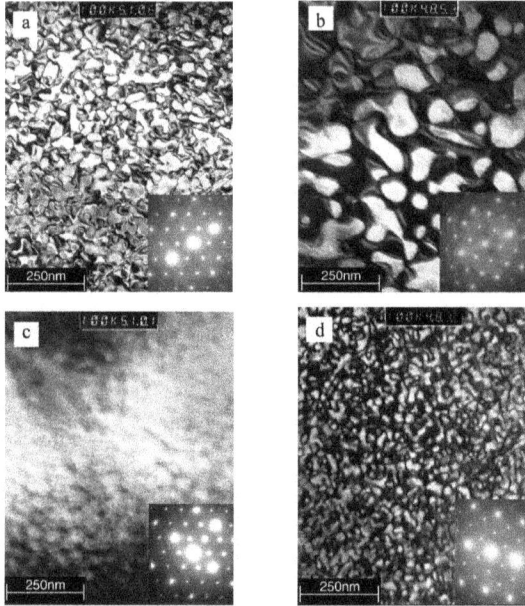

Fig. 5.38 TEM micrographs of the Fe-20Al-5Si alloy: after quenching from 1170K (a), after 100 h annealing at 750K (b), after 100 h annealing at 900K (c) and the Fe-5Al-10Si alloy after 100 h. annealing at 750K (d) (dark field, [110](111)).

### 5.2.3 Fe–Al–(Si, Co, Ge, Mn) Alloys

**Ternary Fe–Al–Si Alloys.** Contrary to the previous group, the addition of Si in Fe-Al neither produces new phases nor trappes C in carbides. Si improves the $D0_3$ order, increasing the transition temperatures of the $D0_3$-to-B2, and B2-to-A2 transitions. We have studies several $Fe_3(Al+Si)$ alloys (Table 5.4). After annealing at 475 °C for 100 h, XRD analyses confirm the $D0_3$ order in all these alloys (Fig. 5.38). The increase of hardness ($HV$) with substitution of Al by Si atoms in $Fe_3(Al+Si)$, and the corresponding decrease of the $D0_3$ lattice parameter ($a$) is presented in Table 5.4, where also the ordering temperature $T_O$ and the Curie temperature $T_C$ are given as determined by DSC and magnetisation measurements. If quenched from 1000 °C the density of vacancies in

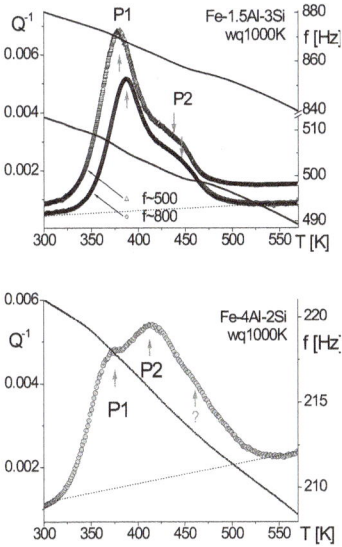

Fig. 5.39. TDIF curves for Fe-1.5Al-3Si (upper) and Fe-4Al-2Si (lower) both water quenched from 725°C. Different frequencies are used (right scale). The P1 peak corresponds to Fe-C-Fe, and the P2 peak corresponds to Fe-C-Me (Me = Al and Si) components of the Snoek-type peaks.

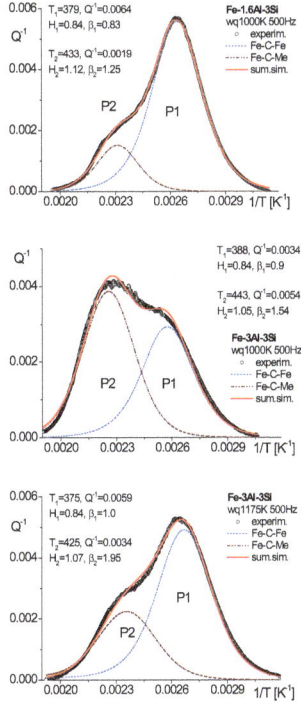

Fig. 5.40. Computer fit for the P1 and P2 components of the Snoek-type peaks: a) Fe-1.5Si-3Al, b) Fe-3Si-3Al (both water quenched from 725 °C, and c) Fe-1.5Si-3Al water quenched from 900 °C.

Fe-25(Si+Al) alloys was detected to be 3-10 times lower than that in Fe-25Al. The $DO_3$ order was detected not only in Fe-25(Al+Si) but also in Fe-15(Al+Si) alloys, with a similar effect of Si substitution on hardness and lattice parameter.

The development of the Snoek peak in binary Fe-3Si and Fe-3Al alloys was discussed in several papers, including [143]. The peak in Fe-Al-Si alloys with Al/Si content (in at.%): 0/2, 0/5, 1.6/3, 3/3, 4/2 clearly consists of two components (Fig. 5.39): a left-side shoulder (P1) corresponds to the "ordinary" Fe-C-Fe Snoek peak in $\alpha$-Fe, i.e., this effect is produced by C atom jumps in the solid solution with local Fe atom surrounding [175]. The second peak (P2) is the result of Snoek-type C atom jumps near Fe-C-Me positions (Me = Al or Si). Temperature positions of both peaks are frequency dependent, i.e. these peaks are caused by thermally activated processes. Al atoms produce a bigger effect as compared with Si on the P2 component of the Snoek-type peak. If the amount of Al/Si in iron increases to 7/2, 6/5, 9/4, 5/10, 8/7 the peak further shifts to higher temperatures due to a high-temperature contribution (P2), and a very small contribution of the "ordinary" Fe-C-Fe (P1) peak can still be seen. The P2 peak becomes dominating, the third peak from the right side appears (a weak peak can be distinguished even in 4/2 composition, see Fig. 5.39). This third peak can be another component of the Snoek-type peak or it can be caused by vacancies.

Table 5.4 Hardness in Fe$_3$(Al+Si) alloy after annealing for 100 h at 750 K, D0$_3$-to-B2 transition temperature, Curie temperature and lattice parameter as a function Al and Si content.

| Alloy | Fe-25Al | Fe-20Al-5Si | Fe-12Al-12Si | Fe-5Al-20Si | Fe-25Si |
|-------|---------|-------------|--------------|-------------|---------|
| HV | 308 | 426 | 459 | 536 | 537 |
| $T_O$ (K) | 819 | 1021 | 1195 | 1389 | |
| $T_C$ (K) | ~780 (in D0$_3$) | 737 | 742 | 789 | 790 |
| $a$ (nm) | 0.5793 | 0.5758 | 0.5713 | 0.5665 | 0.5662 |

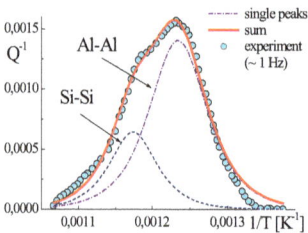

Fig. 5.41. Zener peak fit for the Fe-20Al-5Si alloy by two peaks attributed to reorientation of Al-Al and Si-Si atomic pairs.

This effect was also reported in [305] for the Fe-6wt.%Si (~11.3at.%Si) alloy at 300°C. We can conclude that at least the main contribution to this anelastic effect in ternary Fe-(0-4)Al-(0-5)Si disordered alloys is due to the Snoek-type relaxation, while in Fe-(5-9)Al-(2-10)Si ordered alloys the third component is added to the relaxation process. This third component of the peak decreases with ageing faster than the P1 and P2 peaks supporting the idea of another mechanism of this peak. Computer analyses of the Snoek-type peak are given in Fig. 5.40: increase in % or alloying elements increases the Fe-C-Me component of the peak (Fig. 5.40, a and b) while increase in quenching temperature increase the Fe-C-Fe component of the peak (Fig. 5.40, b and c) due to more random distribution of C atoms.

Apart from the Snoek-type peak, the X, the Zener and the grain boundary peaks are recorded in Fe-Al-Si alloys [199]. An increase in the Si content in $Fe_3(Al,Si)$ from Fe-20Al-5Si to Fe-12.5Al-12.5Si and Fe-5Al-20Si decreases the Snoek peak. The X peak is not well distinguishable in two later compositions, the Zener peak decreases in height and seems to shift slightly to higher temperatures.

The Zener peak shape and position appears to be a compromise between those of the binary Fe–Al and Fe–Si alloys. It can be shown [199] that the Zener peak in ternary Fe-Al-Si alloys with both 20Al-5Si and 5Al-10Si can be well fitted by two peaks (Fig. 5.41) due to reorientation of Al-Al ($H \approx 2.45$ eV, $T_m \approx 808$K) and Si-Si ($H \approx 2.97$ eV, $T_m \approx 837$K) atom pairs.

The relaxation strength of the Zener relaxation ($\Delta = 2Q_{max}^{-1}$) in disordered alloys is proportional to both the amount of substitute atoms and the relaxation strength per atom pair. As Al atoms are bigger and Si atoms smaller than Fe atoms, both Al-Al as well as Si-Si pairs give rise to an increase of the Zener peak height due to changes of the local lattice parameter. Selected data from literature ([267] – open triangles and [306] – closed triangles) are shown in Fig. 5.42 demonstrating the increase in the peak height with both Si and Al content until ordering takes place at ~11 at.%Si or >20 at.%Al in binary alloys.

In ternary alloys, the occurrence of Si-Al pairs may reduce the contribution to the Zener relaxation compared to the binary alloys. Indeed the Zener relaxation in all ternary alloys is significantly lower than in binary alloys. There are at least two reasons for that. Firstly, ordering, which leads to a decrease in the Zener peak height, starts at lower alloying element concentrations in Fe-Si-Al alloys than in Fe-Al alloys. The second reason is that Al and Si atoms in iron at least partly compensate for the elastic distortions produced by each species.

However, as the size differences of Al and Si with respect to Fe are of opposite sign, in the ternary Fe-Si-Al alloys the elastic distortions of Al and Si may compensate each other, leading to an attractive interaction between Al and Si and a preferred formation of Al–Si pairs which are then very weak elastic dipoles. Thus, in most ternary Fe-Si-Al alloys the Zener peak is effectively suppressed as seen in Fig. 5.42, even in the disordered state at $C_{Al+Si} < 15$ at% where, different from the range around 20–25 at%, atomic long-range order cannot be responsible for this effect [254]. Figure 5.43a shows the lattice parameter in dependence on the Si/Al ratio. As expected, Si decreases and Al increases the lattice parameter of $\alpha$-Fe. In the ternary alloys it is lower than in $\alpha$-Fe if Si/Al > 1 and vice versa. Al results in an additional increase of the hardness of the Fe–Si alloys (Fig.

Materials Research Foundations Vol. 30

5.43b). However, the effect of Al on the hardness increase in Fe–Si–Al alloys is weaker than that of Si.

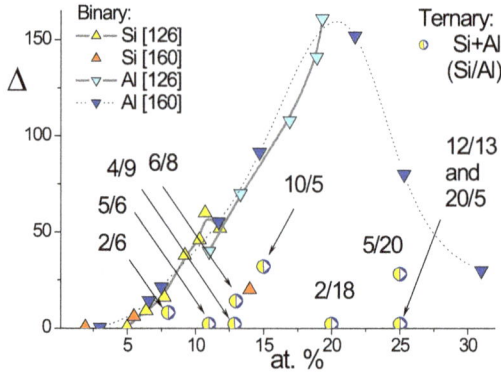

*Fig. 5.42. The relaxation strength of the Zener peak (Δ in binary Fe–Si and Fe–Al alloys, and in ternary Fe–Si–Al alloys. Triangles: up – for Si, down – for Al. Data for ternary alloys [306] are shown by circles. In the left figure, Δ is given as a function of the total amount of Si+Al in at.% and the respective Si and Al contents are indicated at the experimental point as Si/Al.*

The height of the high temperature peak in Fe-Al-Si was studies by Lambri et al [307]. The mechanical spectroscopy response in Fe-25 at.% (Al + Si) and Fe-15 at.% (Al + Si) has been studied in the temperature interval between 525 and 930 °C. An internal friction peak was observed in this temperature range with a maximum close to 730 °C (Fig. 5.44).

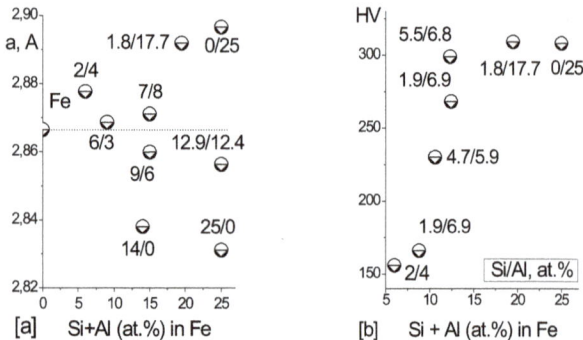

*Fig. 5.43. Lattice parameter (a) and hardness (b) of studied alloys in dependence on Si+Al content in Fe (numbers near experimental points indicate Si/Al contents in at.%.*

224

Peak height depends strongly on the order degree of the Fe-Al-Si alloys as it was judged by *in situ* neutron diffraction studies performed at the D1B powder diffractometer in the Institut Laue Langevin, Grenoble, France [308]. The dislocation structure and solute atoms interacting with grain boundaries control the damping spectrum. The absence of the peak up to 950-1000K during heating is due to a reduced dislocation and grain boundary mobility in the D0$_3$ ordered phase. The appearance of the damping peak during cooling is associated with a higher mobility of dislocations during cooling from disordered state.

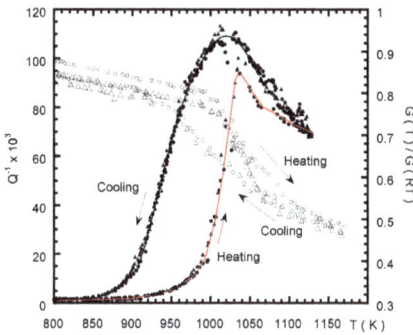

*Fig. 5.44. Internal friction (full symbols, left axis) and relative modulus (empty symbols, right axis) for Fe-6Al-9Si alloy. Circles: first thermal cycle up to 860 °C; triangles: second thermal cycle up to 860 °C.*

Thus, the substitution of Al atoms by Si atoms, forming no carbides and improving the D0$_3$ order (increasing the transition temperature to B2) in Fe-Al-Si alloys, influence on the Snoek-type and Zener peaks' parameters, indicating the difference in C-Si and C-Al interaction in solid solution and contribution of Al-Al, Si-si and Al-Si atom pairs. Si also tends to lower the concentration of vacancies, in agreement with the reduction of the X peak. The anelastic phenomenon at about 1000K (~1 Hz) can be probably attributed to mobility of dislocations in ordered and disordered phases.

**Fe–Al–Co**. The Snoek-type and Zener peaks are recorded in all Fe-Al-Co alloys. The activation energies of Co diffusion in Fe-27Al are 291±4 in A2 and 283±11 kJ/mol in B2 phases [302]. Correspondingly, the Zener peak in Fe-20Al-5Co shifts slightly to higher temperatures, while the Snoek peak very slightly shifts to lower temperature as compared with Fe-25Al because of the lower Al content. XRD shows the B2 order in the specimen both after quenching from 1000 °C (lattice parameter $a$ = 0.2887 nm) and after additional

100 h annealing at 650K ($a$ = 0.2890 nm). The B2 domain structure was detected by TEM, too.

**Fe–Al–Mn.** If Mn is added to Fe-25Al the $T_O$ temperature of the $DO_3$-to-B2 transition is slightly increased: ~560 °C (Fe-25Al-2Mn) and ~575 °C (Fe-25Al-5Mn). Mn is a weak carbide forming element; the activation energy for Mn diffusion in Fe-27Al is 231±5 kJ/mol in the A2 and 234±3 kJ/mol in the B2 phases [300] is close to the activation energy of Al self diffusion in Fe-25Al (236±3 kJ/mol) [153]. Very little influence of ≤ 5% Mn on hardness is detected (HV from 304 to 316). All this is consistent with the small effect of adding 2 or 5% Mn in Fe-25Al on the IF curves. The Snoek, X, and Zener peaks are recorded in the damping curve.

**Fe–Al–Ge.** The repulsive interaction between Al and Ge atoms in iron [309] improves the $DO_3$ order in Fe-Al-Ge alloys, and increases the $DO_3$-to-B2 transition temperature [32, 310]. $T_O$ in Fe-25Al-5Ge is about 652 °C, in Fe-12Al-13Ge about 917 °C; from this we conclude that our IF tests were carried out in the range of the $DO_3$ phase. $T_C$ in Fe-12Al-13Ge is about 487 °C (DSC), and in Fe-20Al-5Ge $T_C \approx 527$ °C (magnetometry). Adding of Ge in Fe-Al as well as substitution of Al by Ge leads to a pronounced increase of hardness, e.g. for Fe-12Al-13Ge $HV = 413$.

The damping curves $Q^{-1}(T)$ for these Fe-Al-Ge alloys are similar to their Fe-Al prototypes: the same Snoek-type and Zener peaks are recorded in all alloys, some indication of the X peak is recorded in the 25% Al and 20% Al containing alloys. Substitution of Al by Ge decreases the X peak which is not observed in the Fe-12Al-13Ge alloy. The Fe-Ge-Al alloys are discussed in the next sub-chapter.

In conclusion, these additional elements exert only little influence on carbide formation, but in case of Ge and Co they affect significantly the ordering temperatures in the opposite directions (Ge increases $T_O$, while Co decreases it). All these alloys can be considered as Fe-Al-based alloys. Thus only small changes in the IF curves are observed in the case of Fe-Al-Mn alloys, while in the case where the 13%Ge or 5%Co are added in the $Fe_3$(Al,Me) compound only the Snoek and Zener peaks can be seen, in agreement with the proposed relaxation mechanisms. In all three systems the Snoek-type and X peaks anneal out after heating to high temperature, contrary to the stable Zener peak confirming the mechanisms proposed for these phenomena.

### 5.3 Fe-Mn based alloys

Structure and anelasticity of Fe-Mn based alloys are significantly different to the discussed above for Fe-Al, Fe-Ga and Fe-Ge alloys. Iron-manganese alloys are known as widely used commercial materials possessing a shape memory (SM) effect, an enhanced damping capacity, and some other attractive mechanical and physical properties. The mechanism of the shape memory effect in these alloys, based on the $\gamma \leftrightarrow \varepsilon$ martensitic transformation (MT), has been thoroughly studied (e.g., [311,312,313,314,315]).

Fe-Mn based alloys do not have structure of intermetallic compounds and, consequently, their chemical composition is given in this section in weight per cents. The main reason to briefly consider Fe-Mn alloys is to study simultaneously the regularities of martensitic transition by in situ neutron diffraction and related anelastic effects.

Three alloys were prepared by melting high-purity components (99.98% Fe; 99.9995% Mn; 99.99% Si) in a Balzers induction furnace in argon atmosphere. According to the spectral analysis data, the alloys contained (I) 22.4%Mn and 2.7%Si (at.%), (II) 25.8%Mn and 3.5%Si, and (III) 25.7%Mn and 6.7%Si. Subsequently, we used the nominal alloy compositions as Fe-22Mn-3Si, Fe-26Mn-4Si, and Fe-26Mn-7Si throughout the research. After casting, the alloys were cold rolled at room temperature until a thickness reduction of about 30% occurred. The typical microstructure of the alloys at room temperature after cold rolling contains a mixture of $\gamma$ (austenite) and $\varepsilon$ (martensite) phases, as shown in Fig. 5.45.

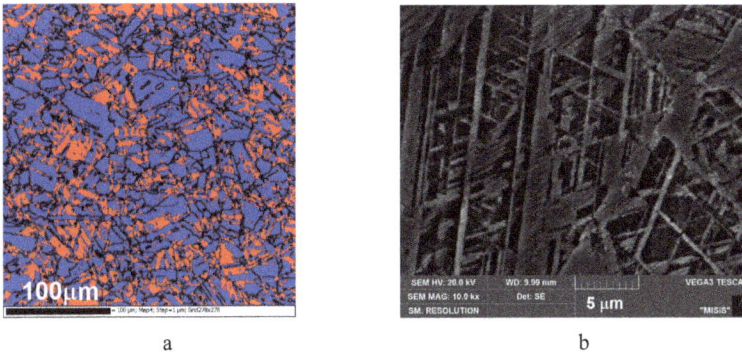

a                                                                b

*Fig. 5.45. Typical microstructure of the alloy after casting and cold rolling at RT: a) EBSD image with martensite in blue and austenite in red; b) structure of $\varepsilon$ martensite observed by SEM in Fe-25.8Mn-3.5Si cold rolled sample (V. Cheverikin, L. Sun).*

Then, the specimens were encapsulated in quartz ampoules under vacuum and annealed at 1000 °C for 1 h, this was followed by cooling the ampoules with specimens in air. Again, a martensitic-austenitic two-phase microstructure was observed at room temperature. The dynamic mechanical analyzer TA Instruments DMA Q800 was used to measure anelastic effects during the martensitic transformation, with a cooling/heating rate of 3 K/min. To determine the phase transformation temperatures and the transformation heat, a differential scanning calorimetry *Perkin Elmer* DSC7 was used in a temperature range of 0 – 350 °C with a heating rate of 10 K/min. Thermal cycling of specimens was carried out directly in the DSC and DMA devices. Microstructural investigations were performed by TEM, in a Hitachi H600 transmission electron microscope with an accelerating voltage of 100 kV, and by neutron diffraction.

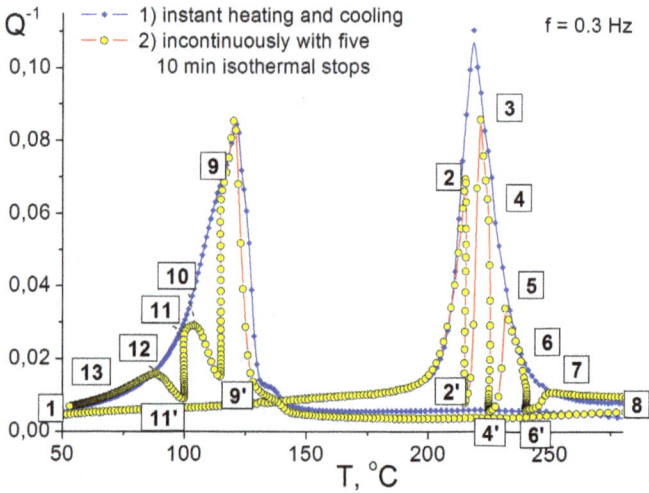

*Fig. 5.46. Temperature dependent IF of the Fe-22Mn-3Si steel (1) during continuous heating (V = 3 K/min, f = 0.3 Hz) and cooling (~2 K/min) and (2) an analogous dependence with five 10_min isothermal holdings during heating at (2–2') 215, (4–4') 225, and (6–6 ') 240°C and during cooling at (9–9 ') 115 and (11–11 ') 100°C at f = 0.3 Hz and $\varepsilon_0 = 3.6 \times 10^{-5}$.*

Figure 5.46 shows the temperature dependence of the internal friction (TDIF) of the Fe-22Mn-3Si alloy upon instant heating and cooling (blue curves, 3 K/min) after several heating and cooling cycles. The TDIF curve exhibits an internal friction peak with a maximum at about 220 °C upon heating and at about 120 °C upon cooling. It is possible to determine the transformation temperature ranges $M_s - M_f$ and $A_s - A_f$; they were found to be $\Delta T_{direct} \approx 70$ °C and $\Delta T_{reverse} \approx 50$ °C, respectively. The internal friction maximum is within the temperature range of the direct martensitic transformation (DMT) upon cooling, and within the range of the reverse martensitic transformation (RMT) upon heating. The hysteresis between the direct and reverse transformations ($\Delta T_{MT}$), which was found from the difference in the temperature positions of the internal friction maxima, is approximately 110 °C.

The second curve (red with circles) illustrates a sharp decrease in the IF values, almost down to the background level, as a result of five 10min isothermal holdings at different temperatures included in the heating and cooling cycle. The sample was heated at a constant rate of 3 K/min in the segments *1–2*, *3–4*, *5–6*, and *7–8*; and was cooled in the regions *8–9*, *10–11*, and *12–13*. The segments *2–2'*, *4–4'*, and *6–6'* correspond to an isothermal holding upon heating; and the segments *9–9'* and *11–11'* to an isothermal holding upon cooling. The regions *2'–3*, *4'–5*, *6'–7*, *9'–10*, and *11'–12* correspond to the transition from the isothermal conditions to heating or cooling.

The internal friction peaks become much higher as the measurement conditions are changed, namely, as the frequency of the forced vibrations decreases and the heating rate increases. The measurements made at different heating rates reveal that the height of the IF maximum for the reverse MT (the frequency and the deformation amplitude being constant) increased with the heating rate (Fig. 5.47). The frequency dependence of the height of the IF maxima for the direct and reverse MTs is shown in Fig. 5.45 at the given heating rates and the deformation amplitude $\varepsilon_0 = 3.6 \times 10^{-5}$.

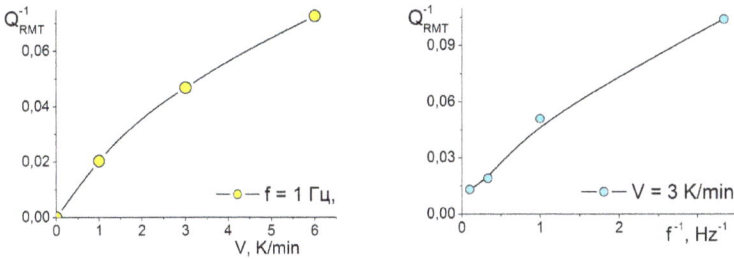

*Fig. 5.47. Variation of the height of the IF peaks during MTs as functions of (a) the heating rate (frequency 1 Hz) and of (b) the inverse frequency of vibrations during heating for the RMT (V = 3 K/min) and during cooling for the DMT (V = 2 K/min). The strain amplitude $\varepsilon_0 = 3.6 \times 10^{-5}$.*

Thermal cycling in the vicinity of the MT temperatures reveals that the position of the peaks and their shape depended on the number of the preceding measurement cycles and change in a regular manner. The temperature dependent internal friction curve in the first heating and cooling cycle has peaks at about 220 °C (heating) and 120 °C (cooling). On cooling the internal friction peak is caused by direct, and on heating by reverse martensitic transformations.

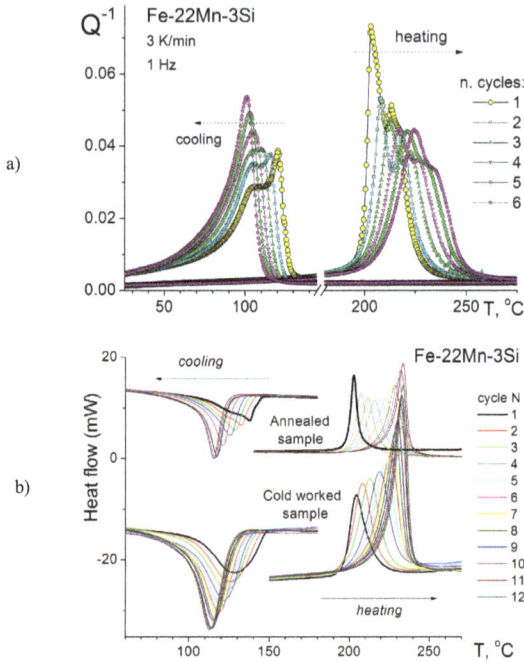

*Fig. 5.48. Effect of thermo-cycling between 20-350 °C on the position and the shape of (a) the internal friction peaks (f = 1 Hz, $\varepsilon_0$ = 3.6 ×10$^{-5}$, T = 3 K/min, n is cycle number); and (b) the DSC peaks during MT in Fe-22%Mn-3%Si.*

The temperature position of peaks and their shape depend on the number of thermo-cycles. During ten-fold thermal cycling (Fig. 5.48a) in the range from 30 to 350 °C the internal friction peak for direct transformation shifts in the direction of lower, and the IF peak for reverse transformation shifts in the direction of higher temperature. There is a change not only in the temperature position but also in peak shape. In the initial state the temperature for the maximum of IF for reverse MT in the range $A_s - A_f$ is significantly closer to the $A_s$ temperature; during thermal cycling the maximum of the IF peak, i.e., the temperature at which the transformation range is at a maximum, shifts closer to $A_f$. This points to formation of an autocatalytic process of phase transformation during cycling.

The shape of the internal friction peak with direct martensitic transformation also changes: there is a tendency towards transition from a "two-headed" to a "single-headed" peak. After homogenizing at 1000 °C its high-temperature component predominates: as a result of thermal cycling it is reduced and moves to a lower temperature. The area beneath the IF peak, i.e., the amount of transformed phase, increases, but the internal friction peak temperature is reduced.

Calorimetric data with thermal cycling (0-350 °C) for annealed and worked specimens of alloy Fe-22%Mn-3%Si entirely confirm the results obtained by the internal friction method (Fig. 5.48b). Insignificant quantitative differences between data obtained by the IF and calorimetric methods are caused by a difference in heating rate during measurements of internal friction (3 K/min) and DSC (10 K/min).

The peak positions for Fe-26Mn-4Si and Fe-26Mn-7Si alloys in the first three cycles from -30 to 350 °C (Table 5.5) confirm this effect – a decrease in the temperature of direct MT and an increase in the temperature of the inverse MT – enhancing the thermal hysteresis $\Delta T$. With an increase in the Si content, this effect becomes stronger lowering the reversibility of the MT.

*Table 5.5    Peak temperatures for direct/reverse MT and corresponding thermal hysteresis values ($\Delta T$) obtained by TDIF and DSC in annealed Fe-22Mn-3Si, cold worked Fe-26Mn-4Si, hot rolled Fe-26Mn-4Si, and hot rolled + annealed Fe-26Mn-7Si alloys (Heating / Cooling Rate: TDIF 3K/min, DSC 10K/min).*

| Alloy | Peak T(°C) TDIF | $\Delta T$ | Peak T(°C) DSC | $\Delta T$ | Transformation heat, J/g |
|---|---|---|---|---|---|
| Fe-22Mn-3Si | 203.1/120.3 208.1/115.6 212.6/110.4 | 82.8 92.5 102.2 | 202.7/132.8 207.3/128.9 211.6/125.8 | 69.9 78.4 85.8 | -9.2/13.5 -11.3/14.9 -13.4/15.8 |
| Fe-26Mn-4Si cold rolled | 212.7/27.8 215.4/24.3 226.0/26.7 | 184.9 191.1 199.3 | 239.8/29.2 244.6/25.9 247.7/23.2 | 210.6 218.7 224.5 | -7.6/5.7 -8.0/5.2 -8.0/4.8 |
| Fe-26Mn-4Si hot rolled | 206.8/19.9 210.6/18.9 226.5/17.4 | 186.9 191.7 209.1 | 233.4/29.3 234.2/26.6 237.4/24.1 | 204.1 207.6 213.3 | -9.3/4.7 -9.0/4.7 -9.1/4.6 |
| Fe-26Mn-7Si hot rolled + annealed | 218.8/-1.0 232.6/-2.8 238.9/-7.2 | 219.8 235.4 246.1 | 232.0/9.4 229.9/8.0 232.5/7.4 | 222.6 221.9 225.1 | -2.2/3.1 -2.9/3.3 -2.7/3.4 |

Thermo-cycling has a pronounced effect on the samples hardness. In the case of annealed samples hardness increases with increase in number of thermo-cycles, while it decreases if cold or hot rolled samples are subjected to thermo-cycling (Fig. 5.49).

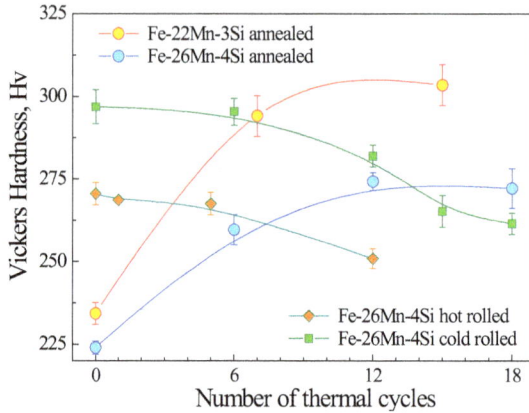

*Fig. 5.49. Change in hardness with thermal cycling for Fe-22Mn-3Si (annealed) and Fe-26Mn-4Si (annealed, cold rolled and hot rolled) samples (L. Sun).*

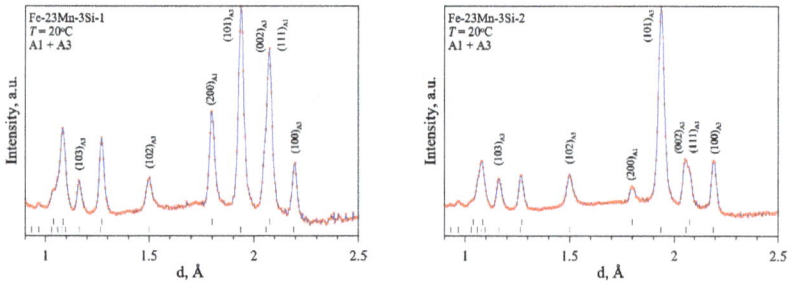

*Fig. 5.50. Neutron diffraction patterns at room temperature of two Fe-23Mn-3Si (sample 1 after one, and sample 2 after twelve heating and cooling thermo-cycles). All recorded peaks correspond either to A1 (Fm3m, upper bars) or A3 (P6₃/mmc, lower bars).*

In situ neutron diffraction tests at instant heating to 400 °C and cooling to 70 °C with the rate of 2.2 K/min (the acquisition time, $t_s$, for each diffraction pattern is 1 min) was carried out after first and twelfth thermo-cycles of the for Fe-22Mn-3Si alloy. Both samples at room temperature have a mixture of A1 (*f.c.c.*) and A3 (hcp) phases (Fig. 5.50):

- the **A1** (*f.c.c.*) has an γ-Fe-type structure with Fe, Mn and Si atoms randomly distributed in (0 0 0) and (1/2 1/2 0) positions, sp. gr. *Fm3m*; $a \approx 3.693$ Å at room temperature,

- the **A3** (hcp) has Mg-type structure with randomly distributed atoms (1/3 2/3 1/4), sp. gr. *P6₃/mmc*; $a \approx 2.534$ Å, $c \approx 4.103$ Å at room temperature.

The amount of A3 phase in twelve times cycled sample is higher (ratio $V_{A3}/V_{A1}$ is about six times higher).

Fig 5.51 exhibits neutron diffraction patterns for Fe-26Mn-4Si samples cold worked in RT (a), after 6 cycles (b) and after 18 cycles (c). At room temperature all studied samples have a two-phase structure: γ austenite (*f.c.c.*, sp. gr. *Fm3m*; $a \approx 3.5970$ Å) and ε martensite (hcp, sp. gr. *P6₃/mmc*; $a \approx 2.5403$ Å, $c \approx 4.1136$ Å) in agreement with the SEM-EBSD results. Intensities of peaks show significant change after six thermal cycles, further thermo-cycling leads only to minor changes.

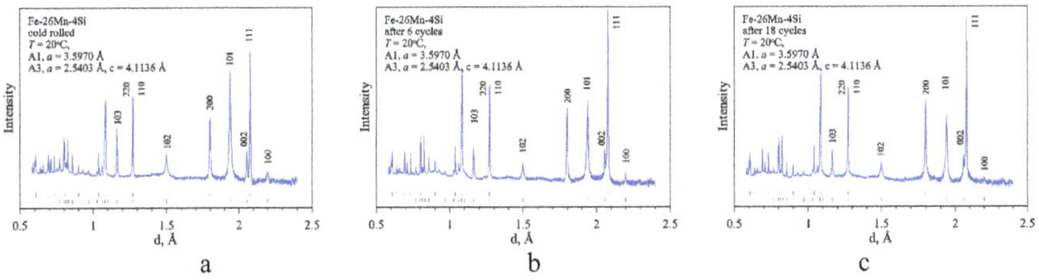

*Fig. 5.51. High resolution neutron diffraction patterns of Fe-26Mn-4Si samples after a) cold rolling at room temperature, (b) followed by six and (c) 18 cycles thermo-cycles. All the recorded peaks correspond either to austenite (Fm3m) or hexagonal martensite (P6₃/mmc). The calculated peak positions are shown in the bottom of the figures.*

The dependence of the peak width, W, for both phases in the form of Williamson–Hall plot ($W^2$ as a function of $d^2$) is presented for the cold worked sample and the sample after 18 thermo-cycles in Fig. 5.52. For the γ phase, the peak broadening is consistent with existence of isotropic microstrains in the sample. The microstrains in the ε martensite are pronouncedly higher and strongly anisotropic. After 18 thermal cycles (350 °C – RT), the $W^2$ of the γ phase has no significant changes, the $W^2$ of the ε phase has very small changes, suggesting that defect density in the structure of the cold worked alloy has not been changed due to thermal cycles.

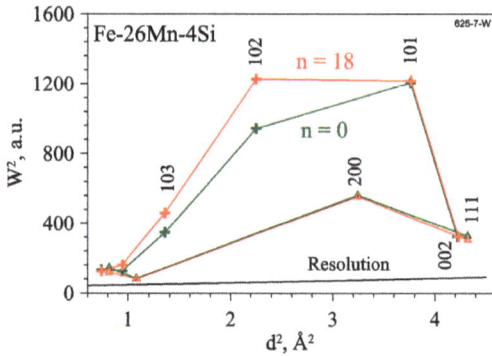

*Fig. 5.52. Williamson-Hall plot: $W^2$ vs. $d^2$ dependences for cold worked and after 18 times thermo-cycled samples. Triangles are for the $\gamma$ phase and crosses for the $\varepsilon$ phase. The bottom black line is the diffractometer resolution function measured with a standard sample.*

Volume faction of athermal and cold worked-induced $\varepsilon$ martensites determined by the *in situ* neutron diffraction against number of thermal cycles are plotted in Fig. 5.53. The volume fraction of the $\varepsilon$ martensite in the cold worked sample is 0.65. The amount of the $\varepsilon$ martensite decreases steeply to 54% after six and to 48% after 18 thermal cycles. The change in the hardness with thermal cycling is explained by a decrease of the ration of $\varepsilon/\gamma$ in the rolled Fe-26Mn-4Si samples.

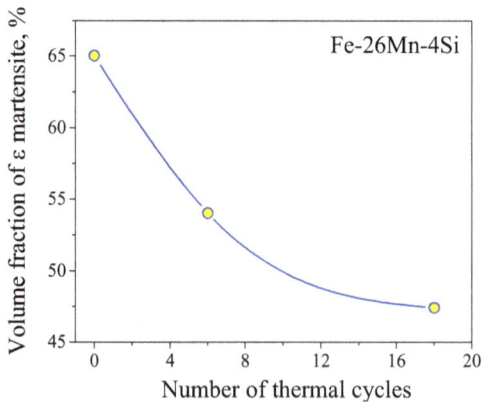

*Fig. 5.53. Change in volume fraction of $\varepsilon$ martensite with number of thermal cycles of the Fe-26Mn-4Si cold rolled sample.*

Neutron diffraction patterns reveal that the *annealed* samples have a mixture of $\gamma$ (for Fe-23Mn-3Si: $a \approx 3.693$ Å, for Fe-26Mn-4Si: $a \approx 3.5970$ Å) and $\varepsilon$ ($a \approx 2.534$ Å, $c \approx 4.103$ Å for Fe-23Mn-3Si and $a \approx 2.5403$ Å, $c \approx 4.11360$ Å for Fe-26Mn-4Si) phases at room temperature, both after the first and twelfth cycles. The dependences from the peak width, $\Delta d$, of both phases in the form of Williamson-Hall plots (($\Delta d)^2$ as a function of $d^2$) are shown in Fig. 5.54 for the first sample. For the $\gamma$ phase, the peak broadening is consistent with isotropic microstrains, which can be estimated as $\delta = \Delta d/d \approx 0.0011$ and $\approx 0.0013$ in the 1 and 12 times thermo-cycled samples, respectively. In turn, the microstrains in the $\varepsilon$ martensite are pronouncedly higher in magnitude and strongly anisotropic. Their mean values are 0.0017 and 0.0025 in the 1 and 12 times thermo-cycled samples, respectively; they are roughly twice higher as compared to the austenite phase. Using the simplest approximation to convert local microstrain into local microstress by the Hooke law: $\sigma = E\delta$, where $E$ is the Young's modulus of the Fe-Mn-Si alloy ($\approx 180$ GPa), the microstress $\sigma$ can be estimated as $\approx 200$ MPa in the $\gamma$ austenite and $\approx 300$-450 MPa in the $\varepsilon$ martensite, which is close to but still below the yield point.

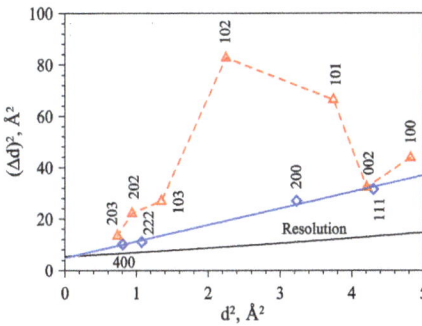

*Fig. 5.54. Williamson-Hall plot: $(\Delta d)^2$ vs. $d^2$ dependences for the same sample. Blue diamonds are for the $\gamma$ phase and red triangles for the $\varepsilon$ phase. The bottom black line corresponds to the diffractometer resolution function measured with a standard sample. The $(\Delta d)^2$ values are multiplied by $10^6$. The lines are guides for the eye.*

Finally, Fig. 5.55 illustrates the intensities of some characteristic diffraction peaks of both phases as a function of temperature for the 1 and 12 times thermo-cycled Fe-22Mn-3Si samples upon heating, showing that the $\varepsilon \rightarrow \gamma$ phase transition is completed at about 190 °C for first sample and at about 220 °C for the second sample. In the same way, the $\gamma \rightarrow \varepsilon$ transition upon cooling is observed at about 120 °C and 100 °C for the 1 and 12 times thermo-cycled samples, respectively. These temperatures for the reversible diffusionless $\varepsilon \leftrightarrow \gamma$ phase transitions are in excellent agreement with the DSC and TDIF results. The figure also shows that the intensity of the $\gamma$ phase peaks does not vanish around room temperature, indicating that a certain volume fraction of the austenite phase remains untransformed.

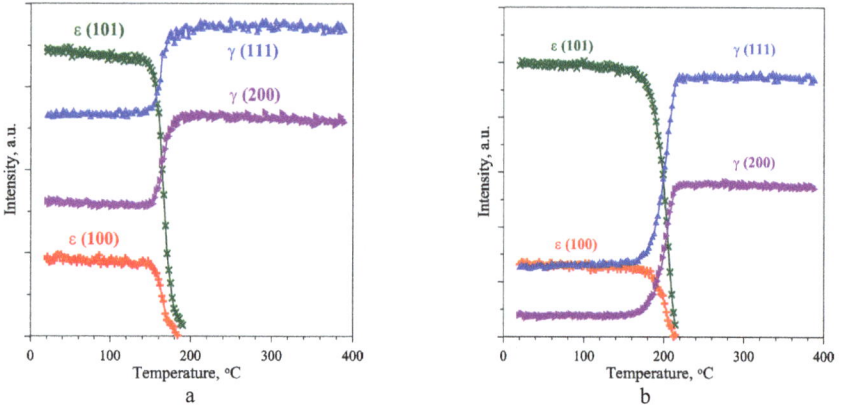

Fig. 5.55. Temperature dependences of the intensity of typical diffraction peaks for $\gamma$ and $\varepsilon$ phases at heating for one time (a) and 12 times (b) thermo-cycled Fe-22Mn-3Si samples.

More details of the $\varepsilon(+\gamma) \rightarrow \gamma$ transition can be retrieved from Fig. 5.56 that shows how the fractions of both phases change in the vicinity of the reverse transition temperature. In this figure, the transition temperatures in the 1 and 12 times thermo-cycled samples are defined as the temperatures where the $\varepsilon$ phase content is reduced by the factor 2. The shift of the reverse MT between the samples is about 34 °C. This is again very similar to the DSC data, where a shift of 31 °C in the peak temperature is observed after 12 cycles. Moreover, it reveals that the kinetics of the transition in the 12 times thermo-cycled sample is somewhat slower than in one time thermo-cycled sample. Upon heating, the content of the $\varepsilon$ phase decreases to zero, and the $\gamma$ phase increases to 100%.

The microstructural observations by TEM are summarized in Fig. 5.57. In the annealed condition, the microstructure at room temperature is composed of martensite plates within the $\gamma$ matrix (Fig. 5.57a), and contains a low amount of defects. Only scarce groups of few dislocations (mostly within the $\varepsilon$ martensite) are visible in some areas (Fig. 5.57b), together with some stacking faults in $\gamma$ phase.

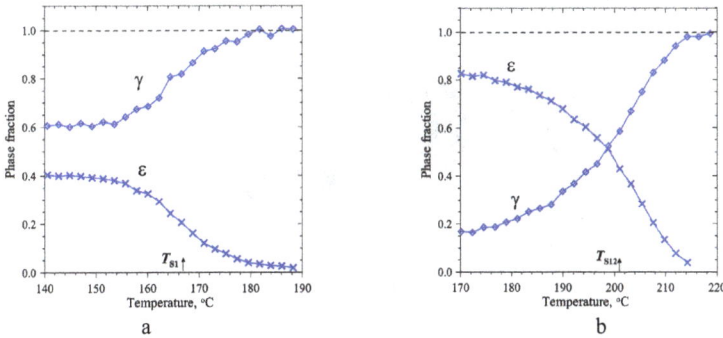

*Fig. 5.56. Temperature dependences of the intensity of typical diffraction peaks for γ and ε phases at heating for one time (a) and 12 times (b) thermo-cycled preliminary annealed samples. The arrows indicate the temperatures at which the ε phase content decreases by 50%: $T_{S1} \approx 167\ °C$, $T_{S12} \approx 201\ °C$.*

Thermal cycling produces a rapid generation of dislocations (see Figs. 5.57c,d after 5 cycles), the density of which increases with the number of cycles (Figs. 5.57e,f after 12 cycles). This result is in agreement with the broadening of the neutron diffraction peaks discussed previously. Dislocations form complex tangles, although a tendency to dislocation alignment in bands is visible in the images. The dislocation bands are parallel to the {111} habit planes between austenite and martensite, as seen in Figs. 5.57d and f. Dislocations are formed as a consequence of the large misfit of the atomic planes at the habit plane between austenite and martensite, due to the large volume change accompanying the transformation in Fe-based alloys. The lattice misfit produces a partial breakdown of coherency at the boundary between the phases, blocking the interface movement and giving rise to an incomplete MT on cooling. In the same way, the dislocation bands make the growth of other martensite variants in the subsequent transformation cycles difficult, which explains the change in the transformation kinetics observed by IF and DSC. Extra overcooling is needed to surpass the dislocation bands that shift the peak temperatures towards the end of the transformation range. At elevated temperature, corresponding to the reverse MT, there is an effect of internal work hardening and strengthening of the austenitic phase due to microplastic deformation, causing additional dislocations and stacking faults formation. The effect of work hardening as a result of direct and reverse martensitic transformation has been observed in different materials [316]. An increase in the internal stresses in the martensitic phase promotes the autocatalytic character of the austenite domains growth during the reverse martensitic transformation [317]. As a consequence, the DSC peak temperature approaches the $A_f$ temperature.

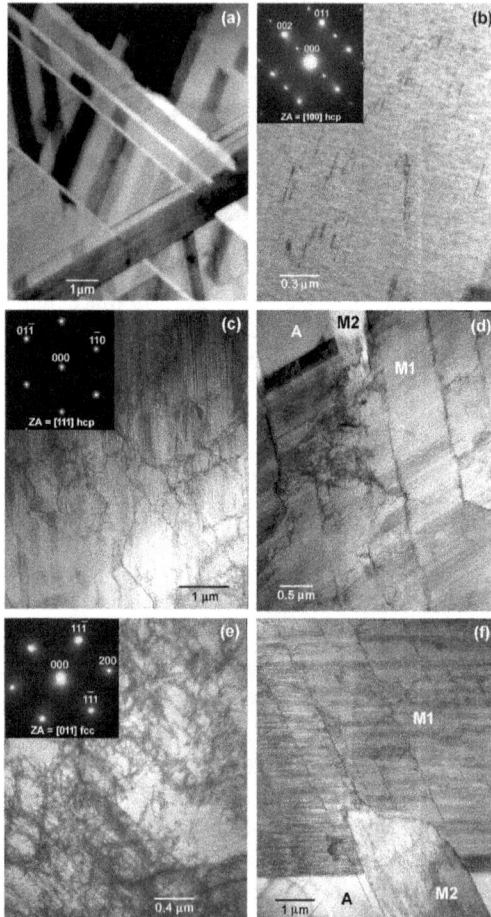

*Fig. 5.57. Microstructure of the alloy Fe-22Mn-3Si in as-annealed condition (a, b); after 5 (c, d); and 12 (e, f) thermal cycles between 0–350°C. a) austenite with some martensite plates; b) dislocations within a martensite plate; c) and e) dislocation tangles with high density; d) and f) martensitic plate (M1) with randomly oriented dislocations and dislocation bands parallel to the {111} habit plane of another martensite plate (M2) with austenite (A) (J. Pons).*

Thus, the combination of in situ neutron diffraction, internal friction, DSC and TEM was used to establish the effect of thermal cycling through the martensitic transformation temperature range in different Fe-Mn-Si alloys. Transmission electron microscopy showed that lattice defects accumulation both in martensitic and austenitic phases in annealed alloy led to an increase in the martensitic transformation hysteresis and promoted the development of autocatalytic reverse transformation processes. From *in situ* neutron diffraction tests we found that microstrains in the $\gamma$ phase can be estimated as $\delta \approx$ 0.0011 and $\approx$ 0.0013 in the one and 12 times cycled samples, respectively, in the annealed Fe-22Mn-3Si alloy. The microstrains in the $\varepsilon$ martensite are pronouncedly higher in magnitude and strongly anisotropic. Their mean values are 0.0017 and 0.0025 in the samples, correspondingly, and they are roughly twice higher as compared to the austenite phase.

Decrease in the hardness in the cold rolled Fe-26Mn-4Si sample is associated with decrease of $\varepsilon$ martensite content due to thermal cycling through the martensitic transformation temperature range. In this case, the effect of thermal cycling is similar to the effect of annealing. In turn, an increase in the hardness in annealed samples (both Fe-22Mn-3Si and Fe-26Mn-4Si) is associated with increase of $\varepsilon$ martensite content and defects accumulation due to direct and inverse martensitic transition during thermal cycling.

Further details of our studies of anelastic effects in steels can be found in [235, 319-320] for Fe-Cr based alloys with magnetomechanical damping, in [321-324] for Fe-Ni based alloys with martensitic transition and in [325-327] for high porous metallic structures.

## References

[270] Kellogg, R.A. Development and modeling of iron–gallium alloys, PhD Thesis Engineering Mechanics, Iowa State University, Ames, Iowa, 2003.

[271] Kellog, R.A.; Flatau, A.B.; Clark, A.E.; Wun-Fogle, M.; Lograsso, T.A. Temperature and stress dependencies of the magnetic and magnetostrictive properties of $Fe_{0.81}Ga_{0.19}$, J. Appl. Phys. 91 (2002) 7821. https://doi.org/10.1063/1.1452216

[272] Kellogg, R.A.; Russel, A.M.; Lograsso, T.A.; Flatau, A.B.; Clark, A.E.; Wun-Fogle, M. Tensile properties of magnetostrictive iron–gallium alloys, Acta Mater. 52 (2004) 5043. https://doi.org/10.1016/j.actamat.2004.07.007

[273] Wu, W.; Liu, J.H.; Jiang, C.B. Tb solid solution and enhanced magnetostriction in Fe$_{83}$Ga$_{17}$ alloys. J. Alloys Compd. 622 (2015) 379-383. https://doi.org/10.1016/j.jallcom.2014.09.151

[274] Yao, Z.Q.; Tian, X.; Jiang, L.P.; Hao, H.B.; Zhang, G.R.; Wu, S.X.; Zhao, Z.Q.; Gerile, N. Influences of rare earth element Ce doping and melt-spinning on microstructure and magnetostriction of Fe$_{83}$Ga$_{17}$ alloy. J. Alloys Compd. 637 (2015) 431–435. https://doi.org/10.1016/j.jallcom.2015.03.009

[275] He, Y.; Jiang, C.; Wu, W.; Wang, B.; Duan, H.; Wang, H.; Zhang, T.; Wang, J.; Liu, J.; Zhang, Z.; Stamenov, P.; Coey, J.M.D.; Xu, H. Giant heterogeneous magnetostriction in Fe-Ga alloys: Effect of trace element doping, Acta Mater.109 (2016) 177–186. https://doi.org/10.1016/j.actamat.2016.02.056

[276] Jin, T.Y.; Wu, W.; Jiang, C.B. Improved magnetostriction of Dy-doped Fe$_{83}$Ga$_{17}$ melt-spun ribbons, Scr. Mater. 74 (2014) 100-103. https://doi.org/10.1016/j.scriptamat.2013.11.010

[277] Ma, T.; Hu, S.; Bai, G.; Yan, M.; Lu, Y. et al., Structural origin for the local strong anisotropy in melt-spun Fe-Ga-Tb: Tetragonal nanoparticles,. 106 (2015) 112401.

[278] Meng, C.; Wang, H.; Wu, Y.; Liu, J.; Jiang, C. Investigating enhanced mechanical properties in dual-phase Fe-Ga-Tb alloys, Sci. Rep. 6, 34258 (2016); doi: 10.1038/srep34258. https://doi.org/10.1038/srep34258

[279] Meng, C.; Wu, Y.; Jiang, C.; Design of high ductility FeGa magnetostrictive alloys: Tb doping and directional solidification, Mater. Des. 130 (2017) 183-189. https://doi.org/10.1016/j.matdes.2017.05.053

[280] Jiang, L.P.; Yang, J.D.; Hao, H.B.; Zhang, G.R.; Wu, S.X.; Chen, Y.J.; Obi, O.; Fitchorov, T.; Harris, V.G. Giant enhancement in the magnetostrictive effect of FeGa alloys doped with low levels of terbium. Appl. Phys. Lett. 102 (2013) 222409. https://doi.org/10.1063/1.4809829

[281] Fitchorov, T.I.; Bennett, S.; Jiang, L.P.; Zhang, G.R.; Zhao, Z.Q.; Chen, Y.J.; Harris, V.G. Thermally driven large magnetoresistance and magnetostriction in multifunctional magnetic Fe-Ga-Tb alloys. Acta Mater. 73 (2014) 19-26. https://doi.org/10.1016/j.actamat.2014.03.053

[282] Golovin, I.S.; Dubov, L.Yu.; Funtikov, Yu.V.; Palacheva, V.V.; Cifre, J.; Hamana, D. Metallurgical and Materials Transactions (A), 46/3 (2015) 1131-1139. https://doi.org/10.1007/s11661-014-2721-3

[283] Golovin, I.S.; Balagurov, A.M.; Emdadi, A.; Palacheva, V.V.; Bobrikov, I.A.; Cheverikin, V.V.; Zanaeva, E.N.; Mari, D. Phase transitions in Fe-27Ga alloys: guidance to develop functionality. Summitted.

[284] Golovin, I.S.; Palacheva, V.V.; Emdadi, A.; Mari, D.; Heintz, A.; Balagurov, A.M.; Bobrikov, I.A. Invited talk at ICIFMS-18, September 12-15, 2017 - Foz do Iguaçu, Brazil, to be published at Materials Research.

[285] Liao, L.-l.; Fang, M.-l.; Zhu, J.; Li, J.-h.; Wang, J. International Journal of Minerals, Metallurgy and Materials, 21/1 (2014) 1-6. https://doi.org/10.1007/s12613-014-0864-2

[286] Coronel V.F., Beshers D.N., Magnetomechanicai damping in iron. Journal of Applied Physics, 64 (1988) 2006-2015. https://doi.org/10.1063/1.341757

[287] Astie B., Degauque J., J. Physique. (Paris) Colloq., 44(C-9) (1983) 461-470.

[288] Astie B., Degauque J., Peyrade, J.P., Proc. of ICIFUAS-10, Sept 6-9, (1993) Italy, 129-134.

[289] Jen, S.U.; Cheng, W.C.; Chiang, F.L. Structural, magneto-mechanical, and damping properties of slowly-cooled polycrystalline $Fe_{81}Ga_{19}$ alloy, J. All. Comp., 651 (2015) 544. https://doi.org/10.1016/j.jallcom.2015.08.153

[290] Jen, S.U.; Liu, C.C.; Magneto-elastic and magnetic domain properties of $Fe_{81}Ga_{19}/Si_{(100)}$ films, J. Appl. Phys. 115 (2014) 013909. https://doi.org/10.1063/1.4861160

[291] Malozemoff, A.P.; Slonczewski, J.C. Magnetic domain walls in bubble materials (New York, Academic Press), 1979.

[292] Dai, L.; Cullen, J.; Wuttig, M.; Lograsso, T.; Quandt, E. Magnetism, elasticity, and magnetostriction of FeCoGa alloys, J. Appl. Phys. 93 (2003) 8627. https://doi.org/10.1063/1.1555980

[293] Clark, A.E.; Restorff, J.B.; Wun-Fogle, M.; Hathaway, K. B.; Lograsso, T. A.; Hung, M.; Summers, E. Magnetostriction of ternary Fe-Ga-X (X=C, V, Cr, Mn, Co, Rh) alloys, J. Appl. Phys. 101 (2007) 09C507. https://doi.org/10.1063/1.2670376

[294] Jen, S. U.; Cheng, W. C.; Lin, Y. C.; Chen, Y. Z.; Golovin, I. S. Magnetic and magneto-mechanical properties of $Fe_{55}Co_{19}Ga_{26}$ alloy, Mater. Lett. 182 (2016) 72. https://doi.org/10.1016/j.matlet.2016.06.080

[295] Chang, F. L. Master Thesis, Mechanical properties of FeNiGa alloys in bulk form, Department of Mechanical Engineering, NTUST, Taiwan.

[296] Schneider, A.; Falat, L.; Sauthoff, G.; Frommeyer, G.: Constitution and microstructures of Fe–Al–M–C (M=Ti, V, Nb, Ta) alloys with carbides and Laves phase Intermetallics 2003, 11, 443-450; Mechanical properties of Fe–Al–M–C (M=Ti, V, Nb, Ta) alloys with strengthening carbides and Laves phase, 13 (2005) 1256-1262.

[297] Stein, F.; Schneider, A.; Frommeyer, G.: Flow stress anomaly and order–disorder transitions in Fe3Al-based Fe-Al-Ti-X alloys with X=V, Cr, Nb, or Mo Intermetallics 11 (2003) 71-82.

[298] Stein, F.; Sauthoff, G.; Palm, M. Phases and phase equilibria in the Fe–Al–Zr system. Z Metallkde, 95 (2004) 469-485. https://doi.org/10.3139/146.017985

[299] Ducher, R.; Stein, F.; Viguier, B.; Palm, M.; Lacaze, J. A Re-Examination of the Liquidus Surface of the Al-Fe-Ti System. Z Metallkde, 94 (2003) 396-410. https://doi.org/10.3139/146.030396

[300] Zuquing, S.; Wangyue, Y.; Yuanding, S. L. H.; Baisheng, Zh.; Jilian, Y. Mater Sci Eng A, 258 (1998) 2557-.

[301] Sima, V. J. All. Comp., 378 (2004) 44-. https://doi.org/10.1016/j.jallcom.2003.11.166

[302] Peteline, S.; Njiokep, E. M. T.; Divinski, S.; Mehrer, H. Def Diff Forum (2003) 216-217, 175. https://doi.org/10.4028/www.scientific.net/DDF.216-217.175

[303] Zhou, Z. C.; Han, F. S. Phys Stat Sol A 2003, 199, 202-206. https://doi.org/10.1002/pssa.200306653

[304] Zhou, Zh.; Gao, T.; Han, F. J Phys Cond Matt 2003, 15, 6809-. https://doi.org/10.1088/0953-8984/15/40/017

[305] Lambri, O. A.; Pérez-Landazábal, J. L.; Cano, J. A.; Recarte, V. Mater Sci Eng A, Mechanical spectroscopy in commercial Fe–6 wt.% Si alloys between 400 and 1000 K. 370 (2004) 459-463.

[306] Golovin, I.S.; Serzhantova, G.V.; Sokolova, O.A.; Semin, V.A.; Jäger, S.; Sinning, H.-R.; Stein, F.; Golovin, S.A. Snoek-type and Zener relaxation in Fe-Si-Al alloys. Solid State Phenomena, 137 (2008) 69-82. https://doi.org/10.4028/www.scientific.net/SSP.137.69

[307] Lambri, O.A.; Pérez-Landazábal, J.I.; Cuello, G.J.; Cano, J.A.; Recarte, V.; Siemers, C.; Golovin, I.S. Mechanical Spectroscopy in Fe-Al-Si alloys at elevated temperatures. J. All. Comp., 468 (2009) 96-10. https://doi.org/10.1016/j.jallcom.2007.12.071

[308] Lambri, O.A.; Pérez-Landazábal, J.I.; Cuello, G.J.; Cano, J.A.; Recarte, V.; Golovin, I.S. Mechanical spectroscopy and neutron diffraction studies in Fe-Al-Si alloys. Solid State Phenomena, 137 (2008) 91-98. https://doi.org/10.4028/www.scientific.net/SSP.137.91

[309] Athanassiadis, G.; Le Gaer, G.; Foct, J.; Rimlinger, L. Phys Stat Sol A 1977, 40, 425-. https://doi.org/10.1002/pssa.2210400208

[310] Gruzin, P. L.; Mkrtchan, V. S.; Rodionov, U. L.; Selisskij, Ia. P.; Khachatrjan, M. Kh. Fiz Metal Metal., 34 (1972) 316-322 (in Russian).

[311] Chowdhury, P.; Sehitoglu, H. Deformation physics of shape memory alloys – Fundamentals at atomistic frontier, Prog. Mater. Sci., 88 (2017) 49-88. https://doi.org/10.1016/j.pmatsci.2017.03.003

[312] Igata, N. Applications for high damping stainless alloys (HIDAS), Key Eng. Mat. 319 (2006) 209-216. https://doi.org/10.4028/www.scientific.net/KEM.319.209

[313] Kim, Y.S.; Han, S.H.; Choi, E.S.; Kim, W.J. Achieving ultrafine grained Fe-Mn-Si shape memory alloys with enhanced shape memory recovery stresses. Mater. Sci. Eng. A. 701 (2017) 285-288. https://doi.org/10.1016/j.msea.2017.06.091

[314] B.C. De Cooman, Y. Estrin, S.K. Kim, Twinning-induced plasticity (TWIP) steels, Acta Materialia, 142 (2018) 283-362. https://doi.org/10.1016/j.actamat.2017.06.046

[315] R. Drevet, Y. Zhukova, P. Malikova, S. Dubinskiy, A. Korotitskiy, Y. Pustov, S. Prokoshkin, Martensitic Transformations and Mechanical and Corrosion Properties of Fe-Mn-Si Alloys for Biodegradable Medical Implants, Metallurgical and Materials Transactions A49, (2018) 3, 1006–1013.

[316] Lobodyuk, V.A.; Estrin, E.I. Martensitic Transformations, Fizmatlit, Moscow, 2009 (in Russian).

[317] Y.K. Lee, Y.S. Seo, W. Jin, C.S. Choi. Effect of thermal cycling ($\gamma \leftrightarrow \varepsilon$) on martensitic transformation kinetics and damping capacity of Fe-17mass%Mn alloy, Key Eng. Mat., 319 (2006) 59-66. https://doi.org/10.4028/www.scientific.net/KEM.319.59

[318] A.A. Bush, K.E. Kamentsev, V.F. Meshcheryakov, Yu.K. Fetisov, D.V. Chashin, L.Yu. Fetisov. Low_Frequency Magnetoelectric Effect in a Galfenol–PZT Planar Composite Structure. Technical Physics, 2009, Vol. 54, n. 9, 1314–1320.

[319] Golovin I.S., Sarrak V.I., Suvorova S.O. Influence of Carbon and Nitrogen on Solid Solution Decay and "475°C Embrittlement" of High-Chromium Ferritic Steels. Metallurgical Transactions, v. 23A, № 9, 1992, p. 2567-2579.

[320] Golovin I.S. Internal friction and modulus defect in α-Fe –based high alloyed hidamet. Journal of Alloys and Compounds. 211/212 (1994), p. 147-151.

[321] Golovin S.A, Golovin I.S., Rodionov Y.L, Seleznev V.N. The influence of chemical composition and thermal treatments on the temperature-dependent internal friction of Fe-Ni alloys in the range of the phase transformation. Journal of Alloys and Compounds, 211/212 (1994), 132, 194-197.

[322] Golovin I.S., Seleznev V.N., Golovin S.A. Isotermal and Athermal Phase Transformation of FeNiMo Alloys. Journal de Physique IV, C8, v.5, (1995), 305-310.

[323] Golovin I.S., Goncharov S.S., Golovin S.A. The contribution of dislocation-impurities interaction to kinetics of martensitic transformation of quenched f.c.c. Fe-Ni-Mo alloys. Journal de Physique, C.8, vol. 6 (1996), 409-412.

[324] Golovin I.S., J.-O. Nilsson, G.V. Serzhantova, S.A. Golovin. Anelastic effects connected with isothermal martensitic transformations in 24Ni4Mo austenitic and 12Cr9Ni4Mo maraging steels. Journal of Alloys and Compounds, v. 310 (1-2), 2000, p. 411-417.

[325] Golovin I.S., Sinning H.-R. Damping in some cellular metallic materials. Journal of Alloys and Compounds, 2003, Vol 355, Iss 1 – 2, p. 2-9.

[326] Arhipov I.K., Golovin I.S., Golovin S.A., Sinning H.-R. Damping caused by microplasticity in porous 316L steels. Philosophical Magazine, 2005, Vol. 85, n. 14, 1557 – 1574

[327] Arhipov I.K., Golovin I.S., Golovin S.A. Damping caused by microplasticity due to fatigue microcrack growth in high porous sintered steel. Philosophical Magazine A. 2006, vol. 86, 2399-2406.

# Keyword Index

## About the Authors

### Professor Dr. Igor S. Golovin

Institution:    National University of Science and Technology "MISIS"
Department:  Physical Metallugry of non-Ferrous Metals
Address:       Leninsky ave. 4, 119049 Moscow, Russia
E-mail:         i.golovin@misis.ru

I.S. Golovin graduated from Tula State University in 1982 and became a Ph.D. student at Moscow Central Research Institute for Physical Metallurgy, Moscow. He got his doctoral (Ph.D.) degree in 1987 and higher D.Sci. degree (Habilitation) in 1998. His research fields are functional alloys, anelasticity, mechanical spectroscopy, internal friction, phase transitions, atomic and magnetic ordering, neutron diffraction, mechanisms of relaxation in crystals, ultrafine grained metals, quasicrystals, hidamets, cellular metallic materials. After several years of working abroad, in 2008 he has accepted a permanent position as a professor at National University of Science and Technology "MISIS".

### Professor Dr. Anatoly M. Balagurov

Institution:    Joint Institute for Nuclear Research
Department:  Frank Laboratory of Neutron Physics, Condensed Matter
                    Department
Address:       JINR, FLNP, Joliot Curie str., 6, 141980 Moscow region,
                    Russia
E-mail:         bala@nf.jinr.ru

A.M. Balagurov graduated from Moscow State University in 1968 and received a research scientist position in Frank Laboratory of Neutron Physics of JINR (Dubna). He was principally responsible for the construction of time-of-flight diffractometers at the pulsed neutron sources in FLNP and development of neutron diffraction as a tool for condensed matter studies. His doctoral (Ph.D.) and higher D.Sci. (Habilitation) degrees are connected with structural studies of complex magnetic oxides (cuprates, manganites, cobaltites), multilayer lipid structures and materials of relevance to nuclear energy sector using neutron beams. In 2000 he was awarded the Russian State Prize for development and realization of neutron diffraction at steady state and pulsed neutron sources. At present he is chief scientist of FLNP, JINR and associated professor at Moscow State University.

9 781945 291647